PROTEIN BIOCHEMISTRY, SYNTHESIS, STRUCTURE AND CELLULAR FUNCTIONS

TROPONIN

REGULATOR OF MUSCLE CONTRACTION

PROTEIN BIOCHEMISTRY, SYNTHESIS, STRUCTURE AND CELLULAR FUNCTIONS

Additional books in this series can be found on Nova's website under the Series tab.

Additional e-books in this series can be found on Nova's website under the e-book tab.

PROTEIN BIOCHEMISTRY, SYNTHESIS, STRUCTURE AND CELLULAR FUNCTIONS

TROPONIN

REGULATOR OF MUSCLE CONTRACTION

J. -P. JIN, M.D., PH.D.
EDITOR

New York

For permission to use material from this book please contact us:
Telephone 631-231-7269; Fax 631-231-8175
Web Site: http://www. novapublishers.com

NOTICE TO THE READER

Additional color graphics may be available in the e-book version of this book.

Library of Congress Cataloging-in-Publication Data

ISBN: 978-1-62948-320-7

Library of Congress Control Number: 2013950459

Published by Nova Science Publishers, Inc. † New York

To my mentors, colleagues, associates, students, and family

Contents

Preface

Muscle contraction is a fundamental biological activity vital to animal mobility and heart function. Skeletal and cardiac muscles are both striated muscles with similar contractile machinery and mechanisms. It is of major scientific and medical importance to understand the molecular mechanism of muscle contraction and regulation, as well as the evolution, development and adaptation of vertebrate striated muscles.

Actin-activated myosin motor function powers the contraction of striated muscles. The contractile machinery in skeletal and cardiac muscle cells are the myofibrils comprised of longitudinally connected sarcomeres. The sarcomere is the primary contractile unit of striated muscle and consists of overlapping thick filaments and thin filaments formed by, respectively, myosin and actin polymers. During contraction, myosin heads interact with actin thin filament to activate myosin ATPase and the hydrolysis of ATP powers conformational changes of myosin. This molecular power stroke displaces the actin thin filament and shortens the sarcomere to result in muscle contraction. The contractile cycles are regulated through the binding of cytosolic Ca^{2+} to the actin filament-associated tropomyosin-troponin system. The troponin complex is, therefore, a central component in the Ca^{2+}-signaling pathway that controls the contraction of cardiac and skeletal muscles.

Since the pioneer work of Professor Setsuro Ebashi and colleagues, more than four decades of intensive research have built a solid basis of knowledge for the molecular structure and the functional mechanism of troponin and its subunit proteins, troponin C, troponin I and troponin T. The comprehensive information revealed from troponin research has made significant contributions to advance the understanding of muscle contraction in health and diseases.

Based on the continuing effort of several generations of many dedicated investigators, this book presents a unique collection of topics broadly selected from for recent advances in troponin research. Although not an all inclusive volume, the chapters of this book cover a broad range of topics including a historical review, the diversities of troponin isoforms, the evolution of troponin genes, the molecular structure of troponin subunits, the biochemical functions of troponin, a kinetic modeling of troponin interaction with the thin filament, the developmental and adaptive regulations, the posttranslational regulations, the role of troponin in length-tension regulation, troponin abnormality in cardiac dysfunction, and troponin mutations in diastolic heart failure.

These integrated topics are intended to represent a core of the current understanding of the regulation and structure-function relationships of the three subunits of troponin. The information will help readers to pursue leading edge inquiries to further research on troponin function and muscle contraction, as well as to pursue translational applications in biology and medicine. We are very proud of the completion of this work that is the very first of its kind and will facilitate progresses in the field. I sincerely thank the contributing authors who are highly regarded experts for their outstanding effort, which made this publication a wonderful success.

J. -P. Jin, M. D., Ph. D.
Professor and Chairman
Department of Physiology
Wayne State University School of Medicine
Detroit, Michigan, USA

List of Contributors

Brandon J. Biesiadecki, PhD
Assistant Professor
Department of Physiology and Cell Biology
The Ohio State University
Columbus, Ohio 43210, USA

Murali Chandra, PhD
Associate Professor
Department of Integrated Physiology and Neuroscience
Washington State University
Pullman, WA 99164, USA

Jonathan P. Davis, PhD
Assistant Professor
Department of Physiology and Cell Biology
The Ohio State University
Columbus, Ohio 43210, USA

Pieter P. de Tombe, Ph. D.
James R. DePauw Professor and Chair
Department of Cell and Molecular Physiology
Loyola University Chicago Stritch School of Medicine
Maywood, IL 60153, USA

Wen-Ji Dong, PhD
Associate Professor
Gene and Linda Voiland School of Chemical Engineering and Bioengineering, Washington State University
Pullman, WA 99164, USA

Han-Zhong Feng, MD
Research Associate
Department of Physiology
Wayne State University School of Medicine
Detroit, MI 48201, USA

Sampath Gollapudi, PhD
Postdoctoral Associate
Department of Integrated Physiology and Neuroscience
Washington State University
Pullman, WA 99164, USA

Xu-Pei Huang, MD, PhD
Professor
Department of Biomedical Sciences
Charles E. Schmidt College of Medicine
Florida Atlantic University
Boca Raton, FL 33431, USA

Pierre-Yves Jean-Charles, PhD
Research Associate
Department of Biomedical Sciences
Charles E. Schmidt College of Medicine
Florida Atlantic University
Boca Raton, FL 33431, USA

J. -P. Jin, MD, PhD
Professor and Chairman
Department of Physiology
Wayne State University School of Medicine
Detroit, MI 48201, USA

Changlong Nan, PhD
Research Assistant Professor
Department of Biomedical Sciences
Charles E. Schmidt College of Medicine
Florida Atlantic University
Boca Raton, FL 33431, USA

Peter J. Reiser, Ph. D.
Professor and Interim Chair
Division of Oral Biology
Ohio State University College of Dentistry
Columbus, OH 43210, USA

Daniel C. Rieck, MSc
Research Assistant
Gene and Linda Voiland School of Chemical Engineering and Bioengineering, Washington State University, Pullman,
WA 99164, USA

Asok K. Sen, PhD
Professor
Richard G. Lugar Center for Renewable Energy and Department of Mathematical Sciences
Indiana University
Indianapolis IN 46202, USA

R. John Solaro, PhD
University Professor and Head
Department of Physiology and Biophysics and Center for Cardiovascular Research
University of Illinois at Chicago College of Medicine
Chicago, IL 60611, USA

Jie Tian, MD
Professor
Division of Cardiology
Children's Hospital
Chongqing Medical University
Chongqing, China

Margaret Westfall, PhD
Associate Professor
Department of Cardiac Surgery
University of Michigan
Ann Arbor, MI 48109, USA

Lei Zhang, MD
Research Associate
Department of Biomedical Sciences
Charles E. Schmidt College of Medicine
Florida Atlantic University
Boca Raton, FL 33431, USA

In: Troponin
Editor: J.-P. Jin

ISBN: 978-1-62948-320-7

Chapter I

A History of Troponin: An Adventure in the Complexities of Calcium-Dependent Regulation

R. John Solaro, PhD *

Department of Physiology and Biophysics and Center for Cardiovascular Research, University of Illinois at Chicago College of Medicine, Chicago, IL

Introduction: Tipping Points in Troponin Research

Troponin and I share a chronological history. My graduate education began in 1966 and troponin was named in 1965. Thus the emergence of understanding of the role of troponin in striated muscle coincided with my own emergence as a scientist focused on troponin under the guidance of my mentor F. Norman Briggs, a pioneer in troponin research. Over the past 45 years or so, I and our lab continue to study troponin. The rationale for this continued interest and excitement about troponin is laid out in the history presented here. As the reader will see, there has been a continuing evolution and expansion of interest in the significance of troponin as a major protein in the control of cardiac and skeletal muscle physiology, as a cause and marker of striated muscle disorders, and as a target for therapies in disorders of both cardiac and skeletal muscle. PubMed searches indicate that in the period 1965-70 there were about 6-7 publications per year with the word "troponin" in the title. In 1980 this number had risen to 43 and by 2000 to 237. Last year (2012) 471 publications had troponin in the title.

I've categorized my take on the milestones in the evolution of troponin research after the naming of troponin in 1965 as "tipping points", but as with all scientific endeavors, the

* Address for Correspondence: R. John Solaro, PhD, Department of Physiology and Biophysics M/C 901, College of Medicine, University of Illinois at Chicago, 835 South Wolcott Avenue, Chicago, IL 60612, 312-996-7620 Fax 312-996-1414, solarorj@uic.edu

discovery and naming of troponin relied heavily on previous work. Reductionist approaches, which determined the nature of the parts of the contractile machine and put the parts back together, while following a read-out reflecting the chemo-mechanical process, formed the beginnings of our understanding of mechanisms in striated muscle contraction and relaxation. Dorothy Needham's *Machina Carnis* published in 1971 provides a comprehensive monograph on muscle history since the 1600s. In the early 1960s, when Ebashi and colleagues were focused on identification of the biochemical mechanism for activation of the actin-myosin interaction by Ca^{2+}, Lyubimova and Engelhart had identified myosin as an ATPase, and actin and myosin were characterized to some extent (see review by Straub in 1950[1]). Moreover, Ebashi's lab had in hand various "tropomyosin" preparations prepared according to Bailey's procedures [2, 3]. They also had in hand read-outs of function including super-precipitation or synaeresis of the actin-myosin proteins in suspension, ATP hydrolysis, and contraction of glycerinated fibers [4]. Moreover, not only was there considerable progress on a role for Ca^{2+} in controlling these reactions, but also Ebashi and colleagues avoided the dead-end of one theory, which proposed that relaxation of muscle occurred through the activity of a "soluble relaxing factor." The idea of a soluble relaxing factor had been promoted by H. H. Weber and later by Fuchs and Briggs [5], who hypothesized that the "grains" (microsomal SR fraction or fragmented SR) identified earlier by Marsh [6] released a factor that induced relaxation. They emphasized in their theory that relaxation is established by the *presence* not the *absence* of a factor, as had been proposed by the work of others in which relaxation was due to the absence of a factor, namely Ca^{2+}.

Evidence for the idea that absence of Ca^{2+} is key to relaxation accumulated in the early and mid-1950s and came to a strong conclusion with a publication By Annemarie Weber in 1959 [7]. Grey and Perry [8] had demonstrated that small amounts of Ca^{2+} were able to activate ATPase activity of myofibrillar proteins. However, they investigated the reversal of a relaxed state induced by high concentrations of ATP, and did not come to the straight-forward conclusion that Ca-binding was involved in the process. Following the discovery of a relaxing factor in muscle by Marsh and by Bendall [6, 9], Hasselbach and H. H. Weber [10] as well as Bendall [9] demonstrated that the Marsh-Bendall factor was able to relax glycerinated fibers. Soon after, Bozler [11] published a seminal paper in which he reported that EDTA could relax fiber bundles induced to contract in the presence of ATP and identified a significant role for Mg addition, thus revealing the importance of MgATP as the myosin substrate inducing relaxation in the absence of Ca^{2+}. He indicated that the "EDTA produces its action by firmly combining with bound Ca" and that EDTA mimicked "all known effects of the relaxation factor...". In a remarkably prescient paper published in 1959, Annemarie Weber [7] speculated that "It might be worthwhile to investigate whether the Marsh-Bendall factor, which is particulate and presumably does not interact directly with the myofibrils, acts by binding Ca^{++}." This speculation was based on a clever set of experiments measuring ATP hydrolysis by myofibrils prepared as originally described by S. V. Perry [12], and use of computations employing equations calculating multi-equilibria of ionic and nucleotide species based on published binding constants. This careful control of Ca^{2+} employing rigorous evaluation of binding constants and *in vitro* conditions continues to be the norm demanded in all the role of Ca^{2+} in all studies of activation and relaxation of contraction.

Interestingly, it was Ebashi working with Lipmann, who reported in 1962[13] that the resealed vesicles of the endoplasmic reticulum were able to accumulate and establish large gradients of Ca^{2+} by an ATP dependent process. They speculated that this ability to accumulate Ca^{2+} indicated the presence of "an energy-requiring system for the transport of Ca^{2+} ions into the tubules and vesicles." As Makoto Endo [14] has written in his vivid description of this period, Ebashi was evidently frustrated with Lipmann's skepticism that something "as simple as Ca^{2+}", could control contraction and relaxation. This resulted in the delay of the publication of the paper. Meanwhile, Ebashi skyrocketed to an appointment as full professor at the University of Tokyo bypassing lower ranks, returned to Japan, and resumed pursuing his passion for determining the Ca^{2+} receptor in the myofilaments. Thus, began an amazing journey in elucidation of the critical reactions switching striated muscle on and off and how adaptive and maladaptive modifications of these reactions are key elements in control of cardiac dynamics in health and disease.

Tipping points I discuss in this journey are the following: 1) The naming, general properties and the elucidation of the hetero-trimeric nature of troponin; 2) Signaling to troponin with changes in mechanical, chemical, and neuro-humoral state , 3) Elucidation of the structure of troponin 4) Troponin as a marker of cardiac disorders and as a drug target, 5) Regulation of troponin activity as an integral mechanism in networks controlling cardiac growth, electro-chemical coupling and metabolism.

The Naming, General Properties and the Elucidation of the Hetero-Trimeric Nature of Troponin

By 1965, Ebashi and colleagues [15] made a seminal conclusion from their studies of control of the activity of *in vitro* systems in which protein fractions identified at that time were mixed to determine their functional role with a read-out of super-precipitation. The mixtures contained myosin, actin, a "tropomyosin" preparation, which was impure and able to induce Ca-responsiveness (unlike Bailey's pure tropomyosin) in the reconstituted system. Ebashi designated the tropomyosin preparation as "native tropomyosin", and set out to determine what other protein component it contained. He named the component, troponin and stated in the paper, "..., the interaction between native tropomyosin and F-actin has a fundamental importance in the function of the thin filament. Troponin is considered to be the key protein in this crucial interaction." As is usually the case, there were skeptics questioning the validity of this assertion. However, studies by Fuchs and Briggs [16] (by now converts to the theory that Ca^{2+} removal from myofilaments induces relaxation) reported a strong correlation between the affinity of troponin for Ca and the Km for activation of the reaction between thin filaments and myosin. These data strongly indicated that troponin is indeed the site controlling activation of contraction.

In July 9-12, 1973, Drabikowski organized a symposium in Jablonna outside Warsaw Poland. The meeting was called "Calcium Binding Proteins" and became famous for the announcement by R. H. Kretsinger of the amino acids forming the "EF" coordination of Ca^{2+} ions in carp muscle, which is the canonical Ca-binding structure in troponin and many other Ca-binding proteins. The proceedings are published in a monograph, *Calcium Binding* Proteins, edited by Drabikowski. Professor Setsuro Ebashi also attended this meeting as did many of the luminaries in field. As a young scientist in awe of the attendees, I asked Ebashi to autograph his abstract of the meeting. He gladly obliged and Figure 1 shows the page and autograph. The abstract predicted the now the well-documented role of Ca-binding to troponin C to alter the troponin I-actin interaction.

Setsuro Ebashi

CA-DEPENDENT INTERACTION OF TROPONIN COMPONENTS AS THE BASIS OF THE CONTROL MECHANISM BY Ca ION

S. Ebashi, S. Ohnishi, S. Abe and K. Maruyama

Dept. of Pharmacol., Facul. of Med., Univ. of Tokyo, Tokyo and Dept. of Biophys., Facul. of Science, Kyoto Univ., Kyoto

It is now widely accepted that troponin exerts its inhibitory action on F-actin through tropomyosin in the absence of Ca ion, and Ca ion removes this inhibition (Ebashi et al., 1968). The chief aim of this investigation is to clarify the interaction among troponin components which underlies the above function of troponin.

Troponin A (A), the Ca-binding factor, was treated by a maleimide label. The spectra of labeled factor (A*) consisted of two components, one (L) moderately immobilized and the other (S) weakly immobilized. Ca ion increased L slightly at the expense of S. This tendency was markedly enhanced by troponin I (I), tropomyosin-aggregating factor. Troponin II (II), the inhibitory factor, showed only a small effect on the spectra of A*. Reconstituted troponin complex showed essentially the same pattern as A* + I (the following table shows an example as to the outermost splitting of ESR spectra of various combinations).

	low Ca	high Ca
A*	57.5^{G}	57.9^{G}
A* + I	60.4	63.7
A* + II	59.5	60.0
A* + I + II	61.4	64.2

These observations are compatible with the suggestion (Ebashi et al., 1972) that Ca ion exerts its function mainly by modifying the A-I interaction and that this modification in turn results in the alteraction of the activity of II. It is conceivable, however, that Ca ion also directly affects the A-II interaction and prepares such a conformational state under which Ca-dependent A-I interaction becomes effective as the control mechanism.

Figure 1. Abstract from the 1973 Meeting "Calcium Binding Proteins" organized by Witold Drabikowski in Jablonna, Poland and signed by Professor Setsuro Ebashi.

By this 1973 meeting in Poland, important steps forward in the understanding of how troponin works had been achieved with studies in the laboratory of David Hartshorne. These studies demonstrated that troponin is a multi-protein complex containing components with different regulatory functions. In 1968 Hartshorne and Mueller [17] reported that, when fractionated, troponin resolved into two fractions distinct in primary structure and in function. One fraction, which they called troponin B induced inhibition and another fraction, troponin A, induced Ca-responsiveness to a reconstituted system. Moreover, troponins A and B could be mixed to restore the function of the parent troponin. Schaub and Perry in 1969 [18] confirmed these results and named troponin B "inhibitory factor" and troponin A calcium-sensitizing factor. By 1971, the multimeric components of skeletal troponin were sorted out in a report by Greaser and Gergely [19]. Although they separated troponin into 4 components as analyzed by the recently developed SDS-PAGE technique [20, 21], they showed that 3 of the components could reconstitute full Ca-regulatory activity in their preparations. The names they chose- troponin C for Ca-binding component, troponin I for inhibitory component, and troponin T for tropomyosin binding component- have stuck and have been indispensable ever since in discourse regarding troponin. Potter and Gergely [22] employed the purified and characterized troponin complex to fully characterize the Ca-binding sites and came to the important conclusions that there are 6 moles of divalent metal binding sites on fsTn, two which are "regulatory", and bind mainly Ca^{2+} with relatively low affinity, and two, which are "structural", of higher affinity and two with ability to also bind Mg^{2+}. Another important conclusion was that the affinity of pure fsTnC regulatory sites for Ca^{2+} was relatively low compared to regulatory sites when fsTnC was combined with in the fast skeletal TnC-TnI or whole fast skeletal troponin complex. These data indicated that when TnI interacts with TnC there is a conformational transition. Potter and Gergely [23] formulated the early models for the interactions among troponin, actin and tropomyosin. Earlier concepts leading to these models are detailed a review by Mannherz and Goody in 1976 [24].

It took another 6-8 years to characterize these components in heart. The first studies by Tsukui and Ebashi [25] produced preparations with the same general components of TnI, TnC, and TnT, but the reconstituted troponin did not regulate very well with addition and removal of Ca^{2+}. A few years later Brekke and Greaser [26] reported full reconstitution of regulatory activity, but the component identified as TnI did not inhibit unless combined with the component identified as TnT. These problems with isolation and reconstitution were likely due to the high amounts of proteases and mitochondrial proteins in heart and the general difficulty of isolating proteins from heart versus fast skeletal muscle. Stull and Buss [27] in 1977 moved this field forward by the use of starting material in the form of myofibrillar fractions extracted with aid of Triton X-100, a method developed by Solaro et al. [28]. The method of Stull and Buss served other studies well in the eventual further detailed identification of Ca-binding properties of cardiac Tn [29] and identification of physiologically significant post-translational modifications [30] discussed below. Thorough determination of the Ca-binding properties of cardiac troponin [29, 30] confirmed the prediction from the structural studies of Van Eerd and Takahashi [31] that cardiac TnC contains a single regulatory site. Differences in Ca-affinity between cardiac TnC alone and in the cTn complex, which also occurs in the cardiac variant, made it important that the cTn preparations contain no free cardiac TnC. Attention to this detail as outlined in the paper ensured accurate determination of the binding constants. Determination of the rates of Ca-exchange with the low affinity site of cardiac TnC alone and in complex with troponin provided definitive evidence that the regulatory site of cardiac TnC is able to bind and release Ca^{2+} over the time course of the contraction relaxation cycle of the heart [30]. The determination of these rates was made possible by methods developed by J. David Johnson [30], who first applied fluorescence labeling of cardiac TnC cysteines in order to provide a reporter of Ca-binding only to the regulatory site. Unfortunately, our colleague David Johnson died suddenly at a young age, at a time when his work was in full bloom. Fluorescence detection of Ca-binding to TnC has been refined in a number of laboratories with work by Jonathan Davis and colleagues [32] as one example.

Signaling to Troponin with Changes in Mechanical, Chemical and Neuro-Humoral State

An exciting advance in our understanding of the role of troponin in contraction and relaxation of muscle occurred with the earliest identifications that troponin is a substrate for kinases and phosphatases [33, 34]. This was particularly exciting in that phosphorylation had been emphasized as a regulator of enzyme activity such as phosphorylase kinase. A role for phosphorylation/dephosphorylation in function of a structural protein such as troponin I was not fully appreciated. Studies in the Stull laboratory [34, 35] were among the first to recognize the potential physiological significance of this mechanism. In their study [34] of phosphorylase kinase dependent phosphorylation of TnI they stated that their results "...suggest that troponin may be interconverted to phosph-dephospho forms in response to physiological stimuli." England reported the first evidence of a correlation between isoproterenol induced increases in contraction of perfused rat hearts and cardiac TnI phosphorylation. The linear correlation was later modified in further studies [36] with other inotropic agents and with dephosphorylation. In 1976 Solaro et al. [37] employed ^{32}P perfused rabbit hearts to demonstrate an increase in cardiac TnI phosphorylation induced by adrenalin. Importantly phospho-peptide analysis indicated that 90-95% of the phosphate incorporation occurred in an N-terminal region unique to the cardiac isoform of TnI. Moreover, the paper reported preliminary data that with increases in cardiac TnI phosphorylation there was a desensitization of regulated acto-myosin to Ca^{2+}. Thus, rather than promoting force generation during adrenergic stimulation this result suggested that cardiac TnI phosphorylation would depress force generation. As predicted in the paper, later work indicated that cardiac TnI phosphorylation at the unique N-terminus plays a role in the "relaxant effect" of adrenergic stimulation a control mechanism important in tuning the contraction/relaxation cycle at the higher frequencies and reduced cardiac cycle time. From these early studies on phosphorylation of cardiac TnI there was an emergence of extensive confirmation of the critical importance of troponin phosphorylation of both cardiac TnI and cardiac TnT. There are a multitude of sites phosphorylated and a variety of kinases and phosphatases controlling these sites in long and short term control of cardiac function. Extensive reviews track the astounding evolution of this field [38-42]. Multiple reaction monitoring has been applied by the Van Eyk laboratory [43, 44] to sort out most of the 14 sites of phosphorylation on cardiac TnI and their potential role in heart failure. A take home lesson is that homeostatic control of cardiac function relies on a balance of cardiac TnI phosphorylations that either promotes or depresses function; with a disturbance in this balance cardiac disorders may occur. Current understanding of post-translational modifications of cardiac TnI other than phosphorylation is in its early stages and offers new and significant challenges. For example, recent studies [45] have identified S-glutathionylation of fast skeletal TnI, most likely at its exposed Cys 133, as a modification affecting myofilament Ca-sensitivity. Post-translational modifications of troponin in the form of proteolytic cleavage are also of interest. Although emphasis has been placed on the pathological implications of cleavage of troponins in myocardial infarction and hypoxia, Jin and colleagues have hypothesized that some cleavages may be physiologically adaptive. One area of their research has been on the restricted deletion of the N-terminal variable region of cardiac troponin T, which demonstrates a novel mechanism by which thin filament regulation adapts to sustain cardiac function under stress conditions [46, 47]. A second related area is the finding that simulated micro-gravity stress induced a restricted deletion of the unique N-terminus of cardiac TnI that is an apparent beneficial response [48]. In the case of troponin proteolysis related to cardiac disorders that appear maladaptive, Westfall and Solaro [49] identified breakdown of cardiac TnI and cardiac TnT in response to global ischemia and indicated that this may be functionally important in depression of function. Gao et. al. [50] reported more definitive evidence on the functional significance of specific cardiac TnI cleavage in a model of cardiac stunning demonstrating a decrease in maximum tension of muscle. Soon after Van Eyk et al. [51] identified a cleavage point at the C-terminus of troponin I. In 2000 Murphy et al. reported stunning in hearts of transgenic mice expressing a truncated cardiac TnI(1-193) missing 18 C-terminal residues. The same paper provided evidence of this cleavage product in human hearts. More studies are required to fully understand the adaptive and maladaptive effects and the mechanisms of proteolysis in relation to modern investigation of the ubiquitin proteasome and autophagy processes.

An early indication that mechanical state might control myofilament sensitivity to Ca and thus involve troponin came from a variety of studies in both skeletal and cardiac muscle. In 1978 Endo [52] reported that skeletal fibers with their sarcolemma removed mechanically showed an increased sensitivity to Ca^{2+} at relatively long versus short sarcomere lengths. At the same time in elegant experiments carried out by Fabiato and Fabiato [53] on 2-7 μm myofibrillar bundles of single cardiac myocytes showed an increase in Ca-sensitivity as a function of increases in sarcomere length. Experiments of Allen and Kurihara [54] supported the idea that length dependent activation is relevant *in situ*. They reported that the imposition of length changes on papillary muscle preparations induced an immediate change in tension, but no change in the peak amplitude of the Ca^{2+} transient as determined with aequorin. A key step forward in this field of research occurred with the publications of Hibberd and Jewell [55, 56], who showed in a high fidelity set of experiments that the length tension relation of cardiac muscle, which occurs at sub-maximal Ca-activation, depends on a length dependence of activation in which the force-pCa relation determined at relatively short sarcomere lengths sits to the right of the force-pCa relation determined at relative long sarcomere lengths. Length dependent activation provided a plausible mechanism for the relatively steep rising phase of the length tension relation in the heart versus skeletal muscle. These observations were recognized as physiologically significant with regard to the Frank-Starling relation and relatively steep relation between length and tension in cardiac muscle compared to fast skeletal muscle.

The role of troponin in length dependent activation was firmly supported by studies demonstrating that switching of troponin isoforms in the heart [57] as well as cardiac troponin phosphorylation [58] is able to modify length dependent activation. The neonatal isoform of troponin I in the heart is the slow skeletal isoform. Expression of slow skeletal TnI in the adult mouse heart induces a significant reduction in the sarcomere length dependence of Ca-sensitivity [59]. More recent experiments have identified cardiac TnI-Thr144, which is localized in the inhibitory peptide and is a Pro in the homologous position in slow skeletal TnI, as a residue significant in the difference between cardiac TnI and slow skeletal TnI with regard to length dependent activation [60]. Experiments with modifications in troponin phosphorylation also indicated a role for troponin in length dependent activation [58]. The length dependence of activation is generally regarded as the basis for steepness of the relation between end systolic volume (ESV) and end systolic pressure (ESP) in the pressure-volume loop analysis of cardiac contractility.

To test this hypothesis, Nowak et al. [61] performed in situ pressure-volume loop analysis in mouse hearts expressing cardiac TnI or slow skeletal TnI with and without adrenergic stimulation with isoproterenol. Slow skeletal TnI is not phosphorylated by cAMP dependent protein kinase, and thus the expectation was that the differential effects of slow skeletal TnI and cardiac TnI together with altered phosphorylation of the cardiac TnI would alter the slope of the ESP-ESV relation. Indeed this was the case and placed mechanisms involving a length sensitivity of cardiac TnI molecular mechanisms as prominent in the Frank-Starling mechanism in control of cardiac output.

Evidence that strong cross-bridge binding to the thin filament affects troponin Ca-binding provides further support for the relation between mechanics and troponin. The numbers of strong cross-bridges reacting with the thin filament may be modified by length, and this forms the basis for one theory of the mechanism of length dependent activation. Mechanical state may also affect troponin Ca-binding via effects of shortening on the number of cross-bridges reacting with the thin filament [62, 63].

A shortening induced deactivation of the myofilaments forms parts of theoretical considerations of control of cardiac dynamics. Although the concept of detailed balance demands that the energies of interaction between troponin and the actin-cross bridge reaction go both ways, the role of force-generating cross bridges as determinant of Ca-binding to troponin remains somewhat controversial. This is discussed in an excellent review by Sun and Irving [64]. The essence of the idea is that cooperative interactions along the thin filament may be more dominant than cooperative Ca-binding induced by strong cross-bridges. Sun et al. [65] show data and argue that only in pathological conditions where rigor complexes may form do strong cross-bridges promote thin filament activation. Indeed the demonstration of effects of strong cross-bridges on thin filament activation has relied almost exclusively on rigor cross-bridges or NEM-modified strong binding, but non-cycling cross-bridges [66].

Studies have been carried out on the effect of force and rigor complexes on Ca-binding to TnC in fast skeletal and cardiac skinned fiber preparations. In studies with glycerol extracted skinned fibers, Fuchs and collaborators [67] and Marston and Tregear [68], pioneered a double isotope labeling technique for the determination of Ca-binding to fsTnC in the myofilament lattice generating force. In general the approaches are technically challenging and the results did not reproduce the titrations that were obtained with soluble Tn complexes. Interference by Ca-binding to myosin light chains [69] may have contributed to the difficulties in addition to the possibility of slowly exchanging sites that needed to be completely exchanged with the ^{45}Ca use to probe bound Ca^{2+}. In studies with skinned fiber bundles from heart muscle, we [70, 71] found a requirement for very long incubations of the fibers for full exchange of Ca-binding sites with ^{45}Ca especially in cardiac TnC even in slow skeletal fibers. With this thorough exchange as well as taking into account the myosin bound Ca^{2+}, we [71] were able to fit the Ca-binding data to a model roughly equivalent to binding data previously obtained with soluble cTn [29].

The data showed only a moderate cooperativity compared to the steepness of the force-pCa relation indicating that force generating cross-bridges may not affect Ca-binding as greatly as theorized. On the other hand, strong binding cross-bridges in the form of rigor complexes promoted an increase affinity of the regulatory sites for Ca^{2+}.

Reactions of troponin I with its neighbors in the thin filament complex are also significantly involved in the effects of acidosis on response of the myofilaments to Ca^{2+}. Clues to the understanding of this important role of troponin came with the studies of Fabiato and Fabiato [72] and Donaldson and Hermanson [73] in 1978. Both studies reported differential effects of lowering pH in skeletal and cardiac myofilaments.

Donaldson and Hermanson [73] reported that there is a direct effect of pH on mechanically skinned fibers, and that the effect on Ca-sensitivity was greatest in heart, less in fast skeletal, and least in slow skeletal muscle fibers. Fabiato and Fabiato [72] also reported a bigger effect of a drop from pH 7.4 to pH 6.2 on the Ca-sensitivity of cardiac skinned fibers compared to skeletal fibers. In our studies of the effect of pH on cardiac and fast skeletal myofilaments, we found that a decrease in pH reduced Ca-binding to cardiac TnC in myofibrillar preparations [74, 75]. This was in contrast to some earlier studies [76, 77], who had reported no change in Ca-binding. The lack of effect of acidic pH appeared to be due to studies of fibers in rigor or problems associated with dissociation of troponin complex or application of EGTA buffers. Although there had been a concern that EGTA, which was employed as a Ca^{2+} buffer in these studies may not be inert and induce artifacts [77], by use of resins to remove contaminant Ca^{2+}, we demonstrated the effect of acidic pH to desensitize cardiac myofibrils in the complete absence of EGTA [74].

The connection of the effects of acidic pH to troponin I began with our finding [78] that neonatal cardiac myofibrils demonstrated a much reduced effect of acidic pH on myofilament response to Ca^{2+} compared to adult heart myofilaments. In comparing the myofilament SDS-PAGE analysis it was clear that cardiac TnI was missing in the neonatal samples, but we did not recognize that the neonates were expressing slow skeletal TnI (slow skeletal TnI). Toyota and Shimada [79] had in 1981 concluded from their immuno-histochemical studies of developing chicken cardiac and skeletal muscles that associated with development there is a synthesis of tissue specific troponin components. Although Humphreys and Cummins [80], for reasons that are not clear, could not detect developmental isoforms of TnI in beef and rat hearts, later studies by Sabry and Dhoot [81] were able to show bands of different mobility identified as different proteins.

They indicated that the mobility of the faster migrating band matched slow skeletal TnI. Critical to the identification of developmental isoforms of both TnI and TnT in the heart was the development of mono-clonal isoform specific antibodies by a number of laboratories [82-85]. Using such antibodies, Saggin et al. [84] positively identified the mobility of slow skeletal TnI as the embryonic, neonatal isoform of TnI in the heart. Molecular cloning and identification of TnI isoforms [86] provided further elucidation of the developmental timing of slow skeletal TnI and cardiac TnI. Importantly, the pattern of TnI expression was also evident in human heart samples [87].

Studies by Cooper and Ordahl [88] revealed developmentally regulated alternate splicing of the TnT gene showing the first evidence that a single cardiac TnT gene is capable of generating multiple isoforms.

The complex expression pattern during cardiac development was eventually worked out by Anderson et al. [82], again with excellent antibodies. These complex patterns of TnI and TnT expression are still not fully understood in terms of functional significance. The importance of interactions between specific isoforms of TnI and TnT in cardiac development has been emphasized in studies of Gomes et al. [89]. For example, they reported that dissociation rates of Ca^{2+} from cardiac TnC in troponin complexes containing slow skeletal TnI dependent on the TnT isoform present.

Elucidation of the Structures of Troponin

In 1985 data reported in two papers [90, 91] revealed the crystal structure of fast skeletal TnC at 2.8 – 3.0 angstrom resolution. I don't think I would be exaggerating to say that for those who had been investigating troponin C for many years seeing the structure for the first time was emotional. I liken it to speaking with someone on the phone for years, and then seeing their face for the first time. Especially impressive was the dumbbell shape with N- and C-terminal lobes connected by a 9 turn alpha-helical stalk. An exciting point of progress in our understanding of troponin came first in 1998, with 2.3 angstrom resolution of the crystal structure of fast skeletal TnC with TnI [92] and then in 2003 with the publication of a seminal paper by Takeda and colleagues [93] revealing the crystal structures of 46 and 52 kDal core domains of cTn in the Ca^{2+} saturated form. The 1998 paper was also historically significant in that it showed that the elongated dumbbell shape of TnC demonstrated in isolation existed as a compact structure in the complex with interaction between the N- and C-terminal lobes. As Vassylyev et al. [92] pointed out, this finding indicated that Ca-receptor proteins may behave differently in complex with near-neighbor proteins. Resolution of the core crystal structure of cTn indicated distinct structural domains with flexible linkers (as the authors suggested) "making the entire molecule highly flexible." The structure also revealed interactions between cardiac TnI and cardiac TnT in the form of a rigid alpha-helical coiled-coil, termed the I-T arm. A model presented in the paper implied that the structure supported the idea that Ca-binding to the regulatory site on cardiac TnC released troponin and tropomyosin from a relatively rigid to more flexible dispositions on the thin filament. Kowelessur and Tobacman [94, 95] have pioneered the application of hydrogen/deuterium exchange studies with troponin in demonstrating the regional flexibility of the cardiac complex and its significance in mutations linked to inherited cardiomyopathies.

Subsequent reports [96] of the crystal structure of fast skeletal troponin complex not only demonstrated Ca^{2+} induced changes in structure, but also identified differences between cTn and fsTn. As expected from previous solution NMR studies [97, 98] and the presence of a defunct Ca-binding domain in cardiac TnC, fsTnC in the complex showed a more open conformation.

The central helix of fsTnC was also shown to be extended in comparison to cardiac TnC. Moreover, unlike the cardiac TnI Ip, which, owing to its flexibility was not visible in the Takeda structure, the fsTnI Ip was resolved. These data revealed an interaction of the inhibitory peptide of fsTnI with the central helix of fsTnC. With this Ca-dependent interaction, the inhibitory peptide undergoes a structural change from a short alpha-helix to an extended loop. On the other hand the equivalent of the I-T arm in the fsTn structure did not change conformation with Ca-binding to fsTnC. Apart from this relevant and important structural information, Vinogradova et al. [96] reported that the detergent Anapoe, when intercalated into the hydrophobic pocket of fsTnC together with the switch peptide induced an increase in the force of contraction of skinned fibers. This is an important observation with regard to the search for small molecules interacting with troponin as considered below. Together with earlier data showing a fundamental difference between cTn and fsTn in the mechanisms for the opening of the hydrophobic pocket, and its interaction with the switch peptide [97, 98], the data indicated the possibility of specific targeting of therapeutic agents.

Troponin As a Marker of Cardiac Disorders and As a Drug Target

A study by Toya-Oka and Ross in 1981 [99] provided early evidence that troponin is released from myofilaments and may be involved in the altered function associated with myocardial infarction. Associated with coronary occlusion in dog hearts was a significant fall in the myofilament content of both cardiac TnC and cardiac TnI and a decrease in Ca-sensitivity. With the discovery that cardiac TnI is a variant unique to the heart, and never expressed in other tissues, to the best of my knowledge Peter Cummins, a biochemist in S. V. Perry's department, at the University of Birmingham in England developed the idea that cardiac TnI in serum should only be there when released from cardiac myocytes. Cummins et al. [100] were the first to carefully evaluate whether cardiac TnI could be useful as a marker of cellular necrosis in myocardial infarction. At the end of the abstract to their paper, they state "Immunoassay of cardiac troponin-I could be a valuable diagnostic aid in the cardiac-specific detection of cell necrosis." Thus began an extensive growth industry in the use of cardiac TnI and cardiac TnT as bio-markers in the diagnosis and prognosis of cardiac damage in myocardial infarction and other conditions such as cardio-renal syndrome etc. A PubMed search "troponin, biomarker" yielded more than 4600 hits. Work in this area continues to evaluate the power of troponin as a biomarker compared to emerging biomarkers. It seems certain than many individuals entering the emergency room made it to timely and successful cardiac catheterization and angioplasty or surgery because of the detection of high serum troponin levels, as a biomarker.

In an era when translational medicine is in vogue, it is gratifying to know that in the history of troponin, a therapeutic application of the acquired knowledge has been applied for over 30 years. To the best of my knowledge the idea that troponin may be a target for inotropic agents began with an early paper published by Herzig and Ruegg [101]. I had the honor of working with them in 1979, when we investigated the effects of vanadate on the actin-myosin reaction [102]. At our breaks at the "kaffe tisch", we had extensive discussions of how realistic it would be to sensitize the myofilaments to Ca^{2+} with small molecules. Eventually Caspar Ruegg came to my lab, and we published a paper [103] indicating strongly that the small molecule ARL-115BS is able to promote submaximal force and increase Ca-bound to cardiac TnC in cardiac myofilaments. ARL-115 BS is a precursor of Pimobendan (UDCG-115 BS). These small molecules were able to induce a leftward shift in the relation between Ca^{2+} isometric force generation and in contrast to troponin I, which S. V. Perry called a Ca-desensitizer, the agents became known as "Ca-sensitizers." As summarized in reviews of the research on inotropic agents acting via direct effects on sarcomeric proteins [104, 105], many companies screened for small molecules with this property. Two compounds made it into clinical trials and clinical use. One is Pimobendan (UDCG 115BS), known as Acardi™ in Japan and the other is Levosimendan also known as Vardax™. Pimobendan demonstrated Ca-sensitization by a mechanism involving enhanced Ca-binding to troponin C. Stereo-selectivity was demonstrated providing early evidence for the feasibility of an intracellular pharmacology and of something that appeared look like a receptor in the troponin. The molecular site of Pimobendan binding was never identified. Boehringer Nippon brought Acardi to market, but it appears to have limited clinical use in humans in Japan. Reports [106] do appear in which it is employed in particularly difficult cases. On the other hand, Pimobendan under the name Vetmedin is widely and successfully used in veterinary medicine for dogs and cats with various familial and acquired cardiomyopathies [107, 108]. Levosimendan was developed by Orion Pharmos in Finland from results of screening of compounds for interaction with cardiac TnC [109, 110]. It is apparent the Levosimendan binds to the hydrophobic pocket induced by Ca-binding to cardiac TnC and prolongs the life-time of the on state. An interesting aspect of this pharmacology is that Levosimendan may undock to cardiac TnC when Ca^{2+} dissociates from cardiac TnC. Both of these Ca-sensitizers also exhibit PDE III inhibition, a property generally considered as negative as elevations in cAMP may increase heart rate and be arrhythmogenic. However, whether the combined properties of Levosimendan as a so-called "ino-dilator" may be beneficial in the context of an appropriate therapeutic window remains unclear [111]. We demonstrated that while Ca-sensitizer drugs were also active as PDE inhibitors; well known PDE III inhibitors such as Milirone were not Ca-sensitizers. Levosimendan has been demonstrated to be of clinical use in alleviating depressed cardiac function following percutaneous coronary interventions [112] and as an inotropic agent improving mortality in cardiac [113] and, where indicated non-cardiac surgery [114]. There is also evidence that Levosimendan may be of use in restoring the depressed function of the diaphragm following prolonged ventilation with a respirator and chronic obstructive pulmonary disease [115].

Despite some setbacks in the use of sarcomeric activators for heart failure, research into the potential application of this idea remains active [116]. The most recent and exciting applications are with agents that target the skeletal Tn complex to increase muscle force in compromised neuromuscular function [117], and in nemaline myopathy in patients with mutations in nebulin [118].

Regulation of Troponin Activity As an Integral Mechanism in Networks Controlling Cardiac Growth, Electro-Chemical Coupling and Metabolism

One of the major challenges in troponin research is the full integration of the properties of troponin in cellular control mechanisms. I have recently reviewed this integration with focus on cardiac TnI phosphorylation [119]. The review stresses that the time in which one may consider troponin simply as a Ca-receptor and Ca-transducing complex is past; investigators are at the brink of new and exciting discoveries of novel roles of troponin. Generally, a major finding linking sarcomeric properties to the integrated activity of the heart came in the form of linkage of sarcomeric mutations to common cardiomyopathies [120]. Linkage of familial hypertrophic cardiomyopathy (FHC) to myosin heavy chain mutations was the first to be identified in 1990, but these prevalent mutations account for less than half of the genetic causes. By 1993, the search for additional genes linked to FHC led initially to a locus at chromosome 1q3, the location of troponin I. Soon after in 1994, the discovery of additional mutations in alpha-tropomyosin and TnT led to the designation by Thierfelder et al. [121] that FHC is a disease of the sarcomere. Since that time there has been identification of a large number of mutations in troponin components linked to both FHC and dilated cardiomyopathy [122]. The question is: How can what appears to be a small change, such as point mutation or deletion of an amino acid, lead to a devastating set of circumstances in which the heart grows and remodels inappropriately with maladaptation and with induction of death causing arrhythmias? The implication in the answer to this question is that troponin operates in a network ensuring homeostatic maintenance of appropriate remodeling of the myocardium and electrical stability. Moreover, as has been argued clearly by Watkins and colleagues [123], the evolution from mutation to hypertrophic phenotype involves mismatching of energy consumed by the myofilaments and by Ca-transport processes. Again the suggestion is that troponin is also a key component in homeostatic control of energy consumption as well as intracellular Ca^{2+}. Better understanding of the mechanisms is demanded with regard to the diagnosis and treatment of these familial disorders [124]. Altered myofilament Ca-sensitivity or cross-bridge cycling kinetics may hold the clue to the triggering of remodeling in as much as diastolic abnormalities may occur. Important clues as the regulatory mechanisms include the demonstration that AMP-activated kinase is able to phosphorylate cardiac TnI [125, 126]. Moreover, we [127] reported that persistent sensitization of cardiac myofilaments to Ca^{2+} by expression of slow skeletal TnI in the adult mouse heart is associated with increased AMP-kinase activity and elevated anaerobic ATP production. Finally, the explicit connection between alterations in myofilament response to Ca^{2+} and triggered arrhythmias has been demonstrated by studies in the laboratory of Bjorn Knollman [128-131]. Mutations in cardiac TnT are of particular interest in that there is a risk of sudden cardiac death with little in the way of hypertrophy and fibrosis [132]. Recent work from this lab [131] indicates that increases in myofilament Ca-binding in turn induce pro-arrhythmic triggered mechanisms by promoting pause dependent increases in Ca-release together with prolongation of the action potential.

I think the integration of troponin into the homeostatic control networks will represent new highlights and milestones of troponin research, when the next history is written.

Acknowledgments

I am grateful to the many colleagues, especially Dr. Tomoyoshi Kobayashi and James Potter, and mentors that have guided our own work on troponin. Our lab is supported by NIH Grants: PO1 HL 062426, RO1 HL 022231, and RO1 064035. I apologize for highlights in the history of troponin that I may have missed.

References

[1] Straub FB. Muscle. *Annu. Rev. Biochem.* 1950;19:371-88.

[2] Bailey K. Tropomyosin: a new asymmetric protein component of the muscle fibril. *Biochem. J.* 1948;43:271-9.

[3] Bailey K. Tropomyosin: a new asymmetric protein component of muscle. *Nature.* 1946;157:368.

[4] Biro NA, Szent-Gyorgyi AE. The effect of actin and physico-chemical changes on the myosin ATP-ase system, and on washed muscle. *Hung. Acta. Physiol.* 1949;2:120-33.

[5] Fuchs F, Briggs FN. The nature of the muscle-relaxing factor. II. Some physicochemical properties. *J. Gen. Physiol.* 1963;46:893-904.

[6] Marsh BB. The effects of adenosine triphosphate on the fibre volume of a muscle homogenate. *Biochim. Biophys. Acta.* 1952;9:247-60.

[7] Weber A. On the role of calcium in the activity of adenosine 5'-triphosphate hydrolysis by actomyosin. *J. Biol. Chem.* 1959;234:2764-9.

[8] Grey TC, Perry SV. A study of the effects of substrate concentration and certain relaxing factors on the magnesium-activated myofibrillar adenosine triphosphatase. *Biochem. J.* 1956;64:184-92.

[9] Bendall JR. Further observations on a factor (The `Marsh' factor) effecting relaxation of ATP-shortened muscle-fibre models and the effect of Ca and Mg ions upon it. *J. Physiol.* 1953;121:232-54.

[10] Hasselbach W, Weber HH. [The effect of the Marsh-Bendall factor on the contraction of a fiber model]. *Biochim. Biophys. Acta.* 1953;11:160-1.

[11] Bozler E. Relaxation in extracted muscle fibers. *J. Gen. Physiol.* 1954;38:149-59.

[12] Perry SV. The adenosinetriphosphatase activity of myofibrils isolated from skeletal muscle. *Biochem. J.* 1951;48:257-65.

[13] Ebashi S, Lipmann F. Adenosine Triphosphate-Linked Concentration of Calcium Ions in a Particulate Fraction of Rabbit Muscle. *J. Cell. Biol.* 1962;14:389-400.

[14] Endo M. Professor Ebashi's journey toward the discovery of troponin: a personal recollection. *Adv. Exp. Med. Biol.* 2007;592:7-9.

[15] Ebashi S, Kodama A. A new protein factor promoting aggregation of tropomyosin. *J. Biochem.* 1965;58:107-8.

[16] Fuchs F, Briggs FN. The site of calcium binding in relation to the activation of myofibrillar contraction. *J. Gen. Physiol.* 1968;51:655-76.

[17] Hartshorne DJ, Mueller H. Fractionation of troponin into two distinct proteins. *Biochem. Biophys. Res. Commun.* 1968;31:647-53.

[18] Schaub MC, Perry SV. The relaxing protein system of striated muscle. Resolution of the troponin complex into inhibitory and calcium ion-sensitizing factors and their relationship to tropomyosin. *Biochem. J.* 1969;115:993-1004.

[19] Greaser ML, Gergely J. Reconstitution of troponin activity from three protein components. *J. Biol. Chem.* 1971;246:4226-33.

[20] Shapiro AL, Vinuela E, Maizel JV, Jr. Molecular weight estimation of polypeptide chains by electrophoresis in SDS-polyacrylamide gels. *Biochem. Biophys. Res. Commun.* 1967;28:815-20.

[21] Weber K, Osborn M. The reliability of molecular weight determinations by dodecyl sulfate-polyacrylamide gel electrophoresis. *J. Biol. Chem.* 1969;244:4406-12.

[22] Potter JD, Gergely J. The calcium and magnesium binding sites on troponin and their role in the regulation of myofibrillar adenosine triphosphatase. *J. Biol. Chem.* 1975;250:4628-33.

[23] Potter JD, Gergely J. Troponin, tropomyosin, and actin interactions in the Ca2+ regulation of muscle contraction. *Biochemistry.* 1974;13:2697-703.

[24] Mannherz HG, Goody RS. Proteins of contractile systems. *Annu. Rev. Biochem.* 1976;45:427-65.

[25] Tsukui R, Ebashi S. Cardiac troponin. *J. Biochem.* 1973;73:1119-21.

[26] Brekke CJ, Greaser ML. Separation and characterization of the troponin components from bovine cardiac muscle. *J. Biol. Chem.* 1976;251:866-71.

[27] Stull JT, Buss JE. Phosphorylation of cardiac troponin by cyclic adenosine 3':5'-monophosphate-dependent protein kinase. *J. Biol. Chem.* 1977;252:851-7.

[28] Solaro RJ, Pang DC, Briggs FN. The purification of cardiac myofibrils with Triton X-100. *Biochim. Biophys. Acta.* 1971;245:259-62.

[29] Holroyde MJ, Robertson SP, Johnson JD, Solaro RJ, Potter JD. The calcium and magnesium binding sites on cardiac troponin and their role in the regulation of myofibrillar adenosine triphosphatase. *J. Biol. Chem.* 1980;255:11688-93.

[30] Robertson SP, Johnson JD, Holroyde MJ, Kranias EG, Potter JD, Solaro RJ. The effect of troponin I phosphorylation on the Ca2+-binding properties of the Ca2+-regulatory site of bovine cardiac troponin. *J. Biol. Chem.* 1982;257:260-3.

[31] van Eerd JP, Takahashi K. The amino acid sequence of bovine cardiac tamponin-C. Comparison with rabbit skeletal troponin-C. *Biochem. Biophys. Res. Commun.* 1975;64:122-7.

[32] Davis JP, Norman C, Kobayashi T, Solaro RJ, Swartz DR, Tikunova SB. Effects of thin and thick filament proteins on calcium binding and exchange with cardiac troponin C. *Biophys. J.* 2007;92:3195-206.

[33] Perry SV, Cole HA. Phosphorylation of troponin and the effects of interactions between the components of the complex. *Biochem. J.* 1974;141:733-43.

[34] Stull JT, Brostrom CO, Krebs EG. Phosphorylation of the inhibitor component of troponin by phosphorylase kinase. *J. Biol. Chem.* 1972;247:5272-4.

[35] England PJ, Stull JT, Krebs EG. Dephosphorylation of the inhibitor component of troponin by phosphorylase phosphatase. *J. Biol. Chem.* 1972;247:5275-7.

[36] England PJ. Studies on the phosphorylation of the inhibitory subunit of troponin during modification of contraction in perfused rat heart. *Biochem. J.* 1976;160:295-304.

[37] Solaro RJ, Moir AJ, Perry SV. Phosphorylation of troponin I and the inotropic effect of adrenaline in the perfused rabbit heart. *Nature.* 1976;262:615-7.

[38] Kobayashi T, Solaro RJ. Calcium, thin filaments, and the integrative biology of cardiac contractility. *Annu. Rev. Physiol.* 2005;67:39-67.

[39] Metzger JM, Westfall MV. Covalent and noncovalent modification of thin filament action: the essential role of troponin in cardiac muscle regulation. *Circ. Res.* 2004;94:146-58.

[40] Solaro RJ. Multiplex kinase signaling modifies cardiac function at the level of sarcomeric proteins. *J. Biol. Chem.* 2008;283:26829-33.

[41] Solaro RJ. Sarcomere control mechanisms and the dynamics of the cardiac cycle. *J. Biomed. Biotechnol.* 2010;2010:105648.

[42] Solaro RJ, van der Velden J. Why does troponin I have so many phosphorylation sites? Fact and fancy. *J. Mol. Cell. Cardiol.* 2010;48:810-6.

[43] Zhang P, Kirk JA, Ji W, dos Remedios CG, Kass DA, Van Eyk JE, et al. Multiple reaction monitoring to identify site-specific troponin I phosphorylated residues in the failing human heart. *Circulation.* 2012;126:1828-37.

[44] Kirk JA, Zhang P, Murphy AM, Van Eyk JE. Troponin I alterations detected by multiple-reaction monitoring: how might this impact the study of heart failure? *Expert Rev. Proteomics.* 2013;10:5-8.

[45] Mollica JP, Dutka TL, Merry TL, Lamboley CR, McConell GK, McKenna MJ, et al. S-glutathionylation of troponin I (fast) increases contractile apparatus Ca2+ sensitivity in fast-twitch muscle fibres of rats and humans. *J. Physiol.* 2012;590:1443-63.
[46] Feng HZ, Biesiadecki BJ, Yu ZB, Hossain MM, Jin JP. Restricted N-terminal truncation of cardiac troponin T: a novel mechanism for functional adaptation to energetic crisis. *J. Physiol.* 2008;586:3537-50.
[47] Zhang Z, Biesiadecki BJ, Jin JP. Selective deletion of the NH2-terminal variable region of cardiac troponin T in ischemia reperfusion by myofibril-associated mu-calpain cleavage. *Biochemistry.* 2006;45:11681-94.
[48] Yu ZB, Zhang LF, Jin JP. A proteolytic NH2-terminal truncation of cardiac troponin I that is up-regulated in simulated microgravity. *J. Biol. Chem.* 2001;276:15753-60.
[49] Westfall MV, Solaro RJ. Alterations in myofibrillar function and protein profiles after complete global ischemia in rat hearts. *Circ. Res.* 1992;70:302-13.
[50] Gao WD, Atar D, Liu Y, Perez NG, Murphy AM, Marban E. Role of troponin I proteolysis in the pathogenesis of stunned myocardium. *Circ. Res.* 1997;80:393-9.
[51] Van Eyk JE, Powers F, Law W, Larue C, Hodges RS, Solaro RJ. Breakdown and release of myofilament proteins during ischemia and ischemia/reperfusion in rat hearts: identification of degradation products and effects on the pCa-force relation. *Circ. Res.* 1998;82:261-71.
[52] Endo M. Length dependence of activation of skinned muscle fibers by calcium. *Cold Spring Harb. Symp. Quant. Biol.* 1973;37:505-10.
[53] Fabiato A, Fabiato F. Myofilament-generated tension oscillations during partial calcium activation and activation dependence of the sarcomere length-tension relation of skinned cardiac cells. *J. Gen. Physiol.* 1978;72:667-99.
[54] Allen DG, Kurihara S. Calcium transients at different muscle lengths in rat ventricular muscle [proceedings]. *J. Physiol.* 1979;292:68P-9P.
[55] Hibberd MG, Jewell BR. Calcium- and length-dependent force production in rat ventricular muscle. *J. Physiol.* 1982;329:527-40.
[56] Hibberd MG, Jewell BR. Length-dependence of the sensitivity of the contractile system to calcium in rat ventricular muscle [proceedings]. *J. Physiol.* 1979;290:30P-1P.
[57] Arteaga GM, Palmiter KA, Leiden JM, Solaro RJ. Attenuation of length dependence of calcium activation in myofilaments of transgenic mouse hearts expressing slow skeletal troponin I. *J. Physiol.* 2000;526 Pt 3:541-9.
[58] Kajiwara H, Morimoto S, Fukuda N, Ohtsuki I, Kurihara S. Effect of troponin I phosphorylation by protein kinase A on length-dependence of tension activation in skinned cardiac muscle fibers. *Biochem. Biophys. Res. Commun.* 2000;272:104-10.
[59] Arteaga GM, Warren CM, Milutinovic S, Martin AF, Solaro RJ. Specific enhancement of sarcomeric response to Ca2+ protects murine myocardium against ischemia-reperfusion dysfunction. *Am. J. Physiol. Heart Circ. Physiol.* 2005;289:H2183-92.
[60] Tachampa K, Wang H, Farman GP, de Tombe PP. Cardiac troponin I threonine 144: role in myofilament length dependent activation. *Circ. Res.* 2007;101:1081-3.
[61] Nowak G, Pena JR, Urboniene D, Geenen DL, Solaro RJ, Wolska BM. Correlations between alterations in length-dependent Ca2+ activation of cardiac myofilaments and the end-systolic pressure-volume relation. *J. Muscle Res. Cell. Motil.* 2007;28:415-9.
[62] Hinken AC, Solaro RJ. A dominant role of cardiac molecular motors in the intrinsic regulation of ventricular ejection and relaxation. *Physiology* (Bethesda). 2007;22:73-80.

[63] Landesberg A. End-systolic pressure-volume relationship and intracellular control of contraction. *Am. J. Physiol.* 1996;270:H338-49.

[64] Sun YB, Irving M. The molecular basis of the steep force-calcium relation in heart muscle. *J. Mol. Cell. Cardiol.* 2010;48:859-65.

[65] Sun YB, Lou F, Irving M. Calcium- and myosin-dependent changes in troponin structure during activation of heart muscle. *J. Physiol.* 2009;587:155-63.

[66] Solaro RJ. Maintaining cooperation among cardiac myofilament proteins through thick and thin. *J. Physiol.* 2009;587:3.

[67] Fuchs F, Bayuk M. Cooperative binding of calcium to glycerinated skeletal muscle fibers. *Biochim. Biophys. Acta.* 1976;440:448-55.

[68] Marston S, Tregear RT. Calcium binding and the activation of fibrillar insect flight muscle. *Biochim. Biophys. Acta.* 1974;347:311-8.

[69] Bremel RD, Weber A. Calcium binding to rabbit skeletal myosin under physiological conditions. *Biochim. Biophys. Acta.* 1975;376:366-74.

[70] Pan BS, Palmiter KA, Plonczynski M, Solaro RJ. Slowly exchanging calcium binding sites unique to cardiac/slow muscle troponin C. *J. Mol. Cell. Cardiol.* 1990;22:1117-24.

[71] Pan BS, Solaro RJ. Calcium-binding properties of troponin C in detergent-skinned heart muscle fibers. *J. Biol. Chem.* 1987;262:7839-49.

[72] Fabiato A, Fabiato F. Effects of pH on the myofilaments and the sarcoplasmic reticulum of skinned cells from cardiace and skeletal muscles. *J. Physiol.* 1978;276:233-55.

[73] Donaldson SK, Hermansen L, Bolles L. Differential, direct effects of H+ on Ca2+ - activated force of skinned fibers from the soleus, cardiac and adductor magnus muscles of rabbits. *Pflugers Arch.* 1978;376:55-65.

[74] Blanchard EM, Solaro RJ. Inhibition of the activation and troponin calcium binding of dog cardiac myofibrils by acidic pH. *Circ. Res.* 1984;55:382-91.

[75] Blanchard EM, Pan BS, Solaro RJ. The effect of acidic pH on the ATPase activity and troponin Ca2+ binding of rabbit skeletal myofilaments. *J Biol Chem.* 1984;259:3181-6.

[76] Fuchs F. The relationship between pH and the amount of calcium bound to glycerinated muscle fibers. *Biochim. Biophys. Acta.* 1979;585:477-9.

[77] Stull JT, Buss JE. Calcium binding properties of beef cardiac troponin. *J. Biol. Chem.* 1978;253:5932-8.

[78] Solaro RJ, Kumar P, Blanchard EM, Martin AF. Differential effects of pH on calcium activation of myofilaments of adult and perinatal dog hearts. Evidence for developmental differences in thin filament regulation. *Circ. Res.* 1986;58:721-9.

[79] Toyota N, Shimada Y. Differentiation of troponin in cardiac and skeletal muscles in chicken embryos as studied by immunofluorescence microscopy. *J. Cell. Biol.* 1981;91:497-504.

[80] Humphreys JE, Cummins P. Regulatory proteins of the myocardium. Atrial and ventricular tropomyosin and troponin-I in the developing and adult bovine and human heart. *J. Mol. Cell Cardiol.* 1984;16:643-57.

[81] Sabry MA, Dhoot GK. Identification and pattern of expression of a developmental isoform of troponin I in chicken and rat cardiac muscle. *J. Muscle Res. Cell. Motil.* 1989;10:85-91.

[82] Anderson PA, Moore GE, Nassar RN. Developmental changes in the expression of rabbit left ventricular troponin T. *Circ. Res.* 1988;63:742-7.

[83] Jin JP, Lin JJ. Rapid purification of mammalian cardiac troponin T and its isoform switching in rat hearts during development. *J. Biol. Chem.* 1988;263:7309-15.
[84] Saggin L, Gorza L, Ausoni S, Schiaffino S. Troponin I switching in the developing heart. *J. Biol. Chem.* 1989;264:16299-302.
[85] Saggin L, Gorza L, Ausoni S, Schiaffino S. Cardiac troponin T in developing, regenerating and denervated rat skeletal muscle. *Development.* 1990;110:547-54.
[86] Murphy AM, Jones L, 2nd, Sims HF, Strauss AW. Molecular cloning of rat cardiac troponin I and analysis of troponin I isoform expression in developing rat heart. *Biochemistry.* 1991;30:707-12.
[87] Hunkeler NM, Kullman J, Murphy AM. Troponin I isoform expression in human heart. *Circ. Res.* 1991;69:1409-14.
[88] Cooper TA, Ordahl CP. A single cardiac troponin T gene generates embryonic and adult isoforms via developmentally regulated alternate splicing. *J. Biol. Chem.* 1985;260:11140-8.
[89] Gomes AV, Venkatraman G, Davis JP, Tikunova SB, Engel P, Solaro RJ, et al. Cardiac troponin T isoforms affect the Ca(2+) sensitivity of force development in the presence of slow skeletal troponin I: insights into the role of troponin T isoforms in the fetal heart. *J. Biol. Chem.* 2004;279:49579-87.
[90] Herzberg O, James MN. Structure of the calcium regulatory muscle protein troponin-C at 2. 8 A resolution. *Nature.* 1985;313:653-9.
[91] Sundaralingam M, Bergstrom R, Strasburg G, Rao ST, Roychowdhury P, Greaser M, et al. Molecular structure of troponin C from chicken skeletal muscle at 3-angstrom resolution. *Science.* 1985;227:945-8.
[92] Vassylyev DG, Takeda S, Wakatsuki S, Maeda K, Maeda Y. Crystal structure of troponin C in complex with troponin I fragment at 2. 3-A resolution. *Proc. Natl. Acad. Sci. U S A.* 1998;95:4847-52.
[93] Takeda S, Yamashita A, Maeda K, Maeda Y. Structure of the core domain of human cardiac troponin in the Ca(2+)-saturated form. *Nature.* 2003;424:35-41.
[94] Kowlessur D, Tobacman LS. Significance of troponin dynamics for Ca2+-mediated regulation of contraction and inherited cardiomyopathy. *J. Biol. Chem.* 2012;287:42299-311.
[95] Kowlessur D, Tobacman LS. Troponin regulatory function and dynamics revealed by H/D exchange-mass spectrometry. *J. Biol. Chem.* 2010;285:2686-94.
[96] Vinogradova MV, Stone DB, Malanina GG, Karatzaferi C, Cooke R, Mendelson RA, et al. Ca(2+)-regulated structural changes in troponin. *Proc. Natl. Acad. Sci. USA.* 2005;102:5038-43.
[97] Li MX, Gagne SM, Spyracopoulos L, Kloks CP, Audette G, Chandra M, et al. NMR studies of Ca2+ binding to the regulatory domains of cardiac and E41A skeletal muscle troponin C reveal the importance of site I to energetics of the induced structural changes. *Biochemistry.* 1997;36:12519-25.
[98] Spyracopoulos L, Li MX, Sia SK, Gagne SM, Chandra M, Solaro RJ, et al. Calcium-induced structural transition in the regulatory domain of human cardiac troponin C. *Biochemistry.* 1997;36:12138-46.
[99] Toyo-Oka T, Ross J, Jr. Ca2+ sensitivity change and troponin loss in cardiac natural actomyosin after coronary occlusion. *Am. J. Physiol.* 1981;240:H704-8.

[100] Cummins B, Cummins P. Cardiac specific troponin-I release in canine experimental myocardial infarction: development of a sensitive enzyme-linked immunoassay. *J. Mol. Cell. Cardiol.* 1987;19:999-1010.

[101] Herzig JW, Feile K, Ruegg JC. Activating effects of AR-L 115 BS on the Ca2+ sensitive force, stiffness and unloaded shortening velocity (Vmax) in isolated contractile structures from mammalian heart muscle. *Arzneimittelforschung.* 1981;31:188-91.

[102] Herzig JW, Peterson JW, Ruegg JC, Solaro RJ. Vanadate and phosphate ions reduce tension and increase cross-bridge kinetics in chemically skinned heart muscle. *Biochim. Biophys. Acta.* 1981;672:191-6.

[103] Solaro RJ, Ruegg JC. Stimulation of Ca++ binding and ATPase activity of dog cardiac myofibrils by AR-L 115BS, a novel cardiotonic agent. *Circ. Res.* 1982;51:290-4.

[104] Endoh M. Cardiac Ca2+ signaling and Ca2+ sensitizers. *Circ. J.* 2008;72:1915-25.

[105] Kass DA, Solaro RJ. Mechanisms and use of calcium-sensitizing agents in the failing heart. *Circulation.* 2006;113:305-15.

[106] Tanaka H, Kawano S, Kozai T, Urabe Y. [Combined therapy with weight loss and amiodarone improved cardiac function in a patient with idiopathic dilated cardiomyopathy complicated with severe obesity: a case report]. *J. Cardiol.* 2007;50:141-8.

[107] Boyle KL, Leech E. A review of the pharmacology and clinical uses of pimobendan. *J. Vet. Emerg. Crit. Care* (San Antonio). 2012;22:398-408.

[108] Gordon SG, Saunders AB, Roland RM, Winter RL, Drourr L, Achen SE, et al. Effect of oral administration of pimobendan in cats with heart failure. *J. Am. Vet. Med. Assoc.* 2012;241:89-94.

[109] Edes I, Kiss E, Kitada Y, Powers FM, Papp JG, Kranias EG, et al. Effects of Levosimendan, a cardiotonic agent targeted to troponin C, on cardiac function and on phosphorylation and Ca2+ sensitivity of cardiac myofibrils and sarcoplasmic reticulum in guinea pig heart. *Circ. Res.* 1995;77:107-13.

[110] Sorsa T, Pollesello P, Solaro RJ. The contractile apparatus as a target for drugs against heart failure: interaction of levosimendan, a calcium sensitiser, with cardiac troponin c. *Mol. Cell. Biochem.* 2004;266:87-107.

[111] Pathak A, Lebrin M, Vaccaro A, Senard JM, Despas F. Pharmacology of levosimendan: inotropic, vasodilatory and cardioprotective effects. *J. Clin. Pharm. Ther.* 2013.

[112] Husebye T, Eritsland J, Muller C, Sandvik L, Arnesen H, Seljeflot I, et al. Levosimendan in acute heart failure following primary percutaneous coronary intervention-treated acute ST-elevation myocardial infarction. Results from the LEAF trial: a randomized, placebo-controlled study. *Eur. J. Heart Fail.* 2013;15:565-72.

[113] Hernandez A, Miranda A, Parada A. [Levosimendan reduces mortality in cardiac surgery:A systematic review and meta-analysis]. *Rev. Esp. Anestesiol Reanim.* 2012;59:6-11.

[114] Rognoni A, Lupi A, Cavallino C, Veia A, Bacchini S, Rosso R, et al. Levosimendan preoperative. *Curr. Pharm. Des.* 2012.

[115] Doorduin J, Sinderby CA, Beck J, Stegeman DF, van Hees HW, van der Hoeven JG, et al. The calcium sensitizer levosimendan improves human diaphragm function. *Am. J. Respir. Crit. Care Med.* 2012;185:90-5.

[116] Meijs MF, Asselbergs FW, Doevendans PA. Omecamtiv mecarbil: a promising new drug in systolic heart failure. *Eur. J. Heart Fail.* 2012;14:232-3.

[117] Russell AJ, Hartman JJ, Hinken AC, Muci AR, Kawas R, Driscoll L, et al. Activation of fast skeletal muscle troponin as a potential therapeutic approach for treating neuromuscular diseases. *Nat. Med.* 2012;18:452-5.

[118] de Winter JM, Buck D, Hidalgo C, Jasper JR, Malik FI, Clarke NF, et al. Troponin activator augments muscle force in nemaline myopathy patients with nebulin mutations. *J. Med. Genet.* 2013.

[119] Solaro RJ, Henze M, Kobayashi T. Integration of troponin I phosphorylation with cardiac regulatory networks. *Circ. Res.* 2013;112:355-66.

[120] Seidman CE, Seidman JG. Mutations in cardiac myosin heavy chain genes cause familial hypertrophic cardiomyopathy. *Mol. Biol. Med.* 1991;8:159-66.

[121] Thierfelder L, Watkins H, MacRae C, Lamas R, McKenna W, Vosberg HP, et al. Alpha-tropomyosin and cardiac troponin T mutations cause familial hypertrophic cardiomyopathy: a disease of the sarcomere. *Cell.* 1994;77:701-12.

[122] Tardiff JC. Thin filament mutations: developing an integrative approach to a complex disorder. *Circ. Res.* 2011;108:765-82.

[123] Ashrafian H, McKenna WJ, Watkins H. Disease pathways and novel therapeutic targets in hypertrophic cardiomyopathy. *Circ. Res.* 2011;109:86-96.

[124] Maron BJ, Maron MS, Semsarian C. Genetics of hypertrophic cardiomyopathy after 20 years: clinical perspectives. *J. Am. Coll. Cardiol.* 2012;60:705-15.

[125] Oliveira SM, Zhang YH, Solis RS, Isackson H, Bellahcene M, Yavari A, et al. AMP-activated protein kinase phosphorylates cardiac troponin I and alters contractility of murine ventricular myocytes. *Circ. Res.* 2012;110:1192-201.

[126] Nixon BR, Thawornkaiwong A, Jin J, Brundage EA, Little S, Davis JP, et al. AMP activated protein kinase phosphorylates cardiac troponin I at Ser-150 to increase myofilament calcium sensitivity and blunt PKA dependent function. *J. Biol. Chem.* 2012.

[127] Pound KM, Arteaga GM, Fasano M, Wilder T, Fischer SK, Warren CM, et al. Expression of slow skeletal TnI in adult mouse hearts confers metabolic protection to ischemia. *J. Mol. Cell. Cardiol.* 2011;51:236-43.

[128] Baudenbacher F, Schober T, Pinto JR, Sidorov VY, Hilliard F, Solaro RJ, et al. Myofilament Ca2+ sensitization causes susceptibility to cardiac arrhythmia in mice. *J. Clin. Invest.* 2008;118:3893-903.

[129] Huke S, Knollmann BC. Increased myofilament Ca2+-sensitivity and arrhythmia susceptibility. *J. Mol. Cell. Cardiol.* 2010;48:824-33.

[130] Knollmann BC, Kirchhof P, Sirenko SG, Degen H, Greene AE, Schober T, et al. Familial hypertrophic cardiomyopathy-linked mutant troponin T causes stress-induced ventricular tachycardia and Ca2+-dependent action potential remodeling. *Circ. Res.* 2003;92:428-36.

[131] Schober T, Huke S, Venkataraman R, Gryshchenko O, Kryshtal D, Hwang HS, et al. Myofilament Ca sensitization increases cytosolic Ca binding affinity, alters intracellular Ca homeostasis, and causes pause-dependent Ca-triggered arrhythmia. *Circ. Res.* 2012;111:170-9.

[132] Maass AH, Ikeda K, Oberdorf-Maass S, Maier SK, Leinwand LA. Hypertrophy, fibrosis, and sudden cardiac death in response to pathological stimuli in mice with mutations in cardiac troponin T. *Circulation*. 2004;110:2102-9.

In: Troponin
Editor: J.-P. Jin

ISBN: 978-1-62948-320-7

Chapter II

Troponin Isoform Diversity

Peter J. Reiser, PhD*
Professor and Interim Chair, Oral Biology, Ohio State University
Columbus, OH, US

Abstract

This chapter provides an overview of the diversity of troponin subunit isoforms among mammals, birds, fish, amphibians and invertebrates. The greatest amount of available information on troponin isoform diversity has been obtained from examinations in mammalian species, and, within mammals, the predominant focus has been on rodents and humans. There, is, however, substantial information from non-mammalian species, including invertebrates, which not only describes interesting isoform expression patterns in individual species, but collectively paints an intriguing evolutionary history of the entire troponin complex.

Introduction

Invariably, the troponin complex consists of three subunits in a 1:1:1 molar ratio. The subunits of the troponin complex orchestrate the coordinated regulation of muscle activation that is controlled by changes in the cytoplasmic free calcium ion concentration in muscle. Multiple isoforms of troponin C (TnC), troponin I (TnI), and troponin T (TnT) are expressed in developmentally regulated and tissue-specific patterns. An underlying premise of much of the study of these patterns is that isoform specific expression of each subunit ideally subserves activation of individual muscles, or individual types of muscle fibers within a muscle, in the execution of specific motor functions. This chapter provides a current view of the understanding of troponin subunit isoform diversity among invertebrates and vertebrates. It had long been believed that a lack of troponin subunits in smooth muscle was a biochemical

* Office: 614-292-8476, E-mail: reiser.17@osu.edu

hallmark that distinguishes it from striated muscle. This review is particularly timely as recent reports from two laboratories have provided strong evidence for not only troponin subunit mRNA, but for troponin subunit protein, as well, in smooth muscle of multiple organs in mice and humans, supporting a report of TnT and TnI in bovine vascular smooth muscle more than twenty years ago (discussed below). There are several other, excellent articles that provide much more detailed information in some areas that can be only briefly mentioned in this chapter. These articles provide either a previous comprehensive review of the diversity in troponin subunit isoforms among animals or an overview of a specific topic (e. g., Hastings, 1997; Gomes et al., 2002; Hooper and Thuma, 2005; Jin et al., 2008; Shaffer and Gillis, 2010; Wei and Jin, 2011).

A brief discussion of homologous proteins, paralagous proteins and orthologous proteins will introduce several important concepts when considering protein isoforms. Inherent in any discussion of the expression of protein isoforms in multiple species is the concept of protein homology. Homologous proteins in multiple species are coded by the same evolutionarily conserved gene, with what are usually multiple, small species-specific differences in the gene coding sequence and in the corresponding amino acid sequence. The amino acid sequence differences are sufficiently small to ensure retention of the same molecular or cellular function, which might be only subtly different between species to best serve the survival of the individual species. For a protein to be homologous among multiple species, the last common ancestor of those species must have expressed that protein. Paralogous proteins are related to each other by gene duplication. Therefore, proteins that are paralogous are expressed in the same given species. Orthologous proteins, on the other hand, are related to each other by a speciation event – that is, two species sharing the same gene. These terms are not used in this chapter, but are nevertheless mentioned for those readers that might want to delve into questions related to evolutionary relationships between specific protein isoforms, that might arise as a result of reading this review.

Mammals

Expression Patterns in Fully Differentiated Muscle

The troponin complex, which is the pivotal site of calcium-regulation of striated regulation, consists of three subunits. TnC is the subunit that binds calcium ions, the event that initiates activation. TnI normally inhibits cross bridge formation, between myosin and actin, when the myoplasmic free calcium ion concentration is low. TnT anchors the trimeric Tn complex to thin filaments through its interaction with tropomyosin. There are two isoforms of TnC expressed in mammals - one isoform in fast muscle fibers (fast TnC) and another isoform that is expressed in slow skeletal muscle fibers and in cardiac muscle (slow/cardiac TnC). Three isoforms of TnI are typically expressed in mammalian muscle - fast, slow and cardiac. The expression of TnT isoforms in mammalian muscle is much more elaborate, with multiple fast isoforms (typically five), slow isoforms (typically four) and cardiac isoforms (typically four). Generally, the expression of cardiac TnT in adult organisms is restricted to the heart, and fast and slow TnT isoforms are restricted to fast and slow muscle fibers, respectively. There are, however, several notable exceptions that will be discussed in this chapter. The same restrictions are also observed for TnC and TnI, with isoforms being expressed in adult cardiac muscle or in either adult fast or slow fibers. It was generally believed that none of the Tn subunits are expressed in smooth muscle, in which activation is dependent on other calcium-binding proteins and mechanisms (reviewed in Berridge, 2008 and Jin et al., 2008). However, there are reports of fast skeletal TnT, TnI and TnC proteins in mouse smooth muscle (Moran et al., 2008) and of cardiac TnT, but not cardiac TnI or cardiac TnC, protein in smooth muscle of several organs in mice and humans (Kajioka et al, 2012). These reports support the initial observation of trace amounts of cardiac TnT and cardiac TnI in bovine vascular smooth muscle (Zanellato et al., 1991).

Tn subunit proteins are expressed by families of genes that code cardiac, slow, and fast isoforms. However, the regulation of the expression of the three subunits differs markedly. The two TnC isoforms, fast and slow/cardiac, are coded by separate genes and only one product is expressed from both genes - fast TnC and slow/cardiac TnC. The three TnI isoforms (fast, slow and cardiac) are coded by separate genes. Only one product is coded by each gene. The regulation of TnT isoform expression is much more complex, consistent with the much greater number of observed TnT proteins. The *TNNT2* gene codes for cardiac TnT isoforms. The *TNNT1* and *TNNT3* genes code for slow and fast TnT, respectively. Multiple TnT isoforms are generated from each of these genes by alternative mRNA splicing (Medford et al. 1984; Cooper and Ordahl, 1984; Breitbart et al., 1985; Huang et al., 1999; reviewed by Wei and Jin, 2011). Generally, only single isoforms of TnC and of TnI are expressed in individual skeletal muscle fibers. However, many muscle fibers individually co-express multiple isoforms of either fast or slow TnT, but not a combination of fast and slow TnT isoforms. Numerous studies have shown that the specific complement of either fast or slow TnT isoforms in a given muscle fiber is associated with differences in the calcium-sensitivity of force generation or of actomyosin ATPase activity (reviewed by Wei and Jin, 2011). TnT and TnI can undergo post-translational modifications, as discussed in another chapter, giving rise to an even greater number of protein species that participate in normal activation of striated muscle.

Adult muscle fibers that stably co-express fast and slow isoforms of contractile proteins are referred to as “hybrid” fibers (reviewed in Stephenson, 2001). Hybrid fibers are normally a minority of the total population of fibers that constitute mammalian limb muscles, although Zhang et al. (2010) reported that fibers that co-express two fast MHC isoforms are fairly common in mouse tibialis anterior and brachioradialis. On the other hand, hybrid fibers represent a large majority of fibers in many craniofacial (extraocular, intrinsic laryngeal, and jaw-closing) muscles (Wu et al., 1998, 2000; Bicer and Reiser, 2004, 2009; Begrin et al., 2006). Most commonly, hybrid fibers express fast and slow isoforms of myosin heavy chains (MHC) and myosin light chains (MLC), but co-expression of fast and slow isoforms of tropomyosin occurs in some extraocular muscle fibers (Bicer and Reiser, 2013). As discussed below, Perry et al. (2009) reported co-expression of multiple TnI isoforms in individual fibers from ghost crab muscle, although the identity of the isoforms (fast, slow or cardiac) was not determined. Identification of these isoforms would be very interesting because normally mammalian muscle fibers individually express only one TnI isoform and only one protein is generated from each of the three mammalian TnI genes.

Wilkinson and Taylor (1986) reported two cardiac-specific TnT isoforms in bovine atria and ventricles and one in rabbit heart. Lee and Tobacman (1987) also reported two cardiac TnT isoforms in bovine hearts and showed that the myosin S-1 MgATPase rate is different, depending on which TnT isoform is present in reconstituted this filaments. More recent studies revealed up to five cardiac-specific TnT isoforms, on one-dimensional protein gels, in adult hearts of several mammalian species (e.g., rabbit: Anderson et al., 1988; rat: Reiser et al., 1994; dog: Bicer and Reiser, 2013; human: Anderson et al., 1995; pig: Korte et al., 2005). It is possible that the greater number reported in some studies is due, at least in part, to different phosphorylation states of some isoforms. However, detection of differently phosphorylated proteins of a given isoform usually requires two-dimensional gel, due to their typically very similar mobility on one-dimensional gels.

Cardiac TnT is also expressed in adult skeletal muscle under abnormal conditions, such as during regeneration following muscle injury, following denervation, in patients with Duchenne muscular dystrophy, and in patients with renal disease (Saggin et al., 1990; Anderson et al., 1991; Jin, 1996; Bodor et al. 1997; McLaurin et al., 1997; Haller et al., 1998; Ricchiuti et al., 1998; Jin et al., 2003). In at least some of these instances, the expression of cardiac TnT in skeletal muscle could be consistent with regeneration of muscle fibers, which normally involves the expression of isoforms of several myofibrillar proteins that are normally expressed during early developmental stages.

The expression pattern of troponin C isoforms in mammalian skeletal and cardiac muscle is relatively simple, compared to that of TnI and TnT. It has been widely observed that a single TnC isoform is expressed in slow skeletal muscle fibers and in cardiac atria and ventricles of many vertebrate species (e.g., Wilkinson, 1980; Putkey et al., 1987). Furthermore, only fast TnC is expressed in fast fibers. Cardiac TnC is highly conserved among vertebrate species, while its intimate partners in muscle activation, TnI and especially TnT, have undergone elaborate evolutionary changes, with respect to isoform sequences and expression patterns. The range in activation-relaxation kinetics in different species is very large. The simplicity of TnC isoform expression, combined with the much more complex patterns of TnI and TnT isoform expression, suggest the importance of events for regulating muscle activation that are downstream from calcium binding to TnC.

Developmental Transitions in Troponin Subunit Expression

Skeletal muscle and the heart express troponin subunit isoforms at early developmental stages that are not detected in the same muscles at the fully differentiated adult stage. Cardiac TnT has been reported to be expressed in skeletal muscle at early developmental stages (Malouf et al., 1992; Toyota and Shimada, 1981 and 1983; Cooper and Ordahl, 1984; Jin, 1996). Cardiac muscle at fetal and early neonatal stages expresses the TnI isoform that is normally expressed in adult slow skeletal muscle (ssTnI), in addition to cardiac TnI (Solaro et al., 1988; Ausoni et al., 1991; Reiser et al, 1994). It appeared that the expression of ssTnI in the period surrounding birth protects the heart from the transient acidosis that normally occurs during the transition from a reliance on maternal blood flow, to support cellular respiration in-utero, to the dependence on air breathing following birth (Solaro et al., 1988; Reiser et al., 1994). This is now strongly supported by the findings of enhanced calcium-sensitivity (Arteaga et al., 2000) in hearts of transgenic mice in which all of the cardiac TnI was replaced with ssTnI. Mouse hearts with forced expression of ssTnI also exhibited elevated anaerobic ATP production during periods of ischemia (Pound et al., 2011).

The complement of either fast or slow TnT isoforms varies among single fibers in mammalian limb muscle. It is still not well understood what regulates which isoforms are expressed in individual fibers. However, there are clear patterns of association between a specific complement of fast TnT isoforms and specific fast MHC and/or MLC isoforms among skeletal muscle fibers (e.g., Greaser et al., 1988; Galler et al., 1997). For example, rat muscle fibers that express predominantly MHC-IIB, the MHC isoform that is associated with the fastest and most powerful contractions in rat limb muscle, also express greater amount of a specific fast TnT isoform compared to other fibers. There appears, therefore, to be coordinated regulation of the expression of specific isoforms of multiple myofibrillar proteins that constitute thick and thin filaments among individual fibers.

Birds

Wilkinson (1978) provided an early description of TnT and TnI isoform expression in chicken breast and leg (apparently fast-type) muscle. He concluded that the patterns in chicken skeletal muscle are very similar to that in rabbit skeletal muscle, and that general description has remained essentially unchanged with similar descriptions from numerous laboratories subsequently. Wilkinson et al. (1984) later described multiple isoforms of TnT in chicken skeletal muscle: one in breast muscle and two lower molecular weight isoforms in leg muscles. The expression of the TnT isoform in pectoralis fibers is associated with enhanced calcium-sensitivity of force generation, compared to fibers that express other fast TnT isoforms (Reiser et al., 1992 and 1996; Ogut et al., 1999). An N-terminal metal binding domain exists in the TnT that is expressed in the pectoralis muscle of all examined species of *Galliformes* (including chickens) and *Craciformes* (Jin and Smillie, 1994), and this domain is not present in TnT of non-avian species (Jin and Samanez, 2001). Zhang et al. (2004) reported that this domain can bind calcium and, thereby, potentially modulate calcium-regulation of muscle. Jin and Smillie (1994) speculated that the unique N-terminal domain of pectoralis TnT in some avian species might facilitate rapid relaxation in species that rely on high intensity, but short duration flight, as these muscles also characteristically are composed of fast glycolytic fibers and fast oxidative-glycolytic fibers (Rosser et al., 1996). The N-terminus of avian pectoralis TnT is also more acidic than that of the fast TnT isoforms expressed in leg muscles. It is possible that this difference also contributes to the increased calcium-sensitivity that is associated with pectoralis TnT, as discussed by Ogut et al. (1999). The multiplicity of TnT isoforms in chicken fast muscle and the association of specific isoforms with greater or lower calcium sensitivity is shared with mammalian skeletal muscle fibers.

Wilkinson and Taylor (1986) reported a single isoform of TnT in chicken atria and ventricles, based upon observations from one-dimensional gels. Analysis of atria and ventricles on two-dimensional gels, in the same study, revealed two TnT spots and the authors concluded that these likely represent different states of phosphorylation. Cardiac TnT isoforms are also expressed in embryonic chicken skeletal muscle (Toyota and Shimada, 1981).

There are two isoforms of chicken TnC, fast and slow/cardiac, as in mammalian species. The expression of these two isoforms in chickens follows the same pattern as in mammals.

One exception is the report of transcript for fast TnC been detected in embryonic quail heart (Stoutamyer and Dhoot, 2005).

Fish

Fish TnT

Multiple TnT isoforms are also expressed in fish skeletal muscle. Transcripts for two slow TnT isoforms were found in red (slow) muscle and expression of a third TnT isoform was detected in fast (white) muscle of sea bream (Campinho et al., 2005). Phylogenetic analysis suggests that one of the slow isoforms is related to other vertebrate slow TnT isoforms, whereas the other sea bream slow isoform and the fast TnT isoform are teleost-specific. Camphino et al. (2005) determined that the slow TnT genes probably arose from duplication of an ancestral gene and that the gene that codes for the fast TnT isoform does not contain introns and, therefore, likely arose from reverse transcription. The complexity of TnT isoform expression in fish muscle is reinforced by the results reported by Feng et al. (2012) which show that there are at least two cardiac TnT isoforms, four TnT isoforms in red muscle and three TnT isoforms in white muscle. The expression of most, perhaps all, of these isoforms appears to be muscle-type specific.

Coughlin and et al. (2005) provide insight into the possible functional significance of fish fast TnT isoforms from a study on rainbow trout. They observed that the ratio of two TnT isoforms in red muscle varies along body length. Isometric activation and relaxation kinetics in fiber bundles that were isolated from multiple locations along the body length axis correlated with the ratio of the two TnT isoforms. Coughlin and co-workers, in the same study, also determined that the TnT ratio does not vary along body length in brook trout and that the activation kinetics also do not vary with length. It is not clear if the identified TnT isoforms are slow or fast isoforms, as red muscle in fish is typically slow-type, but the antibody employed in this study (JLT-12) is normally used for the detection of fast TnT isoforms in mammalian muscle. Nevertheless, this study was the first to show an association of differential TnT isoform expression with contractile properties in fish muscle. Thys et al. (2006) found a similar association between the variation in TnT isoform expression and contraction kinetics in white muscle, along the length of largemouth bass and Thys et al. (1998) found a similar TnT isoform expression pattern in Atlantic cod, which parallels the variation in contraction kinetics in muscle bundles isolated from sites along body length, reported by Davies et al. (1995).

Gallant et al. (2012) compared protein expression patterns between skeletal muscle and the electric organ of mormyrid (*Brienomyrus brachyistius*). Compared to skeletal muscle, the expression of TnT was decreased and that of TnI was increased in the electric organ. This is especially interesting, given that TnT is pivotally involved in calcium-activation of muscle, whereas TnI is important for preventing cross-bridge formation that is required for muscle contraction. Other myofibrillar proteins are expressed in the electric organ of fish and their physiological role in this organ has not been determined.

The expression of troponin I isoforms in fish muscle appears to be much more complex than in mammalian muscle, where one product of each of three genes exists in fast fibers, slow fibers and cardiac muscle. Fu et al. (2009) reported the presence of three TnI genes in zebrafish, with transcripts expressed in fast and slow skeletal muscles and in the heart, very similar to the pattern observed in mammalian species. However, the cardiac TnI isoform in zebrafish is also expressed in craniofacial muscles at least at embryonic stages. Transcripts for as many as seven TnI isoforms in fast and slow skeletal muscle and cardiac (ventricle) muscle were found in rainbow trout (Alderman et al., 2012). Transcripts for all seven TnIs were found in fast, slow and cardiac muscle, but the relative abundance of the seven transcripts was unique in each muscle type. In particular, the most abundant transcript was different in each muscle type. At the protein level, three TnI isoforms were detected in fast muscle and four isoforms in slow muscle, with one isoform being common to both muscle types. Three TnI proteins were detected in cardiac muscle, with one being cardiac-specific and the other two being common to either fast or slow skeletal muscle. Results from mass spectrometry suggested the presence of two TnT isoforms and two TnC isoforms in fast muscle (Alderman et al., 2012). Changes in the abundance of the transcripts of individual TnI isoforms occurred during two months of cold exposure, indicating, as discussed by Alderman and co-workers (2012), adaptation in striated muscle thin filaments that presumably preserves muscle-specific function at lower temperature.

The TnC that is expressed in trout heart is much more sensitive to calcium, compared to mammalian cardiac TnC (Churcott et al., 1994). This difference is thought to preserve cardiac function when trout are normally exposed to low temperatures during seasonal variations. Gillis et al. (2005) identified, using mutational analysis, specific residues in trout cardiac TnC that are responsible for the high calcium-sensitivity. More recently, Kirkpatrick et al. (2011) determined that trout cardiac TnI also contributes to increased myofilament calcium-sensitivity, by substituting TnI in the rat cardiac troponin complex with recombinant trout cardiac TnI. They also determined that trout cardiac TnI lacks the N-terminal extension that exists in cardiac TnI of other vertebrates. This domain contains two phosphorylation sites and the effects of phosphorylation leads to a decrease in calcium-sensitivity. Kirkpatrick and co-workers also showed that phosphorylation of the troponin complex with trout TnI was blunted, compared to the native rat cardiac troponin complex. Comparative studies of this type are extremely valuable by providing new information about non-traditional animal models, in this case, trout heart, to understand how other species have evolved, while also reinforcing what has been determined from more traditional models, such as rat heart.

Amphibians

Amphibian TnI

Warkman and Atkinson (2002, 2006) reported that anuran hearts express cardiac TnI at all stages, unlike mammalian heart, which abundantly expresses slow skeletal TnI at early developmental stages. Little is known about TnI isoform expression in amphibian skeletal muscle.

Amphibian TnC

O'Connell and co-workers report the expression of two TnC isoforms in cane toad muscle. One isoform is expressed in skeletal muscle fibers that also express fast MHC isoforms. The other TnC isoform is expressed in slow (tonic) fibers and in the heart. This pattern is the same as in vertebrate skeletal muscle and, generally, only cardiac TnT is expressed in vertebrate hearts.

Similar to mammalian hearts, there are multiple (four) cardiac TnT isoforms expressed in the Mexican axolotl (Sferrazza et al., 2007). There is, however, at least one notable exception among amphibians. Feng et al. (2012) reported that the heart of the adult American toad (*Bufo americanus*) expresses only slow skeletal TnT isoforms and that this is associated with faster contraction kinetics and the ability to pump blood against a greater afterload. The authors concluded that this is an acquired evolutionary adaptation that increases cardiac output in the American toad heart.

Invertebrates

Invertebrate TnT

Invertebrate TnT isoforms have C-terminal extensions that are not present in vertebrate TnT isoforms. Domingo, et al. (1998) studied TnT isoforms in muscle from *Drosophila melanogaster*, *Lethocerus*, grasshopper and honey bee muscle and reported that these extensions are acidic, due to a high content of glutamic acid and likely contribute to the calcium-binding property of invertebrate TnT that is not observed in vertebrate TnT isoforms. The C-terminal extension increases the molecular mass of invertebrate TnT isoforms, so they migrate markedly slower on protein gels, compared to vertebrate TnT isoforms, the latter generally being ~37 kDa.

Domingo et al. (1998) reported similarly migrating TnT isoforms in *Lethocerus*, grasshopper and honey bee muscle and they showed that there are two TnT isoforms in *D. melanogaster* that are differentially expressed in different muscles – one isoform in head and abdomen and the other in the thorax. Furthermore, a third TnT isoform was detected only at early developmental stages.

Therefore, invertebrate TnT isoforms are distinct from vertebrate isoforms, due to the presence of the acidic C-terminal extension and apparently associated calcium-binding properties, but share the developmental and adult tissue-specific diversity in expression patterns that are pervasive with vertebrate TnT. Domingo and co-workers also showed that *D. melanogaster* TnT is phosphorylated in-vivo, another property shared with vertebrate TnT isoforms.

Marden and co-workers (Fitzhugh and Marden, 1997; Marden et al., 2001) reported evidence for a strong association between TnT isoform expression with contractile properties in the dragonfly, *Libellula pulchella*. Fitzhugh and Marden, 1997) demonstrate a relationship between TnT isoform composition and twitch kinetics and the calcium-sensitivity of force generation at different developmental stages. Marden et al. (2001) reported strong correlations between TnT isoform transcript levels and calcium-sensitivity, peak force generation, work and power generation among adult members of *Libellula pulchella*. The results of these studies suggest that structure-function relationships between TnT isoforms and physiological properties of muscle are likely to be pervasive among vertebrate and invertebrate species.

Invertebrate TnI

There is an isoform of TnI that is expressed specifically in indirect flight muscle (IFM) of some insects. This TnI is referred to as "TnH" and has a long C-terminal extension, compared to vertebrate TnI. This extension has a rich Proline-Alanine content and was believed to participate in a mechanism that underlies stretch-activation, which has been commonly observed in skeletal and cardiac muscle, in general, and is particularly prominent in insect IFM. However, it has recently been demonstrated that removal of the C-terminal extension of TnH in bumblebee IFM does not abolish stretch activation (Iwamoto, 2013). Therefore, the unique contribution of TnH to muscle activation, if any, is not clear. Kobayashi et al. (1989b) reported that crayfish TnI also contains a 30-residue N-terminal extension, similar to vertebrate cardiac TnI.

Perry et al. (2009) reported multiple isoforms, based upon gel electrophoretic mobility, of both TnT and of TnI in individual ghost crab muscle fibers. The authors describe patterns of TnT and TnI isoform among fibers that were similar to those of different myosin chain isoforms in fibers that were either primarily glycolytic (presumably fast) or oxidative (presumably slow), suggesting distinct fast and slow troponin subunits in crab muscle fibers. There were differences in isoform composition of individual fibers between small and large crabs, suggesting the possibility that troponin isoform expression in crabs is developmentally regulated.

Invertebrate TnC

Kobayashi et al. (1989c) reported the expression of two isoforms of TnC, both containing 150 residues, in crayfish tail muscle. Both of these isoforms are believed to bind only two calcium ions, in contrast to fast skeletal TnC of vertebrates. TnC in ascidian body wall muscle apparently binds only two calcium ions, as well (Takagi and Konishi, 1983). Kobayashi et al. (1989a) sequenced TnC in the horseshoe crab and deduced that it has three calcium-binding sites. Therefore, in general TnC among invertebrates binds fewer calcium ions and appears to be variable with respect to calcium-binding capacity, depending on the animal species.

Conclusion

It is very clear that there is a tremendous amount of diversity in troponin subunit isoforms, with respect to the regulation of gene expression, as well with regard to which specific isoforms of TnC, TnI and TnT are expressed in different muscle types, at different developmental stages and among individual fibers of a given muscle. Diversity in the troponin complex exists, not only among vertebrate species, but among invertebrate species, as well. There is overwhelming evidence indicating that diversity in troponin subunit isoforms is a key determinant of major contractile properties in both skeletal and cardiac muscle and an emerging view that troponin subunits might have a role in smooth muscle activation, after all.

References

Alderman, S. L., Klaiman, J. M., Deck, C. A. and Gillis, T. E. Effect of cold acclimation on troponin I isoform expression in striated muscle of rainbow trout. *Am. J. Physiol. Regul. Integr. Comp. Physiol.* 303: R168-R1676, 2012.

Anderson, P. A. W., Greig, A., Mark, T. M., Malouf, N. N., Oakeley, A. E., Ungerleider, R. M., Allen, P. D. and Kay, B. K. Molecular basis of human cardiac troponin T isoforms expressed in the developing, adult, and failing heart. *Circ. Res.* 76: 681-686, 1995.

Anderson, P. A. W., Malouf, N. N., Oakley, A. E., Pagani, E. D. and Allen, P. D. Troponin T isoform expression in humans: a comparison among normal and failing adult heart, fetal heart, and adult and fetal skeletal muscle. *Circ. Res.* 69:1226-1233, 1991.

Anderson, P. A. W., Moore, G. E. and Nassar, R. N. Developmental changes in the expression of rabbit left ventricular troponin T. *Circ. Res.* 63: 742-747, 1988.

Arteaga, G. M. Palmiter, K. A. Leiden, J. M. and Solaro, R. J. Attenuation of length dependence of calcium activation in myofilaments of transgenic mouse hearts expressing slow skeletal troponin I. *J. Physiol.* 526: 541–549, 2000.

Ausoni, S., De Nardi, C., Moretti, P., Gorza, L. and Schiaffino S. Developmental expression of rat cardiac troponin I mRNA. *Development.* 112: 1041-1051, 1991.

Bergrin, M., Bicer, S., Lucas, C. A. and Reiser, P. J. Three-dimensional compartmentalization of myosin heavy chain and light chain isoforms within dog thyroarytenoid muscle. *Am. J. Physiol.* 290: C1446–C1458, 2006.

Berridge, M. J. Smooth muscle cell calcium activation mechanisms. *J. Physiol.* 586: 5047-5061, 2008.

Bicer, B. and Reiser, P. J. Myosin light chain isoform expression among single mammalian skeletal muscle fibers: species variations. *J. Muscle Res. Cell Motil.* 25: 623–633, 2004.

Bicer, S. and Reiser, P. J. Myosin isoform expression in dog rectus muscles: patterns in global and orbital layers and among single fibers. *Invest. Ophthalmol. Vis. Sci.* 50: 157–167, 2009.

Bicer, S. and Reiser, P. J. Complex tropomyosin and troponin T isoform expression patterns in orbital and global fibers of adult dog and rat extraocular muscles. *J. Muscle Res. Cell Motil.* 2013; DOI 10. 1007/s10974-013-9346-9.

Bodor, G. S., Survant, L., Voss, E. M., Smith, S., Porterfield, D. and Apple, F. S. Cardiac troponin T composition in normal and regenerating skeletal muscle. *Clin. Chem.* 43:476-484, 1997.

Campinho, M. A., Power, D. M. and Sweeney, G. E. Identification and analysis of teleost slow muscle troponin T (sTnT) and intronless TnT genes. *Gene* 361: 67-79, 2005.

Churcott, C. S., Moyes, C. D., Bressler, B. H., Baldwin, K. M. and Tibbits, G. F. Temperature and pH effects on Ca^{2+} sensitivity of cardiac myofibrils: a comparison of trout with mammals. *Am. J. Physiol.* 267: R62-R670, 1994.

Cooper, T. A. and Ordahl, C. P. A single troponin T gene regulated by different programs in cardiac and skeletal muscle development. *Science* 226:979-982, 1984.

Coughlin, D. J., Caputo, N. D., Bohnert, K. L. and Weaver, F. E. Troponin T expression in trout red muscle correlates with muscle activation. *J. Exp. Biol.* 208: 409-417, 2005.

Davies, M. L. F., Johnston, I. A. and Van Dewal, J. Muscle fibers in rostral and caudal myotomes of the Atlantic cod (*Gadus morhua L.*) have different mechanical properties. *Physiol. Zool.* 68: 673–697, 1995.

Dhoot, G. K., Frearson, N. and Perry, S. V. Polymorphic forms of troponin T and troponin C and their localization in striated muscle cell types. *Exp. Cell Res.* 122: 339-350, 1979.

Domingo, A., González-Jurado, J., Maroto, M., Díaz, C., Vinós, J., Carrasco, C., Cervera, M. and Marco, R. Troponin-T is a calcium-binding protein in insect muscle: in vivo phosphorylation, muscle-specific isoforms and developmental profile in Drosophila melanogaster. *J. Muscle Res. Cell Motil.* 19: 393-403, 1998.

Feng HZ, Chen X, Hossain MM, Jin JP. Toad heart utilizes exclusively slow skeletal muscle troponin T: an evolutionary adaptation with potential functional benefits. *J. Biol. Chem.* 287: 29753-29764, 2012.

Fitzhugh, G. and Marden, J. Maturational changes in troponin T expression, Ca2+-sensitivity and twitch contraction kinetics in dragonfly flight muscle. *J. Exp. Biol.* 200: 1473-1482, 1997.

Fu, C. Y., Lee, H. C. and Tsai, H. J. The molecular structures and expression patterns of zebrafish troponin I genes. *Gene Expr. Patterns.* 9(5):348-356, 2009.

Gallant, J. R., Hopkins, C. D. and Deitcher, D. L. Differential expression of genes and proteins between electric organ and skeletal muscle in the mormyrid electric fish Brienomyrus brachyistius. *J. Exp. Biol.* 215: 2479-2494, 2012.

Galler, S., Schmitt, T. L., Hilber, K. and Pette. D. Stretch activation and isoforms of myosin heavy chain and troponin-T of rat skeletal muscle fibres. *J. Muscle Res. Cell Motil.* 18: 555-561, 1997.

Gillis, T. . E, Liang, B., Chung, F. and Tibbits, G. F. Increasing mammalian cardiomyocyte contractility with residues identified in trout troponin *C. Physiol Genomics.* 22: 1-7, 2005.

Gomes, A. V., Potter, J. D. and Szczesna-Cordary. D. The role of troponins in muscle contraction. *IUBMB Life.* 54: 323-333, 2002.

Greaser, M. L., Moss, R. L. and Reiser, P. J. Variations in contractile properties of rabbit single muscle fibres in relation to troponin T isoforms and myosin light chains. *J. Physiol.* 406: 85–98, 1988.

Haller, C., Zehelein, J., Remppis, A., Muller-Bardorff, M. and Katus, H. A. Cardiac troponin T in patients with end-stage renal disease: absence of expression in truncal skeletal muscle. *Clin. Chem.* 44: 930-938, 1998.

Hastings, KE. Molecular evolution of the vertebrate troponin I gene family. *Cell Struct. Funct.* 2: 205-211, 1997.

Hooper, S. L. and Thuma, J. B. Invertebrate muscles: muscle specific genes and proteins. *Physiol. Rev.* 85: 1001-1060, 2005.

Iwamoto, H. The long C-terminal extension of insect flight muscle-specific troponin-I isoform is not required for stretch activation. *Biochem. Biophys. Res. Commun.* 431: 47–51, 2013.

Jin, J. -P. Alternative RNA splicing-generated cardiac troponin T isoform switching: a non-heart-restricted genetic programming synchronized in developing cardiac and skeletal muscles. *Biochem. Biophys. Res. Commun.* 225:883-889, 1996.

Jin, J. -P., Brotto, M. A., Hossain, M. M., Huang, Q. Q., Brotto, L. S., Nosek, T. M., Morton, D. H. and Crawford, T. O. Truncation by Glu180 nonsense mutation results in complete loss of slow skeletal muscle troponin T in a lethal nemaline myopathy. *J. Biol. Chem.* 278:26159-26165, 2003.

Jin, J. -P. and Samanez, R. A. Evolution of a metal-binding cluster in the NH_2-terminal variable region of avian fast skeletal muscle troponin T: functional divergence on the basis of tolerance to structural drifting. *J. Mol. Evol.* 52: 103-116, 2001.

Jin, J. -P. and Smillie, L. B. An unusual metal-binding cluster found exclusively in the avian breast muscle troponin T of Galliformes and Craciformes. *FEBS Lett.* 341: 135-140, 1994.

Jin, J. -P., Zhang, Z. and Bautista, J. A. Isoform diversity, regulation, and functional adaptation of troponin and calponin. Crit. Rev. Eukaryot. *Gene Expr.* 18: 93-124, 2008.

Kajioka, S., Takahashi-Yanaga, F., Shahab, N., Onimaru, M., Matsuda, M., Takahashi, R., Asano, H., Morita, H., Morimoto, S., Yonemitsu, Y., Hayashi, M., Seki, N., Sasaguri, T., Hirata, M., Nakayama, S. and Naito, S. Endogenous cardiac troponin T modulates Ca(2+)-mediated smooth muscle contraction. *Sci. Rep.*; 2: 979, 2012. doi: 10. 1038/srep00979.

Kirkpatrick, K. P., Robertson, A. S., Klaiman, J. M. and Gillis, T. E. The influence of trout cardiac troponin I and PKA phosphorylation on the Ca^{2+} affinity of the cardiac troponin complex. *J. Exp. Biol.* 214: 1981-8, 2011.

Kobayashi, T., Kagami, O., Takagi, T. and Konishi, K. Amino acid sequence of horseshoe crab, Tachypleus tridentatus, striated muscle troponin C. *J. Biochem.* 105: 823-828, 1989a.

Kobayashi, T., Takagi, T., Konishi, K. and Cox, J. A. Amino acid sequence of crayfish troponin I. *J. Biol. Chem.* 264: 1551-1557, 1989b.

Kobayashi, T., Takagi, T., Konishi, K. and Wnuk, W. Amino acid sequences of the two major isoforms of troponin C from crayfish. *J. Biol. Chem.* 264: 18247-18259, 1989c.

Korte, F. S., Mokelke, E. A., Sturek, M. and McDonald, K. S. Exercise improves impaired ventricular function and alterations of cardiac myofibrillar proteins in diabetic dyslipidemic pigs. *J. Appl. Physiol.* 98: 461-467, 2005.

Malouf, N. N., McMahon, D., Oakeley, A. E. and Anderson, P. A. A cardiac troponin T epitope conserved across phyla. *J. Biol. Chem.* 267: 9269-9274, 1992.

Marden, J. H., Fitzhugh, G. H., Girgenrath, M., Wolf, M. R. and Girgenrath, S. Alternative splicing, muscle contraction and intraspecific variation: associations between troponin T transcripts, Ca(2+) sensitivity and the force and power output of dragonfly flight muscles during oscillatory contraction. *J. Exp. Biol.* 204: 3457-3470, 2001.

McLaurin, M. D., Apple, F. S., Voss, E. M., Herzog, C. A., Sharkey, S. W. Cardiac troponin I, cardiac troponin T, and creatine kinase MB in dialysis patients without ischemic heart disease: evidence of cardiac troponin T expression in skeletal muscle. *Clin. Chem.* 43:976-982, 1997.

Moran, C. M., Garriock, R. J., Miller, M. K., Heimark, R. L., Gregorio, C. C. and Krieg, P. A. Expression of the fast twitch troponin complex, fTnT, fTnI and fTnC, in vascular smooth muscle. *Cell Motil. Cytoskeleton* 65: 652-661, 2008.

O'Connell, B., Blazev, R. and Stephenson, G. M. Electrophoretic and functional identification of two troponin C isoforms in toad skeletal muscle fibers. *Am. J. Physiol. Cell Physiol.* 290: C515-C5123, 2006.

Ogut, O., Granzier, H. and Jin, J. P. Acidic and basic troponin T isoforms in mature fast-twitch skeletal muscle and effect on contractility. *Am. J. Physiol.* 276: C1162-C1170, 1999.

Perry, M. J., Tait, J., Hu, J., White, S. C. and Medler, S. Skeletal muscle fiber types in the ghost crab, Ocypode quadrata: implications for running performance. *J. Exp. Biol.* 212: 673-683, 2009.

Pound, K. M., Arteaga, G. M., Fasano, M., Wilder, T., Fischer, S. K., Warren, C. M., Wende, A. R., Farjah, M., Abel, E. D., Solaro, R. J. and Lewandowski, E. D. Expression of slow skeletal TnI in adult mouse hearts confers metabolic protection to ischemia. *J. Mol. Cell. Cardiol.* 51: 236–243 2011.

Reiser, P. J., Greaser, M. L. and Moss, R. L. Developmental changes in troponin T isoform expression and tension production in chicken single skeletal muscle fibres. *J. Physiol.* 449: 573-588, 1992.

Reiser, P. J., Greaser, M. L. and Moss, R. L. Contractile properties and protein isoforms of single fibres from the chicken pectoralis red strip muscle. *J. Physiol.* 493: 553-562, 1996.

Reiser, P. J., Westfall, M. V., Schiaffino, S. and Solaro, R. J. Tension production and thin-filament protein isoforms in developing rat myocardium. *Am. J. Physiol.* 267: H1589-H1596, 1994.

Ricchiuti, V., Voss, E. M., Ney, A., Odland, M., Anderson, P. A. W. and Apple, F. S. Cardiac troponin T isoforms expressed in renal diseased skeletal muscle will not cause false-positive results by the second generation cardiac troponin T assay by Boehringer Mannheim. *Clin. Chem.* 44: 1919-1924, 1998.

Rosser, B. W. C., Wick, M., Waldbillig, D. M. and Bandman, E. Heterogeneity of myosin heavy chain expression in fast-twitch fiber types of mature avian pectoralis muscle. *Biochem. Cell Biol.*, 74: 715-728, 1996.

Saggin, L., Gorza, L., Ausoni, S. and Schiaffino, S. Cardiac troponin T in developing, regenerating and denervated rat skeletal muscle. *Development* 110: 547-554, 1990.

Sferrazza, G. F., Zhang, C., Jia, P., Lemanski, S. L., Athauda, G., Stassi, A., Halager, K., Maier, J. A., Rueda-de-Leon, E., Gupta, A., Dube, S., Huang, X., Prentice, H. M., Dube, D. K. and Lemanski, L. F. Role of myofibril-inducing RNA in cardiac TnT expression in developing Mexican axolotl. *Biochem. Biophys. Res. Commun.* 357: 32-37, 2007.

Shaffer, J. F and Gillis, T. E. Evolution of the regulatory control of vertebrate striated muscle: the roles of troponin I and myosin binding protein-C. *Physiol. Genomics.* 42: 406-19, 2010.

Solaro, R. J., Lee, J. A., Kentish, J. C. and Allen, D. G. Effects of acidosis on ventricular muscle from adult and neonatal rats. *Circ. Res.* 63: 779-787, 1988.

Stephenson, G. M. M. Hybrid skeletal muscle fibres: a rare or common phenomenon? *Clin. Exp. Pharmacol. Physiol.* 28: 692–702, 2001.

Stoutamyer, A. and Dhoot, G. K. Transient expression of fast troponin C transcripts in embryonic quail heart. *J. Muscle Res. Cell Motil.* 26: 237-45, 2005.

Takagi, T. and Konishi, K. Amino acid sequence of troponin C obtained from ascidian (Halocynthia roretzi) body wall muscle. *J. Biochem.* 94: 1753-1760, 1983.

Thys, T. M., Blank, J. M., Coughlin, D. J. and Schachat, F. Longitudinal variation in muscle protein expression and contraction kinetics of largemouth bass axial muscle. *J. Exp. Biol.* 204: 4249-4257, 2001.

Thys, T. M., Blank, J. M. and Schachat, F. H. Rostral-caudal variation in troponin T and parvalbumin correlates with differences in relaxation rates of cod axial muscle. *J. Exp. Biol.* 201: 2993-3001, 1998.

Tobacman, L. S. and Lee, R. Isolation and functional comparison of bovine cardiac troponin T isoforms. *J. Biol. Chem.* 262: 4059-4064, 1987.

Toyota N. and Shimada, Y. Differentiation of troponin in cardiac and skeletal muscles in chicken embryos as studied by immunofluorescence microscopy. *J. Cell Biol.* 91: 497-504, 1981.

Toyota, N. and Shimada, Y. Isoform variants of troponin in skeletal and cardiac muscle cells cultured with and without nerves. *Cell* 33: 297-304, 1983.

Warkman, A. S. and Atkinson, B. G. The slow isoform of *Xenopus* troponin I is expressed in developing skeletal muscle but not in the heart. *Mech. Dev.* 115: 143-146, 2002.

Warkman, A. S. and Atkinson, B. G. Amphibian cardiac troponin I gene's organization, developmental expression, and regulatory properties are different from its mammalian homologue. *Dev. Dyn.* 229: 275-288, 2004.

Wei, B. and Jin, J. -P. Troponin T isoforms and posttranscriptional modifications: Evolution, regulation and function. *Arch. Biochem. Biophys.* 505: 144-154, 2011.

Wilkinson, J. M. The components of troponin from chicken fast skeletal muscle. A comparison of troponin T and troponin I from breast and leg muscle. *Biochem. J.* 169: 229-238, 1978.

Wilkinson, J. M., Moir, A. J. and Waterfield, M.D. The expression of multiple forms of troponin T in chicken-fast-skeletal muscle may result from differential splicing of a single gene. *Eur. J. Biochem.* 143: 47-56, 1984.

Wilkinson, J. M. and Taylor, R. D. Expression of troponin T in atria and ventricles of avian and mammalian hearts. *J. Mol. Cell. Cardiol.* 18: 291-297, 1986.

Wilkinson, J. M. Troponin C from rabbit slow skeletal and cardiac muscle is the product of a single gene. *Eur. J. Biochem.* 103: 179-188, 1980.

Wu, Y. Z., Baker, M. J., Crumley, R. L., Blanks, R. H. I and Caiozzo, V. J. A new concept in laryngeal muscle: multiple myosin isoform types in single fibers of the lateral cricoarytenoid. Otolaryngol. *Head Neck Surg.* 118: 86–94, 1998.

Wu, Y. Z., Crumley, R. L. and Caiozzo, V. J. Are hybrid fibers a common motif of canine laryngeal muscles? Single-fiber analyses of myosin heavy chain-isoform composition. Arch. Otolaryngol. *Head Neck Surg.* 126: 865–873, 2000.

Zanellato, A. M., Borrione, A. C., Saggin, L., Giuriato, L., Schiaffino, S. and Sartore, S. Troponin T- and troponin I-like proteins in bovine vascular smooth muscle. *Circ. Res.* 68: 1349-1361, 1991.

Zhang, Z., Jin, J. -P. Root, D. D. Binding of calcium ions to an avian flight muscle troponin T. *Biochem.* 43: 2645-2655, 2004.

Zhang, M. Y., Zhang, W. J. and Medler, S. The continuum of hybrid IIX/IIB fibers in normal mouse muscles: MHC isoform proportions and spatial distribution within single fibers. *Am. J. Physiol. Regul. Integr. Comp. Physiol.* 299(6): R1582–R1591, 2010.

In: Troponin
Editor: J.-P. Jin

ISBN: 978-1-62948-320-7

Chapter III

Evolutionary Lineage of Troponin Isoforms in Vertebrate Species

J.-P. Jin, MD, PhD
Department of Physiology,
Wayne State University School of Medicine
Detroit, Michigan, US

Abstract

The three subunits of the troponin complex are encoded by genes in two separate phylogenetic lineages. While troponin C is a relative of the calmodulin superfamily of EF hand metal ion binding proteins, troponin I and troponin T are striated muscle-specific proteins. Vertebrate troponin I and troponin T have both evolved into three muscle type-specific isoforms. The molecular evolution of troponin I and troponin T genes and their isoforms present valuable data for the understanding of the structure-function relationship of troponin I and troponin T. The novel approach of restoring and detecting evolutionarily suppressed ancestor epitopic structures in troponin I and troponin T has revealed intriguing insights into of the structure-function relationship of troponin in the Ca^{2+}-regulation of muscle contraction and adaptation.

Abbreviations

TnC, troponin C;
TnI, troponin I;
TnT, troponin T.

Introduction

The troponin complex in vertebrate striated muscle consists of three protein subunits: The Ca^{2+}-binding subunit troponin C (TnC), the actomyosin ATPase inhibiting subunit troponin I (TnI), and the tropomyosin-binding subunit troponin T (TnT) (Ohtsuki et al., 1967; Greaser & Gergely, 1971). The physiological function of troponin complex is to transmit cytosolic Ca^{2+} signals in the regulation of the interactions between sarcomeric actin and myosin filaments during the excitation-contraction-relaxation cycles (Tobacman, 1996; Gordon et al., 2000; Jin et al., 2008). Muscle type-specific isoforms of TnC, TnI and TnT have evolved in vertebrate species. These TnC, TnI and TnT isoform genes encode homologous protein variants with conserved but specifically differentiated structure and function. To understand the molecular evolution of the three troponin subunits and their isoforms is essential for the understanding of their structure-function relationship, developmental regulation, posttranslational modification in physiological and pathophysiological adaptations, and mutations that cause myopathies.

The three subunits of troponin are encoded by genes that belong to two separate evolutionary lineages. TnC belongs to the super family of EF-hand Ca^{2+}-binding proteins (Kawasaki et al., 1998; Nakayama et al., 1992) with a close relationship to the calmodulin gene family (Schaub et al., 2008). In contrast, TnI and TnT are encoded by unique genes that are specifically expressed in striated (skeletal and cardiac) muscles (Jin et al., 2008).

The separate phylogenetic lineages of the TnC genes and the TnI and TnT genes are consistent with their distinct functions in the regulation of muscle contraction. TnC is the Ca^{2+} receptor of the troponin complex and TnI and TnT are allosteric adapter proteins that transmit the Ca^{2+} signal in the thin filament regulatory system of striated muscle. Based on the different evolutionary lineages of the three troponin subunits, the troponin complex can be dissected into two functional domains that transmit the Ca^{2+} signaling to the tropomyosin-actin thin filament: the calcium receptor TnC and the molecular relay/adaptor molecules TnI and TnT (Figure 1).

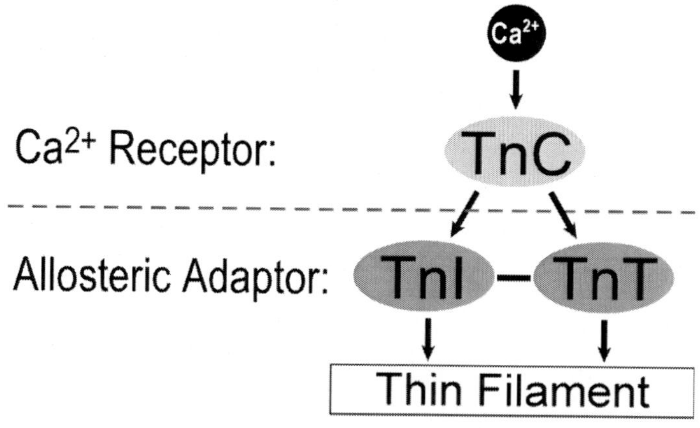

Figure 1. Functional domains of striated muscle troponin complex. Troponin is a protein complex containing three polypeptide subunits: TnC, a member of the calmodulin family, forms the Ca^{2+} receptor domain. TnI and TnT together play an adaptor function to transmit the allosteric signal induced by Ca^{2+} binding to TnC to the tropomyosin-F-actin thin filament.

To provide an integrative outline of the molecular evolution and diversity of the troponin subunits and their isoforms, this chapter will discuss the phylogenetic lineages of TnC, TnI and TnT isoform genes, the molecular evolution of TnI and TnT, and the structural and functional significance of the diverged muscle type-specific isoforms.

Genes Encoding Troponin C Isoforms

Two TnC isoform genes, the fast skeletal muscle TnC (*TNNC2*) and the cardiac/slow skeletal muscle TnC (*TNNC1*), are present in the vertebrate genomes encoding the TnC subunit of troponin in the three types of striated muscle (Parmacek & Leiden, 1989; Parmacek et al., 1990; Gahlmann & Kedes, 1990). The fast skeletal muscle TnC gene is expressed exclusively in fast skeletal muscle fibers of adult animals with a transient expression in embryonic heart (Stoutamyer & Dhoot, 2005). The cardiac/slow skeletal muscle TnC gene is expressed in both cardiac muscle and slow skeletal muscle fibers of adult animals with a transient expression in fast skeletal muscle fibers of embryonic tongue (Prigozy et al., 1997).

The amino acid sequences of fast skeletal muscle TnC and cardiac/slow skeletal muscle TnC in a given species are notably diverged. On the other hand, each of the two TnC isoforms is very much conserved among vertebrate species (Figure 2). This feature indicates that functional adaptation and selection in the differentiated fast versus slow or cardiac muscle fiber environments was the dominant determinant in the molecular evolution of the two TnC isoforms.

Like calmodulin, TnC has two structural domains that contain metal ion-binding sites. Two high affinity metal ion-binding sites (sites III and IV) are present in the C-domain of TnC, which are predominantly occupied by Mg^{2+} in muscle cells under physiological conditions. Two low affinity metal ion-binding sites (sites I and II) are present in the N-domain of fast skeletal muscle TnC to function as the Ca^{2+} receptor sites during the activation of muscle contraction (Johnson et al., 1980; Collins, 1991). The cardiac/slow skeletal muscle TnC differs from the fast skeletal muscle TnC by having only one active Ca^{2+}-binding site (site II) in the N-domain (Johnson et al., 1980; Holroyde et al., 1980). Exchange-replacement of cardiac/slow TnC with fast TnC in skinned cardiac muscle preparations resulted in an increase in the cooperativity of Ca^{2+} activation (Moss et al., 1986; Morimoto & Ohtsuki, 1988).

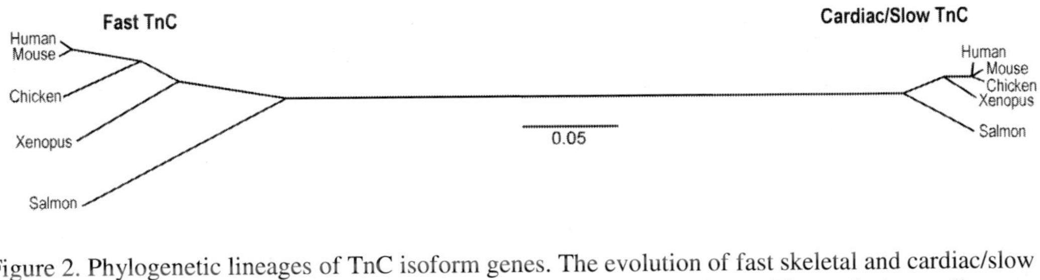

Figure 2. Phylogenetic lineages of TnC isoform genes. The evolution of fast skeletal and cardiac/slow skeletal muscle TnC was analyzed from amino acid sequence comparisons. The evolutionary lineage demonstrated a significant divergence between fast and cardiac/slow TnC isoforms while each of the isoforms has conserved structures across vertebrate species. The evolutionary distances are indicated with a ruler bar for the rate of amino acid substitution.

Therefore, the loss of metal ion-binding site I in cardiac/slow skeletal muscle TnC may have contributed to the lower cooperativity of the Ca^{2+} activation of cardiac and slow skeletal muscle myofilaments as compared with that of fast skeletal muscle.

Considering the presence of four metal ion-binding sites in the closely related EF-hand regulatory protein calmodulin that has a structure very similar to TnC, the fast skeletal muscle TnC may reflect an ancestor structure of the two TnC isoforms. The loss of the metal ion-binding site I in the N-domain of cardiac/slow skeletal muscle TnC was, therefore, a molecular diversity with an evolutionary significance, in which the single regulatory Ca^{2+} binding site in the cardiac/slow skeletal muscle TnC versus the two regulatory Ca^{2+} binding sites in the fast TnC had benefited its evolutionary selection and fixation in heart and slow skeletal muscles.

Genes encoding muscle type-specific TnI and TnT Isoforms are Closely Linked in Physical Pairs in the Vertebrate Genomic DNA

Both TnI and TnT are striated muscle-specific proteins. Distinct isoforms of TnI and TnT are present in the cardiac and skeletal muscles of hagfish (Chong & Jin, 2009), an elementary vertebrate that shared with true vertebrates a common ancestor who lived 500 million years ago (Colbert & Morales, 2001). The long existence of TnI and TnT and their isoform genes has allowed complex and functionally informative evolutionary diversities.

A striking structural feature of the TnI and TnT genes is that they are very closely linked in tandem, with the 3'-end of TnI gene next to the 5'-end of TnT gene, in the chromosomal DNA of vertebrate species (Huang & Jin, 1999; Mullen & Barton, 2000; Cullen et al., 2004). Accordingly, three physical pairs of TnI-TnT genes are present encoding the fiber type-specific isoforms in fast skeletal muscle, slow skeletal muscle and cardiac muscle (Huang & Jin, 1999; Cullen et al., 2004; Baldwin et al., 1985; Wade et al., 1990; Ausoni et al., 1994; Bhavsar et al., 1996; Bucher et al., 1999; Hirao et al., 2004; Jin et al., 2008; Chong & Jin 2009). The closely linked TnI and TnT genes in each pair have their independent promoters and their expressions are regulated separately as well as coordinately in different types of muscle cells, during embryonic and postnatal development, and in physiological or pathophysiological adaptations (Jin et al., 2008; Wei & Jin, 2011).

The close physical link between TnI and TnT genes retained during their hundreds of million years of existence is helpful for understanding the origin and evolution of TnI and TnT and their muscle type-specific isoforms. Among the three TnI-TnT gene pairs, the most closely linked pair is that of cardiac TnI (*TNNI3*) and slow skeletal muscle TnT (*TNNT1*) genes. The physical distance from the transcription termination site of *TNNI3* to the transcription initiation site of *TNNT1* is only ~2.4-2.6-kb as exampled in mouse and human genomes (Huang & Jin, 1999; Cullen et al., 2004).

A mapping of the promoter activity of mouse slow skeletal muscle TnT gene using reporter constructs found that an essential 5'-enhancer activity of the slow TnT gene is integratively present in the upstream cardiac TnI gene (Huang & Jin, 1999). This finding was confirmed *in vivo* in a genetically modified mouse line in which the entire cardiac TnI gene is deleted (Huang et al, 1999). Consistent with the *in vitro* promoter mapping data, the removal of the cardiac TnI gene that contains a portion of the 5'-upstream enhancer region of the slow

TnT gene resulted in a significant knockdown of the expression of slow TnT in soleus and diaphragm muscles (Feng et al., 2009a; 2009b).

These data suggest that the structural integration of TnI and TnT genes has effectively prevented physical separation of their tandem-linkage in the chromosomal DNA during evolution. The preserved close physical linkages indicate directly related origins of the genes encoding TnI and TnT. The close physical linkage also enforces co-evolution of each of the TnI-TnT isoform gene pairs, which helps the investigation of the evolutionary lineage of TnI and TnT isoforms.

Phylogeny of TnI and TnT Isoforms Determined by Sequence Analysis

Sequence alignment of homologous nucleotides or proteins in present-day species, the products of molecular evolution, provides useful information for the phylogenetic relationships of the genes encoding them (Benner et al., 2002). Alignment of the amino acid sequences of TnI and TnT isoforms from representative vertebrate species has demonstrated their evolutionary lineages and distances (Chong & Jin 2009).

An important piece of information from the amino acid sequence comparison of TnI and TnT isoforms (Figure 3) is that the divergence among the three muscle type-specific isoforms in a given species is significantly greater than the divergence of each isoform in different vertebrate species (Jin et al., 1998; 2008; Wei & Jin, 2011).

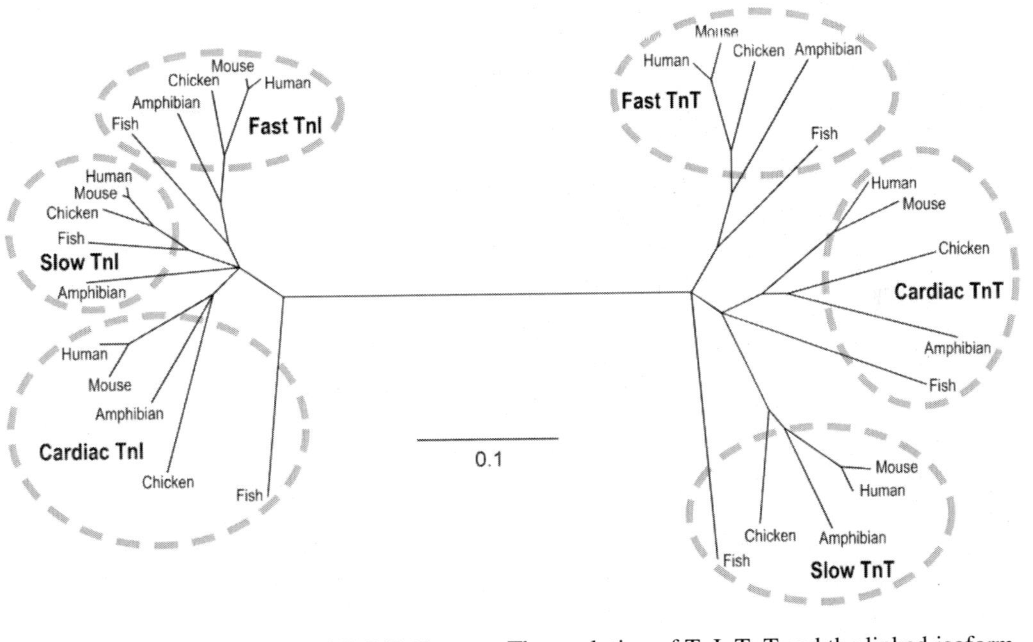

Figure 3. Evolutionary lineages of TnI-TnT genes. The evolution of TnI, TnT and the linked isoform gene pairs was established from amino acid sequence analysis. The evolutionary lineage demonstrated a distal relationship between TnI and TnT, and significant divergence was found between fast, cardiac and slow isoforms while each of the TnI or TnT isoforms has conserved structures across vertebrate species. The evolutionary distances are indicated with a ruler bar for the rate of amino acid substitution. This figure is duplicated with permission from a figure published previously (Chong & Jin, 2009).

Similar to the phylogeny of TnC isoforms (Figure 2), the data indicate that adaptation for functional selection in different muscle types is a driving force in the molecular evolution of TnI and TnT isoforms. This pattern of evolutionary diversity and conservation among TnI and TnT isoforms provided a foundation for understanding the lineage and structure-function relationship of TnI and TnT, and the significance of their diverged muscle type-specific isoforms in the functional properties of cardiac, fast skeletal, and slow skeletal muscles.

Whereas TnI and TnT are significantly diverged proteins, the extremely close physical linkage between TnI and TnT genes suggests their emergence from the duplication of a common ancestor gene. Supporting this hypothesis, phylogenetic analysis of amino acid sequences detected a distant homology between the TnI and TnT gene families (Figure 3). A common structural feature of TnI and TnT proteins is that their N-terminal domain is a regulatory variable structure (Jin et al., 2008; Chong & Jin 2009), also supporting their origination from a common ancestor gene.

To Investigate the Molecular Evolution of TnI and TnT by Detecting Suppressed Ancestral Conformations

Although primary structure/amino acid sequence determines the foundation of protein structure and activity, functional evolution of proteins is very much determined by changes in folded three-dimensional structures. Therefore, the study of molecular evolution solely based on sequence analysis has an intrinsic limitation. To investigate how the folded structure of a protein has evolved and the ancestral relationship with related proteins can significantly enhance our understanding of the evolutionary lineages indicated by sequence analysis.

However, different from the study of gross evolution, which traditionally depends on fossil materials, the remains of ancestor nucleotides and proteins are largely unavailable. This restriction is especially true for directly analyzing the three-dimensional structure of an ancestor protein. Therefore, alternative approaches are needed to extend the studies of protein molecular evolution.

The degree of cross-reactivity of an antibody to homologous proteins reflects the structural similarity and diversity of the epitope(s) (Miller & Cohen, 1991). Based on the concept that the binding affinity of an antibody to an antigen depends on the three-dimensional structural fit between the antibody V region and the epitope on the antigen protein, antibody affinity titrations can detect differences as well as similarities between the molecular structure and conformation of related proteins. This relationship is called the immunological distance and has been employed to evaluate the phylogeny of homologous proteins (Li & Tanimura, 1987; Kaminogawa et al., 1989; Prager, 1993). While the cross-reactivity of polyclonal antibodies reflects the overall structural similarity of related proteins, monoclonal antibodies (mAbs) (Köhler & Milstein, 1975) detect similarities and differences in specific epitopic structures (Chong & Jin 2009).

The three-dimensional structure of a protein may include allosteric conformations that reflect its evolutionary past (Chong & Jin 2009). A protein may also have the potential of restoring ancestral conformations that had been allosterically suppressed by the presence of evolutionarily additive or modified modulator structures. In addition to evaluating static structures, antibodies against allosteric epitopes can be used as probes to detect conformational changes in a protein (34; 36). This concept is illustrated in Figure 4 (Chong & Jin 2009).

The allosteric nature of TnI and TnT proteins allows the use of this novel approach to experimentally study the evolution of the three-dimensional structural relationships between TnI and TnT and among their muscle type-specific isoforms. Employing microtiter plate enzyme-linked immunosorbant assay (ELISA) affinity titrations that effectively detect conformational differences and changes under physiological conditions (Wang & Jin 1998; Jin et al., 2007), we investigated the allosteric structures of TnI and TnT for changes in molecular conformation after modification of evolutionarily additive structures. The approach to experimentally determining the evolutionary lineage of proteins through the analysis of epitope structures using representative antibody probes allows a comparison of protein conformational states to reveal the evolutionary lineages in the TnI-TnT gene family beyond sequence comparisons. The results summarized in the following sections demonstrated informative data for the molecular evolution and structure-function relationships of TnI and TnT and their muscle type specific isoforms.

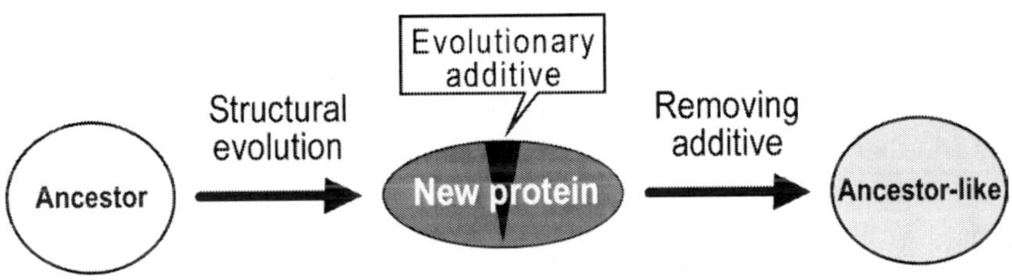

Figure 4. Detection of evolutionarily suppressed epitope structures in present day proteins. A model is proposed to illustrate that the evolution of a protein three-dimensional structure could result from addition and/or alteration of a critical structure that modulates the global conformation and function of the protein. Ancestor-like three-dimensional structures may be suppressed in the new protein and become detectable after removing the modulatory effect. This figure is duplicated with permission from a part of a figure published previously (Chong & Jin, 2009).

This novel approach has a broader significance in the investigation of protein evolution. Although the sequence database in the post-genome era allows extensive analyses of evolutionary relationship of proteins, protein conformational analysis has unique values in understanding the structure-function relationship of proteins. While the protein three-dimensional structure database is rapidly growing, most of the data were based on static structures and the precise correlation between amino acid sequence and folded structure remains to be established. Therefore, an effective and practical method to compare protein epitopic structure and conformation is highly valuable for the study of protein evolution. Our approach to experimentally detecting evolutionarily suppressed ancestor-like conformations in a protein makes it feasible to reveal submolecular fossils and study the origin of protein three-dimensional structures (Chong & Jin, 2009).

Experimental Evidence for the Common Ancestry of TnI and TnT Genes

Gene duplication is a major mechanism to fuel molecular evolution and diversity (Raes & Van de Peer, 2003). The close physical linkage of the genes encoding TnI and TnT in vertebrate genomes (Huang & Jin, 1999; Cullen et al., 2004) and the distant sequence homology (Chong & Jin 2009, Figure 3) suggested their origination from the duplication of a single ancestral gene followed by neofunctionalization.

To investigate whether the distant relationship between TnI and TnT genes indicated by amino acid sequence analysis can be detected as similarities in their three-dimensional protein structure, we explored ancestor-like three-dimensional structure states that are preserved in the TnI and TnT proteins using the epitope analysis approach.

The study first showed that a polyclonal rabbit antiserum RATnI raised by chicken fast skeletal muscle TnI immunization cross-reacts with mouse fast TnT (Chong & Jin, 2009), indicating the presence of TnI-like epitope structures in TnT, likely inherited from a common ancestor of TnI and TnT. This result supports the notion that TnI and TnT genes were evolved by duplication, other than splitting, of an ancestor gene.

In contrast, a polyclonal antiserum RATnT generated from fast skeletal muscle TnT immunization did not show any detectable cross-reactivity to TnI (Chong & Jin, 2009). This unidirectional immunological cross-reactivity of anti-TnI and anti-TnT polyclonal antibodies indicates that the dominant immunogenic epitopes in the TnT immunogen are not present in TnI, reflecting evolutionarily diverged structures that were absent in the common ancestor of TnI and TnT.

The N-terminal variable region of TnT is an evolutionarily added structure (Jin & Samanez, 2001) that plays a role in conformational and functional modulations (Wang & Jin, 1998; Jin & Root, 2000; Biesiadecki et al., 2007). While the polyclonal antiserum RATnI raised against fast skeletal muscle TnI has no detectable cross-reactivity with intact cardiac TnT, the deletion of the N-terminal segment of cardiac TnT resulted in significant binding of RATnI (Chong & Jin 2009).

The C-terminal end segment of TnI is evolutionarily conserved among isoforms and across species (Jin et al., 2001). An mAb, TnI-1 that specifically recognizes the C-terminal end segment of all three TnI isoforms in vertebrate species from fish to human also showed a significant cross-reaction to N-terminal truncated cardiac TnT whereas no cross-reaction with intact cardiac TnT was detectable (Chong & Jin, 2009).

Distinct from destructive effects in which deleting a portion of a protein often destroys the integrity of antibody epitopes, the selective removal of the N-terminal segment of cardiac TnT based on its structural and functional domains revealed the presence of a TnI-like molecular conformation in cardiac TnT, which is suppressed in intact cardiac TnT but can be restored by removing the evolutionarily added modulatory effect of the N-terminal segment.

The data that TnT contains static and allosteric TnI-like conformational states provide the first three-dimensional structure evidence for the genes encoding TnI and TnT to share a common ancestor. The conserved primary function of troponin is an inhibitor of actin-myosin interaction and TnI is the inhibitory subunit of troponin complex (Perry, 1999). Therefore, the experimental data indicating the presence of evolutionarily suppressed TnI-like three-dimensional ancestor structures in TnT logically suggest a hypothesis that the ancestor of TnI and TnT was likely a TnI-like inhibitory protein (Chong & Jin, 2009).

TnI and TnT have diverged prior to the emergence of vertebrates (Chong & Jin, 2009). The data that after the three-dimensional structure of TnT evolved to become significantly distinct from that of TnI, removal of the evolutionarily additive and conformationally modulating N-terminal segment of TnT restores an ancestor-like conformation as recognizable by the anti-TnI antibodies. Despite permanent changes in amino acid sequences during the evolution of TnI and TnT, the reversibility of the allosteric states of TnT to restore an ancestral conformation is a novel mechanism to increase functional variability in the regulation of muscle contractility.

Evolution of Muscle Type-Specific TnT Isoforms

Vertebrate TnI and TnT have each evolved into three isoforms specific for fast skeletal, cardiac, and slow skeletal muscles (Barton et al., 1997; Tiso et al., 1997; Huang & Jin, 1999; Perry, 1998; 1999; Jin et al., 2008). As discussed above, these TnI and TnT isoforms are encoded by three linked pairs of genes, suggesting their parallel evolution through a series of duplication events from an ancestral TnI-TnT gene pair.

The first conformational evidence supporting this hypothesis is that the polyclonal anti-TnI antiserum RATnI cross-reacts to fast skeletal muscle TnT but not cardiac or slow skeletal muscle TnT (Chong & Jin 2009), indicating a higher similarity between TnI and fast TnT. Accordingly, fast skeletal muscle TnT may be the first TnT directly evolved from the duplication of the TnI-like ancestor gene. Therefore, the first TnI-TnT gene pair emerged during evolution may be a fast TnI-fast TnT-like pair and the present-day fast TnI and fast TnT genes have preserved their structural and functional features.

The primary structures of the three muscle type-specific TnT isoforms differ mainly in the N-terminal variable region (Jin et al., 2008; Wei & Jin, 2011). The N-terminal region of TnT does not bind to any known proteins in the myofilaments but functions as a regulatory structure modulating the conformation and function of other regions of TnT. In contrast, the middle region and C-terminal region of TnT are highly conserved, corresponding to the fundamental functions of these two regions that directly interact with TnC, TnI and tropomyosin (Jin et al., 2008; Wei & Jin, 2011). This feature is consistent with the regulatory function of the N-terminal variable region of TnT that provides a structural basis for adaptations to the functional demands in different types of muscle, at different stages of development, or in the muscle of different species.

We first examined the evolutionary relationship among the muscle type-specific TnT isoforms by determining the static state immunological distance between fast skeletal muscle TnT and cardiac or slow TnT using antibody affinity titration (Chong & Jin 2009). The results showed that the polyclonal antiserum RATnT raised by immunization with fast skeletal muscle TnT cross-reacts more strongly to cardiac TnT than that to slow skeletal muscle TnT. Similarly, an mAb T12 generated from fast skeletal muscle TnT immunization cross-reacts with cardiac TnT but not slow skeletal muscle TnT. The closer immunological distance of fast TnT to cardiac TnT than that to slow TnT suggest that the slow TnI-cardiac TnT gene pair was directly derived from the duplication of the fast TnI-fast TnT-like ancestor gene pair (Chong & Jin 2009).

This direct lineage between fast TnT and cardiac TnT was further demonstrated by the restoration of suppressed fast skeletal muscle TnT-like epitope structures in cardiac TnT after experimentally removing the N-terminal segment of cardiac TnT. Deletion of the evolutionarily added N-terminal region significantly increased the affinity of mAb T12. In contrast, N-terminal truncation of slow skeletal muscle TnT did not result in such change (Chong & Jin 2009). The restoration of evolutionarily suppressed fast TnT-like conformational state in cardiac TnT after removal of the evolutionary effect of the N-terminal domain of cardiac TnT indicates an ancestral position of fast TnT, supporting the notion that the slow TnI-cardiac TnT gene pair was originated from duplication of the ancestral fast TnI-fast TnT-like gene pair.

The data that mAb T12 generated from fast TnT immunization cross-reacts to cardiac TnT but not to slow TnT support that the cardiac TnI-slow skeletal muscle TnT gene pair had evolved from duplication of a slow TnI-cardiac TnT-like ancestral gene pair other than the original fast TnI-fast TnT-like gene pair. This evolutionary relationship between cardiac TnT and slow TnT was confirmed by affinities of an mAb 1G9 raised by cardiac TnT immunization. mAb 1G9 exhibited a weak cross-reaction to slow TnT with no cross reaction to fast TnT, indicating a presence of cardiac TnT-like epitope in slow TnT but not in fast TnT.

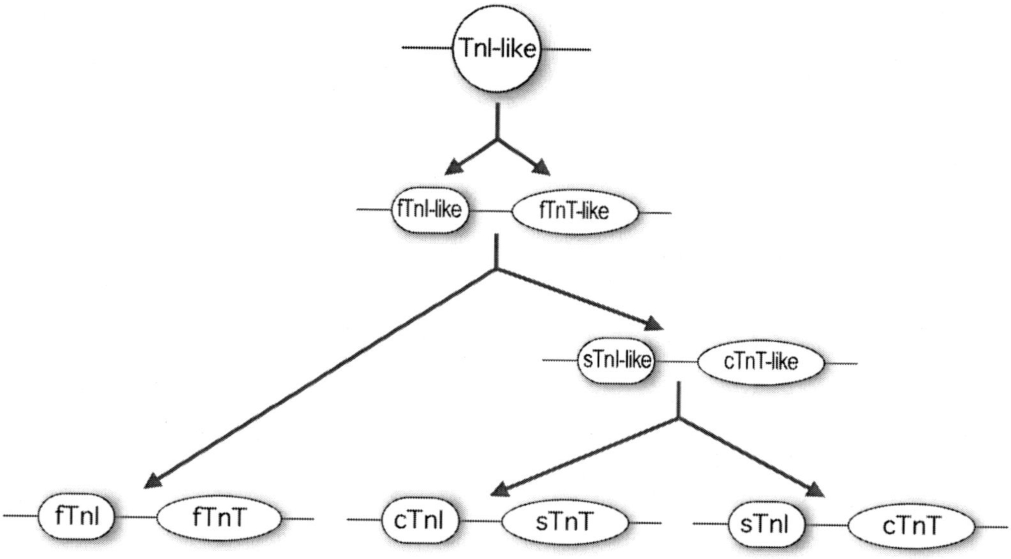

Figure 5. Evolution of TnI and TnT isoforms. The evolution of TnI, TnT and the isoform gene pairs is summarized from data of sequence analysis, immunological distance, and experimental determination of evolutionarily suppressed conformational states. This evolutionary lineage began from a TnI-like ancestor that was duplicated to form a fast TnI-fast TnT-like gene pair. A later duplication event resulted in the emergences of a slow TnI-cardiac TnT-like gene pair that was further duplicated to form the present-day slow TnI-cardiac TnT and cardiac TnI-slow TnT gene pairs. This figure is duplicated with permission from a part of a figure published previously (Chong & Jin, 2009).

Serial N-terminal truncations of slow skeletal muscle TnT resulted in progressively increased affinity to mAb 1G9 (Chong & Jin 2009). Therefore, the cardiac TnT-like 1G9 epitope structure located in the C-terminal T2 domain, which had been evolutionarily suppressed in intact slow TnT, can be more clearly detected when the conformational modulating effect of the N-terminal region is removed. This cardiac TnT-like three-dimensional structure in slow TnT is further restored when both the N-terminal and middle regions were removed as shown by the more increased affinity of mAb 1G9 to the isolated C-terminal T2 domain of slow TnT (Chong & Jin 2009). In contrast, removing the N-terminal and middle regions of fast skeletal muscle TnT did not produce any binding of mAb 1G9. The complete absence of 1G9 epitope structure in fast TnT suggests its emergence after the divergence of slow TnI-cardiac TnT and fast TnI-fast TnT gene pairs.

Summarized in Figure 5, the evolutionary lineage of TnI-TnT isoform genes determined from protein epitopic structures is consistent with the [fast skeletal muscle (slow skeletal muscle, cardiac muscle)] phylogenetic relationship indicated by sequence analyses of TnT and TnI (Figure 3, Chong & Jin, 2009) and other muscle proteins (Oota et al., 1999).

The undifferentiated utilization of the cardiac/slow isoform of TnC in cardiac and slow skeletal muscles supports the hypothesis that the evolutionary emergence of the cardiac and slow TnI-TnT gene pairs was a relatively recent event and the cardiac TnI-slow TnT gene pair is the newest pair among the three muscle type-specific isoforms (Figure 5) (Chong & Jin 2009). The latest emergence of the cardiac TnI-and slow TnT gene pair is also supported by the presence of a unique N-terminal extension in cardiac TnI (Jin et al., 2008; Akhter et al., 2012).

Muscle type-specific TnI Isoforms

Same as the TnT isoforms, three muscle type-specific TnI genes are present in vertebrate species encoding cardiac, slow and fast skeletal muscle isoforms (Hastings, 1997; Jin et al., 2008). Vertebrate TnI is highly conserved among the three isoforms and across species in majority of the amino acid sequences. In addition to the ~180 amino acids conserved structure, cardiac TnI has a unique N-terminal extension of ~30 amino acids encoded by an cardiac TnI-specific exon (exon 3) (Jin et al., 2008), which is absent in the fast and slow skeletal muscle TnIs (Perry, 1999). The N-terminal extension of cardiac TnI contains two protein kinase A (PKA)-phosphorylated residues, Ser_{22} and Ser_{23}, and therefore is a target site downstream of the β-adrenergic signaling. Two more Ser residues in the N-terminal region of cardiac TnI (Ser_5 and Ser_6) may also play a regulatory role upon protein kinase C-ε-catalyzed phosphorylation (Henze et al., 2013).

In adult skeletal muscle, the expression of the two skeletal muscle TnI isoform genes, *TNNI1* (slow TnI) and *TNNI2* (fast TnI), are coupled with the expression of slow and fast skeletal muscle TnT isoform genes, respectively, in a fiber type-specific manner (Brotto et al., 2006; Yu et al., 2007). During the slow to fast fiber type transition in muscle adaptation to simulated microgravity unloading, a parallel slow to fast isoform switching occurs for both TnI and TnT (Stevens et al., 2002; Yu et al., 2007).

The early embryonic heart expresses only slow skeletal muscle TnI (Sasse et al., 1993). The expression of cardiac TnI begins in mouse heart at around embryonic day 10 and the level gradually increases during perinatal development while the expression of slow TnI is down-regulated. This switch completes in mouse heart within two weeks after birth and the adult heart expresses solely cardiac TnI (Saggin et al., 1999). The N-terminal extension of cardiac TnI can be cleaved via restrictive proteolysis in cardiac adaptations (Yu et al., 2001; Feng et al., 2008), and transgenic mouse heart expressing 100% N-terminal truncated cardiac TnI remain functional (Feng et al., 2009a). Therefore, the N-terminal extension specific to cardiac TnI is not a part of the TnI core structure but an evolutionarily added regulatory domain for unique functions of the adult cardiac muscle.

The linkage between TnI and TnT isoform genes (Figure 5) reflects their functional relationships. Although the cardiac TnI-slow TnT and the slow TnI-cardiac TnT gene pairs appear as scrambled linkages of isoform genes, the functionally paired expression of slow TnI and cardiac TnT in embryonic heart indicates that they represent functionally meaningful pairing as well.

Functional Implications

The expression of the three TnI and three TnT isoform genes in adult cardiac and skeletal muscles is strictly controlled in a muscle type-specific manner. Knockout of the cardiac TnI gene (Huang et al., 1999) or cardiac TnT gene (Nishii et al., 2008) in mice resulted in postnatal or embryonic lethality. A nonsense mutation in human slow TnT gene resulting in complete loss of slow TnT protein causes a recessive nemaline myopathy with infantile lethality (Jin et al., 2003). Therefore, the presence of multiple TnI and TnT isoforms is not a redundancy, but essential for the function of the different types of striated muscle.

The different TnI isoforms contribute to the Ca^{2+} sensitivity, cooperativity, and pH tolerance of force development in different types of muscle fibers (Westfall et al., 1999). In transgenic mice over-expressing slow TnI in the heart to replace the endogenous cardiac TnI, the cardiac muscle displayed an increased Ca^{2+} sensitivity in force development (Fentzke et al., 1999). Slow TnI confers a higher resistance to the acidosis-caused inhibition in embryonic cardiac muscle than that in adult heart muscle that contains cardiac TnI (Westfall et al., 2000). Cardiac myocytes transfectively expressing slow TnI exhibited increased Ca^{2+} sensitivity of tension development with decreased cooperativity (Westefall & Metzger, 2001). As an adult cardiac muscle-specific regulatory structure, the evolutionarily added N-terminal extension of cardiac TnI is a target for physiological and pharmacological regulations of cardiac function, such as the modifications via PKA-catalyzed phosphorylation (Solaro, 2003) and the restrictive proteolytic deletion (Yu et al., 2001; Barbato et al., 2005; Feng et al., 2008).

Similarly, investigating the evolutionary lineage of the three TnT isoform genes helps to understand the structure-function relationship of TnT as well as the physiological or pathological significance of the muscle type-specific and developmental isoforms. The effect of removing the evolutionarily added N-terminal segment of TnT on revealing evolutionarily suppressed epitopic structures inherited from an ancestor isoform demonstrates another convincing example for the role of the N-terminal variable region as a conformational modulator (Wang & Jin, 1998; Jin & Root, 2000; Biesiadecki et al., 2007).

The observation that ancestor-like conformations may be retained in a suppressed state in the folded protein structures and be restored when the evolutionarily modulating structure is removed provides a novel insight into the structure-function relationship of TnI and TnT. A practical significance of the critical role of evolutionarily determining structures is that when the stabilizing effect of the conformational modulator is removed, the suppressed ancestor-like three-dimensional structure or conformation will be exhibited (Chong & Jin, 2009). The broader meaning of this mechanism may explain the molecular basis of atavism, in which a mutation that causes the loss of evolutionary modulation in a protein would result in the expression of an ancestral trait rather than a radical new trait. Similarly, a mutation of TnI or TnT affecting the function of an evolutionary modulator site or responding site would result in the devolution of troponin function to that in an ancestor species.

In addition to strengthening the sequence analysis evidence that TnI and TnT arose from gene duplication, the conformational studies provided data for the presence of allosteric conformational structures shared by TnI and TnT, which might have been retained for a functional benefit. It would be interesting to investigate the nature of these structures and their role in regulating muscle function. For example, removing the N-terminal segment of TnT by restricted proteolysis during acute functional adaptation (Zhang et al., 2006; Feng et al., 2008b) restores an epitope structure recognized by mAb TnI-1 against the C-terminal end segment of TnI (Chong & Jin 2009). The C-terminal end segment of TnI is the most conserved region in the TnI polypeptide chain, which undergoes Ca^{2+}-regulated conformational changes in the troponin complex (Jin et al., 2001). Three-dimensional structure reconstruction (Pirani et al., 2006) and protein binding (Akhter et al., 2012) studies suggested a role for the C-terminal end segment of TnI in trapping tropomyosin during myofilament regulation. Therefore, the presence of suppressed but restorable TnI C-terminus-like structure in TnT not only indicates that TnI and TnT arose from a TnI-like ancestor protein, but also may have a functional significance in muscle adaptation to stress conditions.

Further studies are merited to explore the presence and functional significance of structural similarities preserved in both TnI and TnT. In addition to identifying structural modifications that were critical to the diversity of TnI and TnT isoforms, to understand the role of conformational modulator sites that determined the evolutionary divergence and the consequence of manipulating these sites will help to predict the structural and functional potentials of the protein and guide the design of structural modifications and the search for effective drug targets. On the other hand, mutations in such evolutionarily determining structure of TnI and TnT might have high potency to modify functions and to cause diseases.

Conclusion

TnI and TnT have evolved into three muscle fiber type-specific isoforms with diverged functions. In addition to revealing the evolutionary lineages in the TnI-TnT gene family, recent studies of TnI and TnT isoform evolution provided data to better understand the structure-function relationship of the TnI and TnT isoforms. It remains to be investigated that as evolutionarily added modulatory structures, how the N-terminal extension of cardiac TnI regulates the function of TnI and how the N-terminal variable region of TnT functions as an evolutionary determinant in the molecular conformation and function of TnT. Future work is needed and answers to these questions will ultimately lead to fundamental advancement of our understanding of the molecular mechanisms of muscle contraction and the physiological and pathophysiological adaptations of striated muscles.

Acknowledgments

Many of my past and present lab members participated in our studies reviewed in this chapter. Their names are presented in the references cited and I sincerely thank everyone for their outstanding contribution. This work was supported by grants from the National Institutes of Health AR048816 and HL098945.

References

Akhter, S., Zhang, Z., and Jin, J.-P. (2012) The heart-specific NH_2-terminal extension regulates the molecular conformation and function of cardiac troponin I. *Am. J. Physiol. Heart and Circ. Physiol.* 302:H923-H933.

Ausoni, S., Campione, M., Picard, A., Moretti, P., Vitadello, M., De Nardi, C., Schiaffino, S. (1994) Structure and regulation of the mouse cardiac troponin I gene. *J. Biol. Chem.* 269:339-46.

Barbato, J.C., Huang, Q.-Q., Hossain, M.M., Bond, M., and Jin, J.-P. (2005) Proteolytic N-Terminal Truncation of Cardiac Troponin I Enhances Ventricular Diastolic Function. *J. Biol. Chem.* 280:6602-6609.

Baldwin, A.S., Jr., Kittler, E.L., Emerson, C.P., Jr. (1985) Structure, evolution, and regulation of a fast skeletal muscle troponin I gene. *Proc. Natl. Acad. Sci. U S A* 82:8080-4.

Barton, P.J., Townsend, P.J., Brand, N.J., and Yacoub, M.H. (1997) Localization of the fast skeletal muscle troponin I gene (TNNI2) to 11p15.5: genes for troponin I and T are organized in pairs. *Ann. Human. Genet.* 61: 519-523

Benner, S.A., Caraco, M.D., Thomson, J.M., and Gaucher, E.A. (2002) Planetary biology--paleontological, geological, and molecular histories of life. *Science* 296: 864-868

Bhavsar, P.K., Brand, N.J., Yacoub, M.H., Barton, P.J. (1996) Isolation and characterization of the human cardiac troponin I gene (TNNI3). *Genomics* 35:11-23.

Biesiadecki, B.J., Chong, S.M., Nosek, T.M., and Jin, J.-P. (2007) Troponin T core structure and the regulatory NH2-terminal variable region. *Biochemistry* 46:1368-1379

Brotto, M.A., Biesiadecki, B.J., Brotto, L.S., Nosek, T.M., Jin, J.-P. (2006) Coupled expression of troponin T and troponin I isoforms in single skeletal muscle fibers correlates with contractility. *Am. J. Physiol. Cell Physiol.* 290:C567-76.

Bucher, E.A., Dhoot, G.K., Emerson, M.M., Ober, M., Emerson, C.P., Jr. (1999) Structure and evolution of the alternatively spliced fast troponin T isoform gene. *J. Biol. Chem.* 274:17661-70.

Chong, S.M., and Jin, J.-P. (2009) To Investigate Protein Evolution by Detecting Suppressed Epitope Structures. *J. Mol. Evol.* 68:448-60.

Colbert, E.H. & Morales, M. (2001) Colbert's Evolution of the Vertebrates: A History of the Backboned Animals Through Time. 4th edition. John Wiley & Sons, Inc., New York - ISBN 978-0-471-38461-8.

Collins, J.H. (1991) Myosin light chains and troponin C: structural and evolutionary relationships revealed by amino acid sequence comparisons. *J. Muscle Res. Cell. Motil.* 12:3-25.

Cullen, M.E., Dellow, K.A., Barton, P.J.. (2004) Structure and regulation of human troponin genes. *Mol. Cell. Biochem.* 263:81-90.

Feng, H.-Z., Chen, M., Weinstein, L.S. and Jin, J.-P. (2008) Removal of the N-terminal extension of cardiac troponin I as a functional compensation for myocardial b-adrenergic deficiency. *J. Biol. Chem.* 283:33384-93.

Feng, H.-Z., Biesiadecki, B.J., Yu, Z.-B., Hossain, M.M., and Jin, J.-P. (2008b) Restricted N-terminal truncation of cardiac troponin T: A novel mechanism for functional adaptation to energetic crisis. *J. Physiol.* (London) 586:3537-3550.

Feng, H.-Z., Hossain, M.M., Huang, X.-P., and Jin, J.-P. (2009a) Myofilament Incorporation Determines the Stoichiometry of Troponin I in Transgenic Expression and the Rescue of A Null Mutation. *Archives Biochem. Biophys.* 487:36-41.

Feng, H.-Z., Wei, B., and Jin, J.-P. (2009b) Deletion of a genomic segment containing the cardiac troponin I gene knocked down the expression of slow troponin T gene and impaired fatigue tolerance of diaphragm muscle. *J. Biol. Chem.* 284:31798–31806.

Fentzke, R.C., Buck, S.H., Patel, J.R., Lin, H., Wolska, B.M., Stojanovic, M.O., Martin, A.F., Solaro, R.J., Moss, R.L., Leiden, J.M. (1999) Impaired cardiomyocyte relaxation and diastolic function in transgenic mice expressing slow skeletal troponin I in the heart. *J. Physiol.* (London) 517:143-57.

Gahlmann, R., Kedes, L. (1990) Cloning, structural analysis, and expression of the human fast twitch skeletal muscle troponin C gene. *J. Biol. Chem.* 265:12520-8.

Gordon, A.M., Homsher, E., Regnier, M. (2000) Regulation of contraction in striated muscle. *Physiological. Rev.*, 80:853-924.

Greaser, M.L., Gergely, J. (1971) Reconstitution of troponin activity from three protein components. *J. Biol. Chem.* 246:4226-4233.

Hastings, K.E. (1997) Molecular evolution of the vertebrate troponin I gene family. *Cell. Struct. Funct.* 22:205-11.

Henze, M., Patrick, S.E., Hinken, A., Scruggs, S.B., Goldspink, P., de Tombe, P.P., Kobayashi, M., Ping, P., Kobayashi, T., Solaro, R.J. (2013) New insights into the functional significance of the acidic region of the unique N-terminal extension of cardiac troponin I. *Biochim. Biophys. Acta.* 1833:823-32.

Hirao, C., Yonemura, I., Miyazaki, J. (2004) Genomic structure of the chicken slow skeletal muscle troponin T gene. *Gene* 338:243-56.

Holroyde, M.J., Robertson, S.P., Johnson, J.D., Solaro, R.J., Potter, J.D. (1980) The calcium and magnesium binding sites on cardiac troponin and their role in the regulation of myofibrillar adenosine triphosphatase. *J. Biol. Chem.* 255:11688-93.

Huang, Q.-Q., and Jin, J.-P. (1999) Preserved close linkage between the genes encoding troponin I and troponin T, reflecting an evolution of adapter proteins coupling the Ca(2+) signaling of contractility. *J. Mol. Evol.* 49:780-788.

Huang, X.-P., Pi, Y.Q., Lee, K.J., Henkel, A.S., Gregg, R.G., Powers, P.A., Walker, J.W. (1999) Cardiac troponin I gene knockout: A mouse model of myocardial troponin I depletion. *Circ. Res.* 84:1-8.

Jin, J.-P., Chen, A., Huang, Q.-Q. (1998) Three alternatively spliced mouse slow skeletal muscle troponin T isoforms: conserved primary structure and regulated expression during postnatal development. *Gene* 214:121-129.

Jin, J.-P., & Root, D.D. (2000) Modulation of troponin T molecular conformation and flexibility by metal ion binding to the NH_2-terminal variable region. *Biochemistry* 39:11702-11713.

Jin, J.-P., & Samanez, R. (2001) Evolution of a metal-binding cluster in the NH(2)-terminal variable region of avian fast skeletal muscle troponin T: functional divergence on the basis of tolerance to structural drifting. *J. Mol. Evol.* 52:103-116.

Jin, J.-P., Yang, F.-W., Yu, Z.-B., Ruse, C.I., Bond, M., and Chen, A. (2001) The highly conserved COOH-terminus of troponin I forms a Ca^{2+}-modulated allosteric domain in the troponin complex. *Biochemistry* 40:2623-2631.

Jin, J.-P., Brotto, M.A., Hossain, M.M., Huang, Q.-Q., Brotto, L.S., Nosek, T.M., Morton, D.H., Crawford, T.O. (2003) Truncation by Glu180 nonsense mutation results in complete loss of slow skeletal muscle troponin T in a lethal nemaline myopathy. *J. Biol. Chem.* 278:26159-26165.

Jin, J.-P., Chong, S.M., and Hossain, M.M. (2007) Microtiter plate monoclonal antibody epitope analysis of Ca^{2+}- and Mg^{2+}-induced conformational changes in troponin C. *Arch. Biochem. Biophys.* 466:1-7.

Jin, J.-P., Zhang, Z., and Bautista, J.A. (2008) Isoform diversity, regulation and functional adaptations of troponin and calponin. *Crit. Rev. Eukar. Gene Expr.* 18:93-124.

Johnson, J.D., Collins, J.H., Robertson, S.P., Potter, J.D. (1980) A fluorescent probe study of Ca2+ binding to the Ca2+-specific sites of cardiac troponin and troponin C. *J. Biol. Chem.* 255:9635-40.

Kawasaki, H., Nakayama, S. & Kretsinger, R. H. (1998) Classification and evolution of EF-hand proteins. *BioMetals* 11:277–295.

Kaminogawa, S., Shimoda, M., Kurisaki, J., and Yamauchi, K. (1989) Application of a monoclonal antibody to a comparative study of alpha-lactalbumins from various species. *J. Dairy Sci.* 72:1124-1129.

Köhler, G., & Milstein, C. (1975) Continuous cultures of fused cells secreting antibody of predefined specificity. *Nature* 256:495-497.

Li, W.H., and Tanimura, M. (1987) The molecular clock runs more slowly in man than in apes and monkeys. *Nature* 326:93-96.

Miller, E.J., & Cohen, A.B. (1991) Use of antibodies in the study of protein structure and function in lung diseases. *Am. J. Physiol.* 260:L1-12.

Morimoto, S., Ohtsuki, I. (1988) Effect of substitution of troponin C in cardiac myofibrils with skeletal troponin C or calmodulin on the Ca2+- and Sr2+-sensitive ATPase activity. *J. Biochem.* (Tokyo) 104:149-54.

Moss, R.L., Lauer, M.R., Giulian, G.G., Greaser, M.L. (1986) Altered Ca2+ dependence of tension development in skinned skeletal muscle fibers following modification of troponin by partial substitution with cardiac troponin C. *J. Biol. Chem.* 261:6096-9.

Mullen, A.J., Barton, P.J. (2000) Structural characterization of the human fast skeletal muscle troponin I gene (TNNI2). *Gene* 242:313-20.

Nakayama, S., Moncrief, N.D., Kretsinger, R.H. (1992) Evolution of EF-hand calcium-modulated proteins. II. Domains of several subfamilies have diverse evolutionary histories. *J. Mol. Evol.* 34:416-48.

Nishii, K., Morimoto, S., Minakami, R., Miyano, Y., Hashizume, K., Ohta, M., Zhan, D.Y., Lu, Q.W., Shibata, Y. (2008) Targeted disruption of the cardiac troponin T gene causes sarcomere disassembly and defects in heartbeat within the early mouse embryo. *Dev. Biol.* 322:65-73.

Ohtsuki, I., Masaki, T., Nonomura, Y., Ebashi, S. (1967) Periodic distribution of troponin along the thin filament. *J. Biochem.* 61:817-819.

Oota, H., Saitou, N., Matsushita, T., Ueda, S. (1999) Molecular genetic analysis of remains of a 2,000-year-old human population in China-and its relevance for the origin of the modern Japanese population. *Am. J. Hum. Genet.* 64:250-258.

Parmacek, M.S., Leiden, J.M. (1989) Structure and expression of the murine slow/cardiac troponin C gene. *J. Biol. Chem.* 264:13217-25.

Parmacek, M.S., Bengur, A.R., Vora, A.J., Leiden, J.M. (1990) The structure and regulation of expression of the murine fast skeletal troponin C gene. Identification of a developmentally regulated, muscle-specific transcriptional enhancer. *J. Biol. Chem.* 265:15970-6.

Perry, S.V. (1998) Troponin T: genetics, properties and function. *J. Muscle Res. Cell Molti.* 19:575-602.

Perry, S.V. (1999) Troponin I: inhibitor or facilitator. *Mol. Cell. Biochem.* 190:9-32.

Pirani, A., Vinogradova, M.V., Curmi, P.M., King, W.A., Fletterick, R.J., Craig, R., Tobacman, L.S., Xu, C., Hatch, V., Lehman, W. (2006) An atomic model of the thin filament in the relaxed and Ca2+-activated states. *J. Mol. Biol.* 357:707-17.

Prager, E.M. (1993) The sequence-immunology correlation revisited: data for cetacean myoglobins and mammalian lysozymes. *J. Mol. Evol.* 37:408-416.

Prigozy, T.I., Dalrymple, K., Shuler, C., Kedes, L. (1997) Differential expression of troponin C genes during tongue myogenesis. *Dev. Dyn.* 209:36-44.

Raes, J. and Van de Peer, Y. (2003) Gene duplication, the evolution of novel gene functions, and detecting functional divergence of duplicates in silico. *Appl. Bioinformatics.* 2:91-101.

Saggin, L., Gorza, L., Ausoni, S., Schiaffino, S. (1989) Troponin I switching in the developing heart. *J. Biol. Chem.* 264:16299-302.

Sasse, S., Brand, N.J,, Kyprianou, P., Dhoot, G.K., Wade, R., Arai, M., Periasamy, M., Yacoub, M.H., Barton, P.J. (1993) Troponin I gene expression during human cardiac development and in end-stage heart failure. *Circ. Res.* 72:932-8.

Schaub, M.C., Heizmann, C.W. (2008) Calcium, troponin, calmodulin, S100 proteins: From myocardial basics to new therapeutic strategies. *Biochemical and Biophysical Research Communications* 369:247–264.

Solaro, R.J. (2003) The special structure and function of troponin I in regulation of cardiac contraction and relaxation. *Adv. Exp. Med. Biol.* 538:389-401; discussion 401-2.

Stevens, L., Bastide, B., Kischel, P., Pette, D., Mounier, Y. (2002) Time-dependent changes in expression of troponin subunit isoforms in unloaded rat soleus muscle. *Am. J. Physiol. Cell. Physiol.* 282:C1025-30.

Stoutamyer, A., Dhoot, G.K. (2005) Transient expression of fast troponin C transcripts in embryonic quail heart. *J. Muscle Res. Cell. Motil.* 26:237-45.

Tiso, N., Rampoldi, L., Pallavicini, A., Zimbello, R., Pandolfo, D., Valle, G., Lanfranchi, G., and Danieli, A. (1997) Fine mapping of five human skeletal muscle genes: alpha-tropomyosin, beta-tropomyosin, troponin-I slow-twitch, troponin-I fast-twitch, and troponin-C fast. *Biochem. Biophys. Res. Commun.* 230:347-350.

Tobacman, L.S. (1996) Thin filament-mediated regulation of cardiac contraction. *Annu. Rev. Physiol.* 58:447-481.

Wade, R., Eddy, R., Shows, T.B., Kedes, L. (1990) cDNA sequence, tissue-specific expression, and chromosomal mapping of the human slow-twitch skeletal muscle isoform of troponin I. *Genomics* 7:346-57.

Wang, J., & Jin, J.-P. (1998) Conformational modulation of troponin T by configuration of the NH2-terminal variable region and functional effects. *Biochemistry* 37:14519-14528.

Wei, B., and Jin, J.-P. (2011) Troponin T Isoforms and Posttranscriptional Modifications: Evolution, Regulation and Function. *Arch. Biochem. Biophy.* 505:144-154.

Westfall, M.V., Albayya, F.P., Metzger, J.M. (1999) Functional analysis of troponin I regulatory domains in the intact myofilament of adult single cardiac myocytes. *J. Biol. Chem.* 274:22508-16.

Westfall, M.V., Albayya, F.P., Turner, I.I., Metzger, J.M. (2000) Chimera analysis of troponin I domains that influence Ca(2+)-activated myofilament tension in adult cardiac myocytes. *Circ. Res.* 86:470-7.

Westfall, M.V., Metzger, J.M. (2001) Troponin I isoforms and chimeras: tuning the molecular switch of cardiac contraction. *News Physiol. Sci.* 16:278-81.

Yu, Z.-B. Zhang, L.-F. and Jin, J.-P., (2001) A proteolytic NH_2-terminal truncation of cardiac troponin I up-regulated in simulated microgravity. *J. Biol. Chem.* 276:15753–15760.

Yu Z.-B, Gao, F., Feng, H.-,Z., Jin, J.-P. (2007) Differential regulation of myofilament protein isoforms underlying the contractility changes in skeletal muscle unloading. *Am. J. Physiol. Cell Physiol.* 292:C1192-203.

Zhang, Z., Biesiadecki, B.J., and Jin, J.-P. (2006) Selective removal of the N-terminal variable region of cardiac troponin T in ischemia-reperfusion by myofibril-associated m-calpain cleavage. *Biochemistry* 45:11681-94.

In: Troponin
Editor: J.-P. Jin

ISBN: 978-1-62948-320-7

Chapter IV

Biochemical Functions of Troponin

Sampath K. Gollapudi, PhD and Murali Chandra, PhD*

Department of Integrated Physiology and Neuroscience,
Washington State University, Pullman, Washington, US

4.1. Introduction

Ever since Ebashi discovered troponin (Ebashi 1963), great advances have been made in our understanding of how the troponin (Tn) complex regulates striated muscle (cardiac and fast skeletal muscle) contraction. It is now well established that the triggering event for striated muscle contraction comes from the binding of Ca^{2+} to the Tn complex and subsequent structural changes within the Tn complex. Structural changes in the Tn complex trigger a cascade of allosteric/cooperative interactions within the thin filament, culminating in the activation of thin filaments caused by a shift in the position of tropomyosin (Tm) on the thin filament (Haselgrove 1972; Huxley 1972). Although Ca^{2+}-activation of the thin filament is essential for muscle contraction, the force-bearing ('open') state of the thin filament depends on the crossbridge (XB)-mediated effects on cooperative/allosteric processes operating in the thin filament. During this XB-mediated activation of the thin filament, strongly-bound XBs promote the recruitment of other strong XBs by inducing further movement of Tm-Tn on the thin filament (Figure 1).

Whether it is Ca^{2+}- or XB-mediated activation of the thin filament, it is now well appreciated that the movement of Tm-Tn on the actin filament is fundamental to the magnitude of contractile activation and deactivation. Since such movement of Tm-Tn on the actin filament requires tightly coordinated protein-protein interactions, any alterations in the composition of Tn—as it occurs in different tissues—is expected to modify tissue function. Therefore, our discussion will place more emphasis on the biochemical role of Tn subunits in the activation of thin filaments and how their activity is altered by tissue-specific expression of different genes.

* Correspondence: Murali@vetmed.wsu.edu.

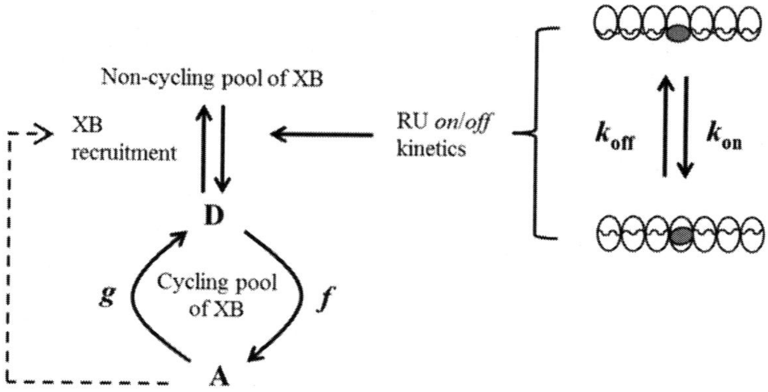

Figure 1. A simplified scheme for the activation of thin filaments by Tn and subsequent force development in striated muscles. Tm is shown as a wavy line and Tn is shown as a filled ellipse. '*off-on*' state transition of the regulatory unit, RU (Tm-Tn), depends on Ca^{2+} binding to Tn. The Ca^{2+}-mediated movement of RU permit XBs to transition from the non-cycling to the cycling pool. In the cycling pool, force-bearing XBs are represented by 'A' and non-force-bearing XBs are represented by 'D'. XBs in the cycling pool may undergo transition from D→A or A→D according to the rate constants, *f* and *g*, respectively. The transition between the cycling and non-cycling XBs is not only influenced by the '*off-on*' state of RU, but is also influenced by strong XB-mediated cooperative feedback on the thin filament, as shown by the dotted line (model scheme was adapted from Campbell 1997).

Because Ca^{2+}-induced structural changes in the Tn complex is obligatory for the '*off-on*' state transition of striated thin filaments, it is imperative to understand the biochemical nature of the Tn complex so that its role in muscle contraction could be fully recognized. The Tn complex comprises three individual subunits: 1) troponin C (TnC), which binds Ca^{2+}; 2) Troponin I (TnI), which inhibits strong actin-myosin interactions by blocking the myosin-binding site on actin in the absence of Ca^{2+}; and 3) troponin T (TnT), which binds strongly to Tm and helps anchor TnC-TnI on the actin filament. It is now well understood that the process of muscle contraction begins when Ca^{2+} ions, which are released from the sarcoplasmic reticulum, bind to TnC in a reversible manner. This initial binding of Ca^{2+} to TnC leads to a cascade of structural changes within the Tn complex, leading to two fundamental structural events that are cardinal to muscle contraction: 1) the release of TnI-induced inhibition of actin-myosin interactions; 2) alterations in the spatial disposition of Tm on the actin filament. Tn-aided movement of Tm allows the initial binding of XBs to actin and subsequent transition to their force-bearing state. Thus, Ca^{2+}-triggered changes in the Tn complex permit Ca^{2+}-mediated regulation of striated muscle contraction.

4.2. TnC

The letter 'C' in TnC originates from its capacity to bind Ca^{2+} selectively. The ability of TnC to bind Ca^{2+} and initiate contraction are experimentally proven using a variety of techniques such as site-directed mutagenesis of Ca^{2+} binding sites in TnC (Gulati et al. 1992; Chandra et al. 1994; Smith et al. 1994; Rao et al. 1995), selective removal of TnC from muscle fibers (Moss et al. 1986; Morimoto and Ohtsuki 1987; Babu et al. 1989), *in vitro* reconstituted assays (Dotson and Putkey 1993; Sorenson et al. 1995; Fuchs and Grabarek 2011), NMR solution measurements (Brito et al. 1993; Sia et al. 1997; Mercier et al. 2000), and X-Ray crystallography studies (Herzberg and James 1985; Herzberg and James 1988; Satyshur et al. 1988). The most well studied TnC isoform is that of fast skeletal TnC (fsTnC), which contains two globular domains, each carrying two helix-loop-helix metal binding loops (Herzberg and James 1985). A solvent exposed α-helical peptide (central helix) separates the two domains: 1) the amino (N-) terminal domain, which binds Ca^{2+} weakly; and 2) the carboxyl (C-) terminal domain, which binds both Ca^{2+} and Mg^{2+} with a relatively higher affinity.

There are two isoforms of TnC: 1) fsTnC; and 2) slow skeletal/cardiac muscle TnC (ssTnC/cTnC). fsTnC and ssTnC/cTnC isoforms are encoded by two distinct genes in mammalian species. Because of their centrality in muscle contraction, these isoforms share many similar features. For example, when the cTnC isoforms of different species (human, bovine, porcine, rabbit, rat, and mouse) are aligned, a high degree of similarity is evident, as revealed by an identity of ~98-100% (Figure 2A).

Similarly, a comparison of fsTnC across these species also reveals an identity of ~97-100% (Figure 2B), suggesting an evolutionarily conserved, tissue-specific functional roles for fsTnC and cTnC isoforms. In contrast, a sequence comparison between human cTnC and fsTnC reveals a striking difference (only 65% similarity; Figure 3), leading to the expectation that molecular aspects of Ca^{2+}-mediated activation is different in fast skeletal and cardiac muscles.

4.2.1. Structural Differences between cTnC and fsTnC

The amino acid sequence of a protein is tightly correlated to its folded structure. Thus, the heterogeneous nature of cTnC and fsTnC isoforms raises two important questions: 1) are the tertiary structures of these TnC isoforms significantly different?; 2) would such structural differences shed some light on physiologically relevant functional differences between muscle types?

At least at the gross structural level, the overall shape of cTnC (Sia et al. 1997) and fsTnC (Herzberg and James 1985) are similar. For example, both cTnC and fsTnC comprise two globular domains—the N- and C-domains—connected by a flexible central α-helix (Figure 4). In both isoforms, the N- and C-domains contain two helix-loop-helix metal binding loops, each of which in turn consists of a 12 to 14-amino acid Ca^{2+}-binding loop flanked by α-helices on either side. In addition to the four α-helices, the N-domain contains an extra α-helix at the very N-Terminus termed the "N-helix" (Herzberg et al. 1986). The four α-helices flanking the N-helix are named A-D in the N-domain and are followed by the central helix, while the four helices in the C-domain are named E-H. The Ca^{2+}-binding loops are numbered I-IV in the increasing order of the sequence. This is where the similarity between cTnC and fsTnC ends and significant differences between them begin to appear upon closer inspection.

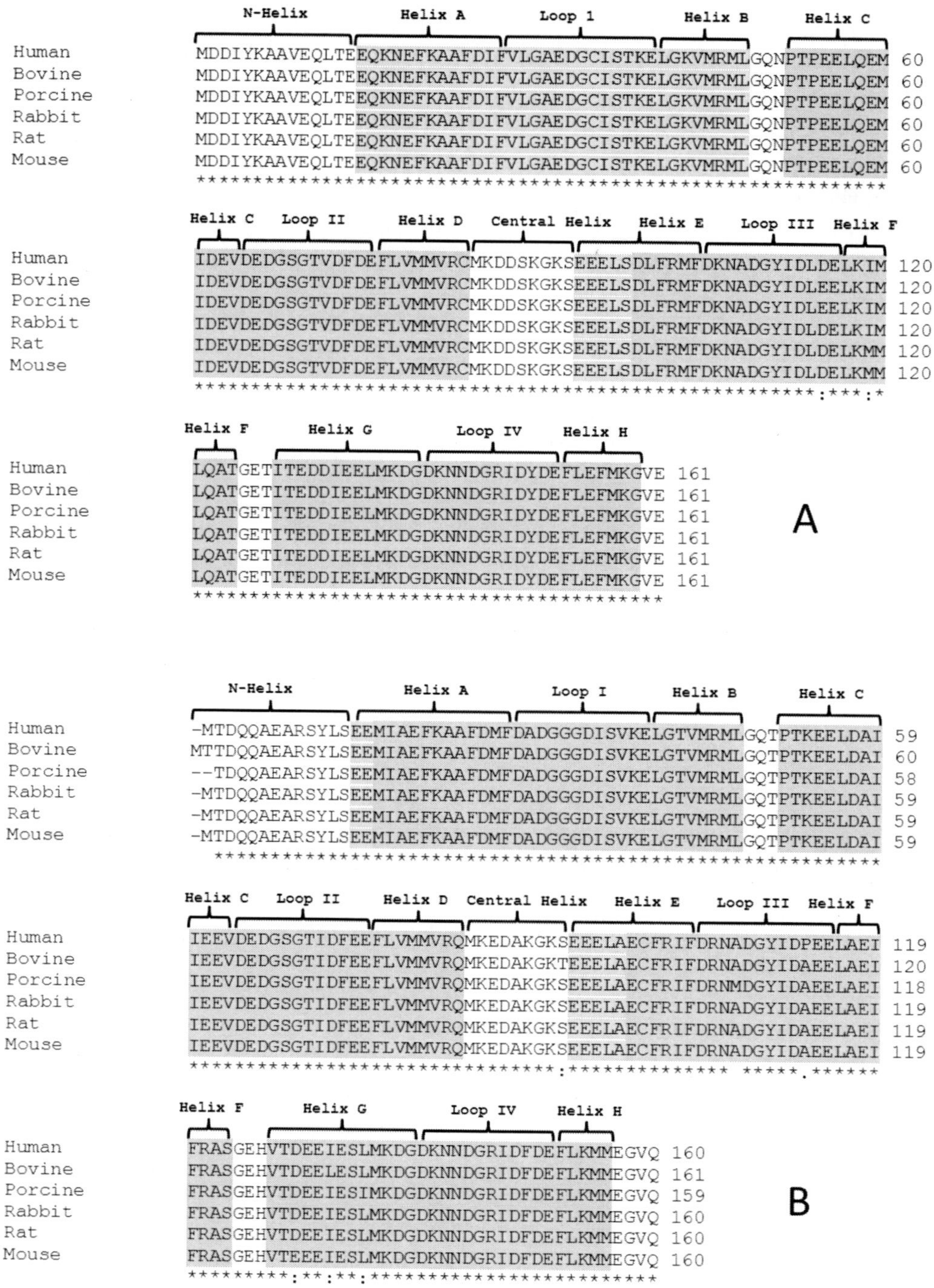

Figure 2. Comparison of striated muscle TnC sequences from different species using Clustal W. (A) cTnC comparison, and (B) fsTnC comparison. "*" indicates fully-conserved residues, "." indicate conservation between groups of weakly similar properties, ":" indicates conservation between groups of strongly similar properties, and "-" indicates deletion introduced to maximize sequence similarities. Shaded portions represent EF-hands 1-4 in the increasing order of the protein sequence, each consisting of helix-loop-helix motif.

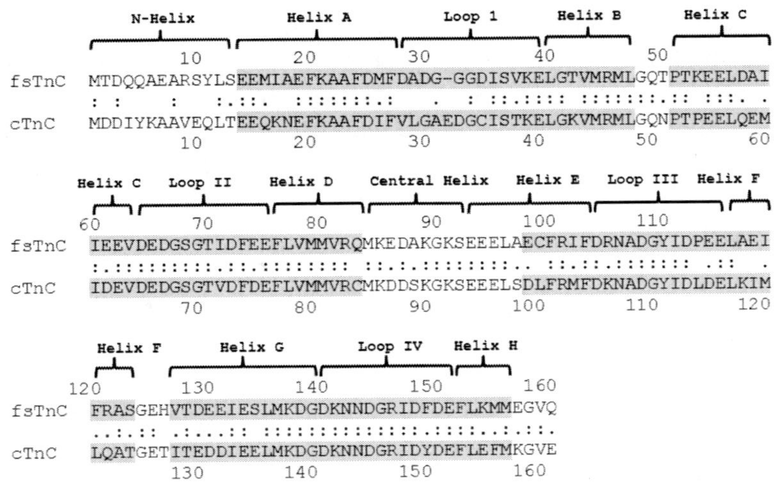

Figure 3. Comparison of human cTnC and fsTnC using LALIGN program (Huang and Miller 1991). "." indicates conservative replacement, ":" indicates identity, "·" indicates non-conservative replacement, and "-" indicates deletion introduced to maximize sequence similarities. Shaded portions represent EF-hands 1-4, each consisting of helix-loop-helix motif.

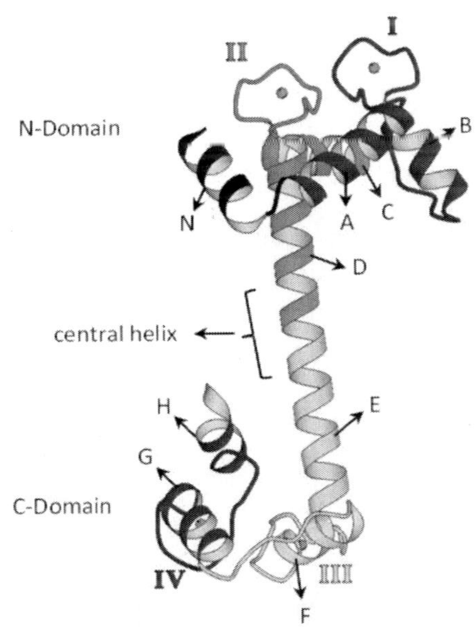

Figure 4. Ribbon model of a 4 Ca^{2+}-bound rabbit fsTnC (reproduced with permission from Houdusse et al. (1997)). The N-domain contains 5 helices (N, A, B, C, and D) and two Ca^{2+} binding loops (I and II). The C-domain contains 4 helices (E, F, G, and H) and two metal binding loops (III and IV). The N- and C-domains are connected by a flexible "central α-helix".

One of the major differences between fsTnC and cTnC is that loops I and II of fsTnC contain functional Ca^{2+} binding sites; whereas loop I is defunct in cTnC. Thus, cTnC binds Ca^{2+} only at loop II. Because Ca^{2+} binds weakly to loops I and II in fsTnC and only to loop II in cTnC, these loops participate in the regulatory function of TnC (Potter and Gergely 1975; Leavis and Kraft 1978). On the other hand, loops III and IV of both fsTnC and cTnC contain high-affinity Mg^{2+}- and Ca^{2+}-binding sites that do not directly participate in the regulatory processes, but are thought to be important for anchoring TnC on the thin filament. Because of its role in contractile regulation, most of the studies have focused on the Ca^{2+}-mediated structural alterations in the N-domain of fsTnC and cTnC.

4.2.2. N-Domain of TnC

The N-helix (1-14 residues) of TnC does not contain Ca^{2+} binding sites nor is known to interact with any other thin filament protein. However, the N-helix is thought to be important for the proper alignment of the N-domain of TnC and for intra-molecular communications between helices A-D in the N-domain (Herzberg et al. 1986). For example, the replacement of 1-41 residues of fsTnC by the corresponding region of cTnC imparts a cTnC-type function to fsTnC in rat cardiac trabeculae (Gulati, Babu et al. 1992). However, the replacement of just the N-helix of fsTnC with that of cTnC or the inactivation of site I in fsTnC significantly reduces the ability of rat cardiac trabeculae to recover tension, suggesting that interactions between the cardiac-specific N-helix and loops I-II,are important for cTnC function.

Crystallographic structure of turkey fsTnC demonstrates that Arg at site 11 forms salt bridges with E16 in helix A, E76 in helix D, and E127 in helix G (Herzberg and James 1985; Herzberg and James 1988; Satyshur et al. 1988; Gulati et al. 1995). These structural studies postulated that breaking salt bridge connections, via structural alterations in the N-helix, would affect TnC function by altering intra-molecular communications within the N-domain. The deletion of 1-14 residues in avian fsTnC or 1-11 residues in chicken fsTnC markedly reducesCa^{2+} sensitivity of force development in regulated actin-myosin system, supporting the idea that the N-helix is important for structural integrity of the N-domain of TnC (Chandra et al. 1994; Smith et al. 1994). Interestingly, there are several hydrophobic contacts between Ala8, Val9, and Leu12 of the N-helix with Leu78, Val79, and Val82 of the D-helix of chicken cTnC (Sia et al. 1997), the heart-specific significance of which still remains to be fully resolved. The structural/functional significance of the N-helix may go beyond that of the N-domain because structural alterations in the N-helix affects metal binding loops III and IV in the C-domain of TnC (Smith et al. 1999).

4.2.3. C-Domain of TnC

The C-domain of both cTnC and fsTnC contain two high affinity Ca^{2+}/Mg^{2+}loops, termed III and IV; these loops bind $Ca^{2+}$and Mg^{2+} with binding constants of $2x10^7$ M^{-1} and $5x10^3$ M^{-1}, respectively(Potter and Gergely 1975). In a relaxed muscle, both of these loops are saturated with Mg^{2+} and their rate constant of dissociation is low. When the muscle is activated, Ca^{2+} competes with Mg^{2+} for loops III and IV, but the replacement of $Mg^{2+}$by Ca^{2+} is rather slow because Mg^{2+}dissociatesslowly from loops III and IV. Therefore, the primary trigger for the muscle contraction is thought to be the binding of Ca^{2+} to the low affinity regulatory loops I and II in the N-domain of fsTnC and loop II in cTnC. Thus, in both tissues, loops III and IV are thought to play a structural role in anchoring TnC on the thin filament. This is supported by the observation that extraction of TnC from the muscle is relatively easy when Ca^{2+}/Mg^{2+} are not bound to loops III and IV, but the metal bound loops III and IV render it difficult to remove TnC from the thin filament (Zot and Potter 1982).

If the metal binding loops III and IV are purely meant for structural stabilization, why is it that the C-domain of fsTnC and cTnC show substantial sequence heterogeneity? To understand the structure/function relationship of Ca^{2+}/Mg^{2+} binding loops III and IV in the C-domain, many studies have inactivated these metal binding loops by replacing the first Asp residue of these sites with Ala in both cTnC (Negele et al. 1992;Dotson and Putkey,1993)and in fsTnC (Sorenson et al. 1995; Szczesna et al. 1996). The inactivation of either loop III or loop IV in cTnC has no effect onCa^{2+}-mediated activation of bovine cardiac myofibrillar ATPase activity. However, the inactivation of both loops III and IV requires ~10-100-fold greater TnC concentrations to achieve the same level of Ca^{2+}-regulation, suggesting that only one of the two C-terminal metal binding sites is required for proper binding and strong activation characteristics of cTnC (Negele et al. 1992; Dotson and Putkey, 1993). Structural consequences of Ca^{2+} binding to loops III and IV of cTnC have been studied using NMR techniques (Brito et al. 1993). In the absence of Ca^{2+}, the C-domain of cTnC is suggested to be compact and highly disordered. Upon Ca^{2+} binding, a short β-sheet, connecting loops III and IV, is observed in the NMR structure; this β-sheet is considered to be important for reorganizing the C-domain into a more compact and ordered structure (Brito et al. 1993). However, the formation of this β-sheet requires Ca^{2+}-binding to only one of the two high affinity sites in the C-domain, which reinforces the earlier conclusions that only one metal binding site in the C-domain is required for anchoring cTnC on the thin filament. Studies on fsTnC have demonstrated that the inactivation of loop III or IV of the C-domain significantly increases the Ca^{2+} sensitivity of force development in fast skeletal myofibers (Szczesna et al. 1996) and enhances the dissociation of fsTnC (~10 fold) from the thin filament in the presence of Mg^{2+}(Sorenson et al. 1995). Thus, in contrast to cTnC, both loops III and IV of fsTnC are considered important for normal function of the C-domain, suggesting that sequence heterogeneity in the C-domain of TnC may have some correlation to divergent functions in different muscle tissues.

Several biochemical and muscle mechanical studies have suggested a role for the C-domain of TnC in the Ca^{2+}-mediated regulation of muscle contraction (Calvert et al. 2000; Burkart et al. 2003; Sun et al. 2006; Pinto et al. 2008; Fuchs and Grabarek 2011).

For example, a microcalorimetry study shows that the affinity of the C-domain of cTnC for cTnI increases by 8-fold in the presence of Ca^{2+} versus Mg^{2+} (Calvert et al. 2000). The affinity of recombinant chicken fsTnC for the thin filament increases with the number of strongly-bound XBs in rabbit skeletal myofibers (Pinto et al. 2008). Pinto et al. suggest that the C-domain of fsTnC is important for strong XB-mediated effects on the strength of fsTnC binding to the thin filament. Divergent effects on the cooperativity of ATPase activity are also observed when the C-domain of cTnC is occupied by Ca^{2+} versus Mg^{2+}. (Fuchs and Grabarek 2011).

Other studies of both cTnC and fsTnC have also suggested that the Ca^{2+} affinity for the N-domain of TnC is also mediated by the C-domain (Burkart et al. 2003; Sun et al. 2006). Collectively, these studies not only demonstrate the complex functions of the C-domain, but also suggest a modulatory role for the C-domain of both cTnC and fsTnC in the activation of thin filaments.

4.2.4. The Physiological Significance of the Hydrophobic Pocket

Based on the crystal structure model of turkey fsTnC (Herzberg et al. 1986), it is suggested that, in the absence of Ca^{2+}, helices B and C move away from the N, A, and D helices in the N-domain and orient in a direction parallel to the central helix; whereas the helices F, G, and H in the C-domain are oriented perpendicular to the central helix. When Ca^{2+} binds to the N-domain of TnC, the N-domain changes its orientation such that its conformation is now similar to that of the C-domain. This structural transition exposes the hydrophobic pocket in the N-domain of fsTnC, which is important for stronger contacts with fsTnI. Because of this stronger interaction with fsTnI, the Ca^{2+}-bound fsTnC is now able to pull fsTnI away from actin, leading to the release of actin-myosin interactions. Because such reorganization of helices is essential to expose the TnI binding site on TnC, any structural modifications that interfere with the exposure of hydrophobic pocket should alter the way the muscle is activated by Ca^{2+}. Interestingly, naturally occurring structural modification in the N-domain of cTnC prevents this Ca^{2+}-induced reorganization of helices. For example, the N-domain of cTnC exists in a "relatively closed" conformation in the Ca^{2+}-saturated state(Sia et al. 1997); this eventuates because both B and C helical structures move only marginally away from the N, A, and D helices (Figure 5). Thus, the hydrophobic pocket is only partially exposed and the cTnI binding interface on cTnC is substantially reduced. Despite this lack of exposure of the hydrophobic pocket, Ca^{2+}-bound cTnC is still able to participate in triggering cardiac muscle contraction, suggesting that Ca^{2+}-mediated activation of thin filaments in fast skeletal and cardiac muscle is different. A reduced cTnI binding interface on cTnC also suggests that cTnC-cTnI interactions are weaker when compared to fsTnC-fsTnI interactions. Such relatively weaker interactions may provide a molecular basis for the 4-fold decrease in the free energy of Ca^{2+} binding to the cTnC-cTnI complex (Liao et al. 1994),as compared to the fsTnI-fsTnC complex (a *DG*of 2. 7 $kcal{\cdot}mol^{-1}$ versus 0. 65 $kcal{\cdot}mol^{-1}$ for cTnC-cTnI). These weaker interactions between cTnC and cTnI may relate to a three-fold increase in the *off* rate of Ca^{2+} from cTnC-cTnI, as compared to fsTnC-fsTnI (Sundaralingam et al. 1985). Differences in the binding interface for cTnI on cTnC, lower free energy of Ca^{2+} binding to cTnC, and faster dissociation rate of Ca^{2+} from cTnC could make cTnC a more dynamic calcium sensor (Sia et al. 1997).

As an integral part of a tightly-linked allosteric system, cTnC does not work in isolation. Thus, its interaction with other proteins, mainly cTnI, is likely to modulate its regulatory activity. This is where the interplay between protein structures and internal dynamics begins to unravel some unique aspects of cardiac thin filament activation. Biochemical work done in several independent laboratories demonstrates that a complete Ca^{2+}-induced opening of the N-domain of cTnC occurs only in the presence of cTnI (Li et al. 1999), whereas Ca^{2+} binding alone is sufficient to induce a large opening of the hydrophobic pocket in fsTnC (Mercier et al. 2000). cTnI-induced opening of the N-domain in cTnC provides a much larger binding interface for stronger interaction with cTnI. Although both cTnC and fsTnC adopt similar open conformations in the end (Li et al. 1999; Mercier et al. 2000), the mechanistic path by which they reach this important conformational milestone differs significantly. An interesting angle here is that the full activation of cTnC-cTnI interaction requires the participation of strong XBs (Grabarek et al. 1983; Tobacman and Sawyer 1990; Moss et al. 2004). These observations are strongly suggestive of tissue-specific differences in kinetics of Tn structural transitions, differences that are likely to have divergent effects on dynamics of contractile regulation in fast skeletal and cardiac muscle.

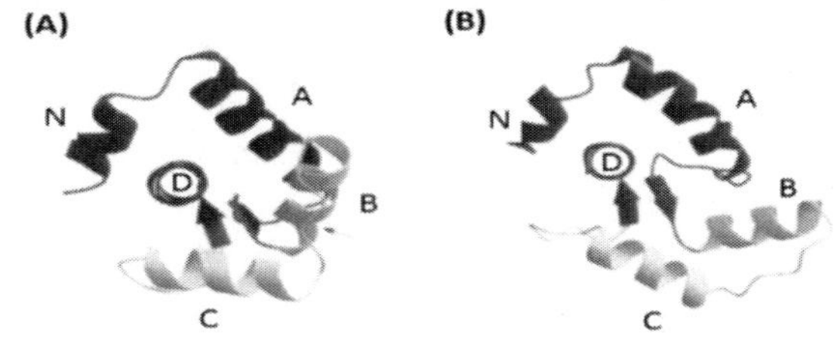

Figure 5. Comparison of the N-domains of cTnC and fsTnC in the Ca^{2+}-bound state (reproduced with permission from (Sia et al. 1997)). (A) a ribbon model of the Ca^{2+}-bound N-domain of cTnC; (B) a ribbon model of the Ca^{2+}-bound N-domain of fsTnC. The naming convention of the helices is similar to that of Figure 4.

4.2.5. Functional Differences between cTnC and fsTnC

Although the Ca^{2+}-activation of force development in fast skeletal and cardiac muscle is highly cooperative, there are significant differences in the way the cooperativity is expressed in these two types of muscles. Even though the evidence bearing on the question of cooperativity between Ca^{2+} binding to loops I and II of fsTnC has been contradictory (Johnson and Potter 1978; Potter et al. 1981; Guth and Potter 1987), it is likely that the cooperativity of Ca^{2+}-activated force development does not reside solely in the binding of Ca^{2+} to TnC because cooperativity of cardiac muscle is high despite the presence of just one Ca^{2+} binding site in cTnC (Dobesh et al. 2002). Thus, the source of this apparent thin filament cooperativity must be external to the Ca^{2+} binding to the N-domain of cTnC and fsTnC. Because fsTnC binds 2 Ca^{2+} and cTnC binds one Ca^{2+}inthe regulatory domain, it was speculated earlier that the differences in cooperative activation of fast skeletal and cardiac muscle resulted from differences in the effects of 2 Ca^{2+}-bound vs 1 Ca^{2+}-bound TnC on near-neighbor interactions in the thin filament.

Findings from two independent studies show that the presence of non-functional fsTnC greatly decreases myofilament Ca^{2+} sensitivity and cooperativity of rabbit skeletal muscle fibers, while the presence of non-functional cTnC shows no such effects in rat cardiac trabeculae (Gillis et al. 2007; Regnier et al. 2002). These observations may also relate to a larger functional unit size (number of actin monomers turned on per Ca^{2+} bound to Tn) of 10-12 in skeletal muscle versus 7 in cardiac muscle (Gillis et al. 2007; Regnier et al. 2002). Thus, unlike the cardiac muscle, the activation of thin filaments in skeletal muscle appears to depend primarily on the near-neighbor regulatory unit (Tm-Tn) interactions, a notion supported by several studies. Partial replacement of rabbit fsTnC with bovine cTnC attenuatesCa^{2+}-activated tension in rabbit skeletal muscle fibers within the pCa range 5.0-6.5 (Moss et al. 1986). Reconstitution of porcine cTnC into rabbit fast skeletal myofibrils significantly attenuates myofilament Ca^{2+} sensitivity and cooperativity of force development (Morimoto and Ohtsuki 1987). Likewise, the replacement of porcine cTnC with rabbit fsTnC in porcine cardiac fibers increases myofilament cooperativity but attenuates myofilamentCa^{2+}

sensitivity (Morimoto and Ohtsuki 1988). Although TnC-based effects alone cannot explain differences in cooperativity of fast skeletal and cardiac muscles, different structural changes associated with Ca^{2+} binding to cTnC and fsTnC must have some impact on thin filament cooperativity. Strong XBs are also known to variably affect the activation of thin filaments in cardiac versus skeletal muscle. For example, the treatment of detergent-skinned muscle fibers with non-force generating derivatives of myosin S1 (NEM-S1 heads) potentiates force at all sub-maximal Ca^{2+} activations in both cardiac and skeletal muscle preparations (Fitzsimons et al. 2001a, 2001b). However, this potentiation of force is much higher in cardiac muscle, suggesting that it is more responsive to strong XBs (Fitzsimons et al. 2001a, 2001b). Inferences drawn from these observations suggest that the activating effects of both Ca^{2+} and XBs are different in these two muscle types.

Another major difference between fast skeletal and cardiac muscle is that in the physiological range of muscle length, the cardiac muscle responds more steeply to an increase in muscle length. In other words, the cardiac muscle produces more tension for a given length change unlike that of the skeletal muscle. This has important implications for the heart muscle because the length-mediated increase in the binding of Ca^{2+} to Tn—also known as the Frank-Starling law of the heart—is more pronounced in the heart muscle (Gulati et al. 1991; Konhilas et al. 2002; Martyn and Gordon2001). According to the Frank-Starling law of the heart, cardiac myofilaments are equipped with a muscle length-sensing mechanism that tunes the activity of XBs to the volume of blood in the ventricle (Konhilas et al. 2002; de Tombe et al. 2010). With respect to the steeper length-dependent activation (LDA) of cardiac muscle, the presence of only one active Ca^{2+} binding site in cTnC poses a structural conundrum: what is the source of cooperativity in the cardiac muscle that is important for LDA? Studies have shown that cTnC is not the length sensor in cardiac muscle (Moss et al. 1991; McDonald et al. 1995), and that changes in the interfilament spacing alone cannot explain effects of increasing muscle length on Ca^{2+}-responsiveness in cardiac myofilaments (Konhilas et al. 2002). For example, at longer sarcomere lengths, bovine ventricular cardiac muscle preparations show a change in the affinity of $Ca^{2+}$for cTnC, whereas slow-twitch rabbit skeletal muscle preparations do not (Wang and Fuchs 1994). However, both ventricular myocardium and slow-twitch muscle types contain the same isoforms of cTnC and myosin heavy chain (MHC), substantiating earlier observations that cTnC is not the length sensor. Thus, this unique effect of LDA in cardiac muscle is likely to involve complex molecular mechanisms that have its origin in tissue-specific expression of other sarcomeric proteins.

4.3. TnI

The letter 'I' in TnI originates from its capacity to inhibit actin-myosin interactions in the absence of Ca^{2+} (Greaser et al. 1972). Because the inhibitory activity of TnI is weaker in the absence of Tm, there is a strong correlation between TnI-Tm interactions and the inhibition of actin-myosin interactions. For example, TnI's inhibitory activity is exerted along the entire length of the actin filament and its periodicity of 38 nm corresponds to the effective length of the filamentous Tm molecules (Ohtsuki and Shiraishi 2002). Another major feature of TnI is that it interacts extensively with TnC and the strength of this interaction increases several fold in the presence Ca^{2+}. Therefore, the Ca^{2+}-bound TnC is able to pull TnI away from actin and thus release the inhibition of TnI on actin-myosin interactions. There are three known isoforms of TnI: cardiac TnI (cTnI); fast skeletal TnI (fsTnI); and slow skeletal TnI (ssTnI), which are encoded by three distinct genes in mammalian species (Syska et al. 1974). When the cTnI from different species (human, bovine, porcine, rabbit, rat, and mouse) are aligned, there is a high degree of similarity (~89-99%; Figure 6A). Likewise, both fsTnI and ssTnI show remarkable sequence identity when compared across species (Figure 6B and 6C). However, when sequence comparisons are made between cTnI, fsTnI, and ssTnI isoforms within species, sequence heterogeneity becomes dramatic. For example, human cTnI and fsTnI isoforms share a sequence homology of only 55%, cTnI and ssTnI share a sequence homology of only 60%, and fsTnI and ssTnI share a sequence homology of only 58% (Figure 7). Given how point mutations and minor alterations in protein sequences affect biological functions, an inescapable consequence is that such major heterogeneity between different TnI isoforms must have significant relevance to the tissue-specific function.

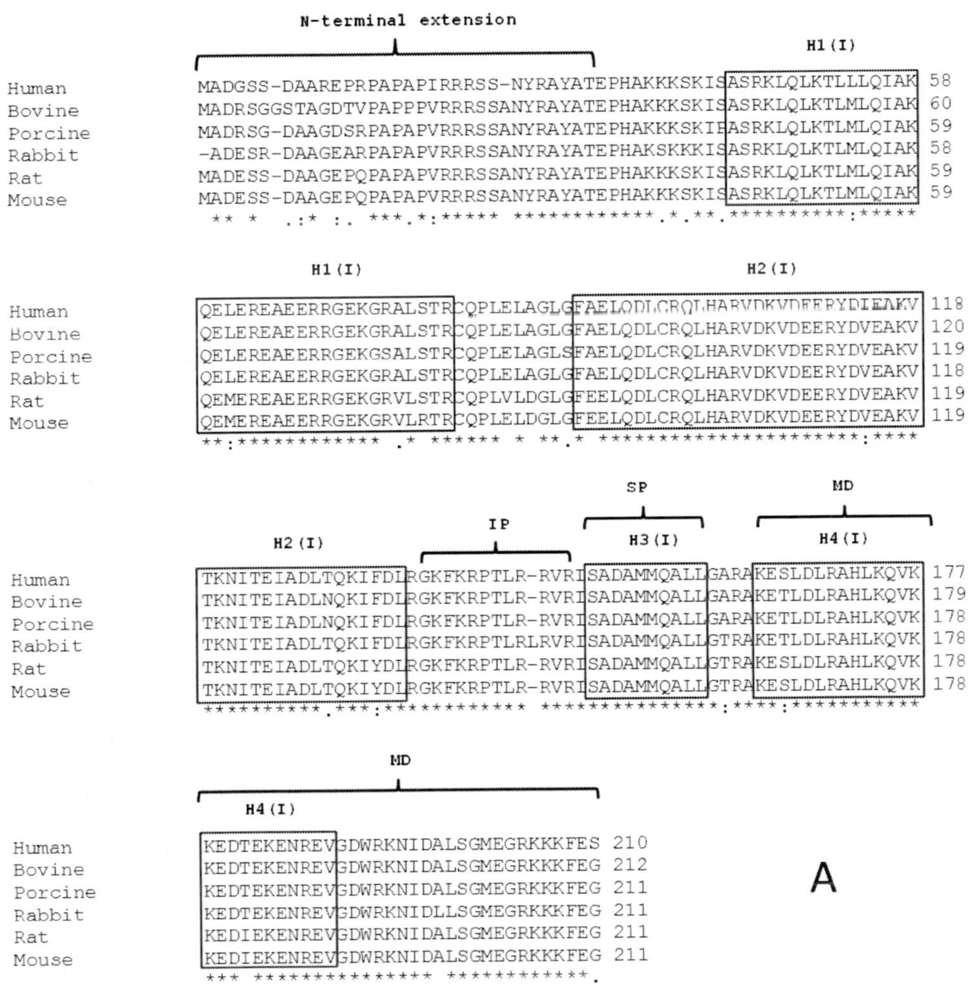

Figure 6. (Continued).

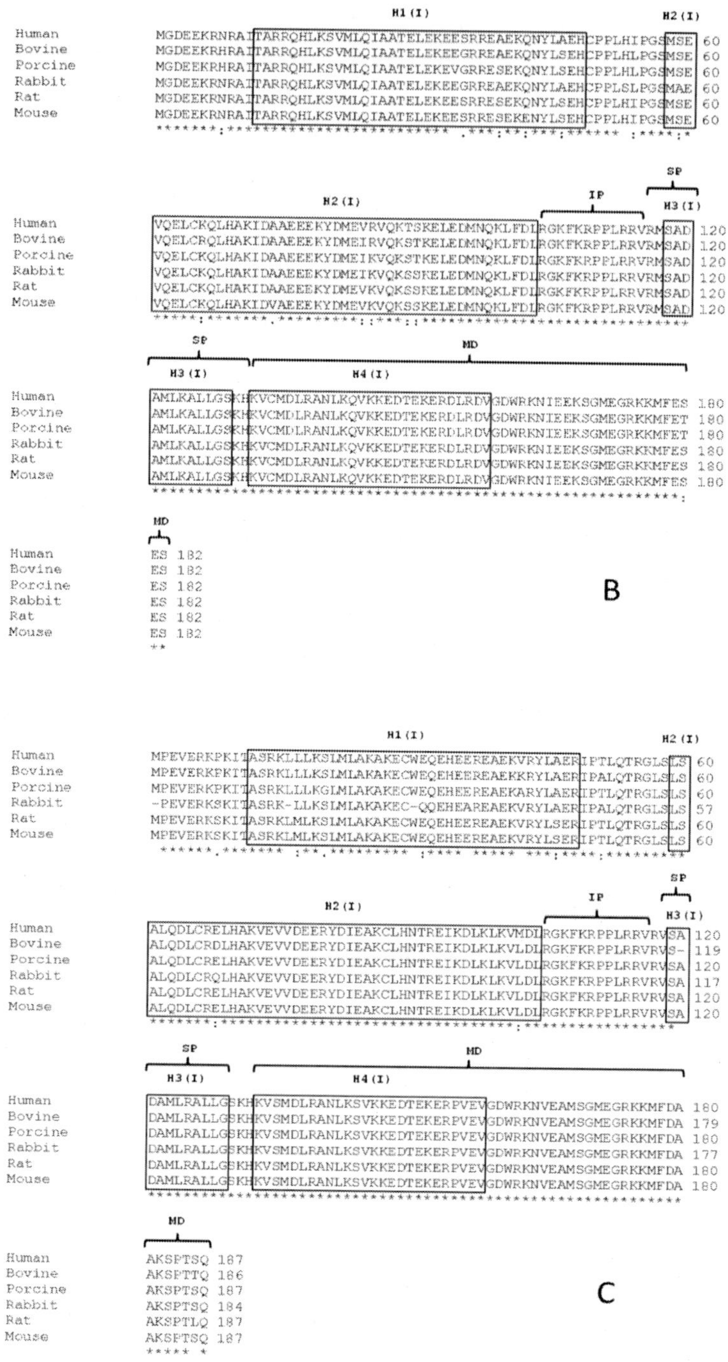

Figure 6. Sequence comparison of striated muscleTnI isoforms from different species using ClustalW. (A) cTnI comparison (B) fsTnI comparison, and (C) ssTnI comparison. "*" indicates fully conserved residues, "." indicate non-conservative replacements, ":" indicates conservative replacements and "-" indicates deletion introduced to maximize sequence similarities. IP, inhibitory peptide; SP, switch peptide; MD, mobile domain.

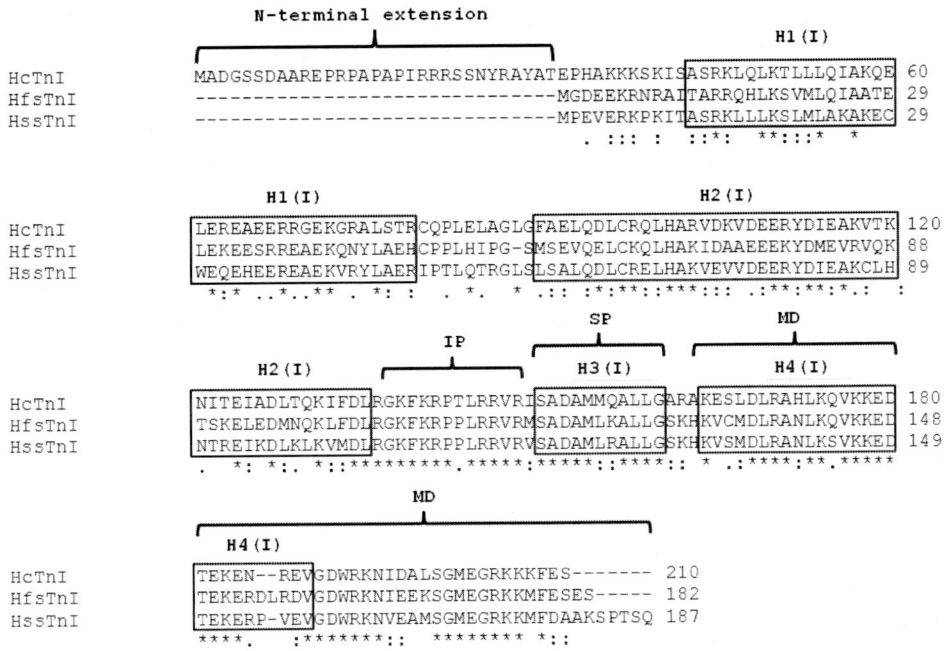

Figure 7. Comparison of human cTnI, fsTnI, and ssTnI isoforms using ClustalW. "*" indicates fully conserved residues, "." indicate conservation between groups of weakly similar properties, ":" indicates conservation between groups of strongly similar properties, and "-" indicates deletion introduced to maximize sequence similarities.

4.3.1. Structural Details of fsTnI

TnI's ability to inhibit actin-myosin interactions is due mainly to its ability to interact with actin in the absence of Ca^{2+}. In rabbit fsTnI, this inhibition was originally shown to be accomplished by the inhibitory peptide (IP, see Figs 6 and 7) comprising residues 96-117(Syska et al. 1976), although a later study showed that the minimal IP (residues 104-115) was sufficient for the normal inhibitory activity of TnI (Talbot and Hodges 1981). In the presence of Ca^{2+}, both cTnC and fsTnC reverse the inhibition of actin-myosin ATPase activity caused by cTnI and fsTnI, demonstrating that the IP of TnI interacts with the Ca^{2+}-bound TnC. Although this biochemical description of TnI is generally accepted, recent studies have further enhanced our understanding of TnI's role in regulating muscle contraction. In addition to the IP (first actin binding region), fsTnI also contains a second actin binding region (secondary IP) comprising residues 140-148 and a regulatory switch peptide (SP) comprising residues 116-131, whose strong interaction with the hydrophobic pocket of fsTnC is important for the Ca^{2+}-mediated activation of thin filaments (Tripet et al. 1997).

High resolution structures of intact TnI are not available. However, inferences made from biochemical studies and structural interpretations suggest that TnI can be broadly divided into two regions: 1) the N-terminal region of TnI (1-104 residues of human fsTnI or 1-135 residues of human cTnI), which not only anchors TnC to the thin filament, but also aids in transmitting the Ca^{2+}-induced structural changes to the rest of the thin filament; and 2) the C-terminal region (105-182 residues of human fsTnI or 136-210 residues of human cTnI), which acts as a Ca^{2+}-sensitive switch by aiding the movement of TnI between actin and TnC. Although both regions of TnI appear to be involved in the Ca^{2+}-mediated activation of thin filaments, distinct segments within each region are known to perform unique function by interacting with various thin filament regulatory proteins (TnC, TnT, Tm and actin).

4.3.2. Structural Details of cTnI

More details on the structure of cTnI are provided by the high resolution crystal structures of the Ca^{2+}-bound human cardiac Tn complex, which reveal that the N- and C-termini of cTnI contain two α-helices each (Takeda et al. 2003). Residues 43-79 of human cTnI make up the amphiphilic H1 α-helix, of which residues 43-65 interact with the C-domain of cTnC (first binding site for cTnC). Residues 90-135 of human cTnI make up the H2 α-helix that forms a coiled-coil with the C-terminal residues 226-271of cTnT. Residues 137-148 of human cTnI comprise the IP that switches between cardiac actin and cTnC in the absence and presence of Ca^{2+}, respectively. This is followed by another amphiphilic H3 α-helix, also termed the SP (residues 150-159 of human cTnI) that interacts strongly with the hydrophobic pocket of the N-domain of cTnC in a Ca^{2+}-dependent manner(second binding site for cTnC). Further downstream is the mobile domain (MD) of human cTnI, which includes residues 164-188 (H4 α-helix) and 189-210 that interact with cardiac actin and Tm.

4.3.3. The Role of TnI in the Ca^{2+}-Mediated Activation

In light of the structural details of cTnI and fsTnI, the mechanism of Ca^{2+} sensitive switch is detailed as follows. In the absence of Ca^{2+}, the minimal IP (137-148 of cTnI or 104-115 of fsTnI) binds strongly to actin-Tm to block actin-myosin interactions. While this general description of the IP function is similar in cTnI and fsTnI, the mechanism of Ca^{2+}-mediated structural changes are different in fast skeletal and cardiac muscle. For example, minor structural changes caused by the binding of first Ca^{2+} to site II in fsTnC prepares it for the binding of Ca^{2+} to site I (Strynadka et al. 1997). When Ca^{2+} binds to site I, there is a large structural change in the N-domain of fsTnC; this is accompanied by a large free energy change that helps fsTnC to overcome the energy barrier to expose the hydrophobic pocket to the solvent (McKay et al. 2000). On the other hand, a large structural change in the N-domain of cTnC does not eventuate because Ca^{2+} cannot bind to site I in cTnC. As a consequence, the large free energy change—which is cardinal for overcoming the energy barrier to expose the hydrophobic pocket—is unavailable. Therefore, minor structural changes that accompany the binding of $Ca^{2+}$to site II leads to partial opening of the hydrophobic pocket in the N-domain of cTnC, as shown in Figure 5A(Sia et al. 1997; Spyracopoulos et al. 1997). However, the binding of the SP of cTnI to this partially exposed hydrophobic pocket in cTnC releases sufficient free energy to cause major structural changes in the Ca^{2+} defunct site I of cTnC (Li et al. 1999); free energy released during this structural change then leads to a large exposure of the hydrophobic pocket in cTnC that stabilizes cTnC-cTnI interactions (McKay et al. 2000).

Although both cTnC-cTnI and fsTnC-fsTnI achieve more or less the same configuration at the end, diverse conformational states of these proteins at various stages of activation and differences in strengths of interaction between them may have implications for divergent function in different tissues. For example, the binding of the SP of fsTnI (residues 115-131) to the hydrophobic pocket of fsTnC was found to be ~6-fold stronger than the binding of the SP of cTnI (residues 147-163) to the hydrophobic pocket of cTnC (Li et al. 1999). A stronger fsTnI-fsTnC interaction may make skeletal muscle thin filaments more sensitive to Ca^{2+}, implying that the strength of the Ca^{2+} signaling process is tightly modulated by tissue-specific TnC-TnI interactions (Ball et al. 1994; Strauss et al. 1996; Konhilas et al. 2002). Because there is coupling between the binding of Ca^{2+} to TnC and TnI-TnC interactions in a tissue-specific manner, differences in strengths of interaction between them will have a significant impact on the dynamics of Ca^{2+} binding to the cardiac and skeletal muscle Tn complex. For example, it is now well known that Ca^{2+} dissociates more rapidly from cTnC when compared to fsTnC. The dissociation rate of $Ca^{2+}$is 30 s^{-1} in rat cardiac Tn complex (Kreutziger et al. 2011) versus 4.57 s^{-1} in rabbit skeletal Tn complex (Kreutziger et al. 2007); the dissociation rate of $Ca^{2+}$is 25.3±0.7 s^{-1} in rabbit cardiac myofibrils reconstituted with the human cardiac Tn complex (Little et al. 2012) versus 15±3 s^{-1} in rabbit psoas rigor myofibrils reconstituted with the chicken fast skeletal Tn complex (Little et al. 2011).

4.3.4. Inhibitory Sites of TnI Isoforms and their Function

Because the inhibition of actin-myosin interactions requires the positioning of Tm on the thin filament in the "off" state, it is believed that the IP may interact directly with Tm and actin. Synthetic peptide studies have demonstrated that the minimum sequence required for inhibition consists of residues 104-115 in fsTnI (Talbot and Hodges 1981; Grabarek and Gergely 1987); this region corresponds to residues 137-148 in cTnI. In attempts to understand the differences between cardiac and skeletal muscle TnI isoforms, several biochemical studies have employed deletion mutants of TnI in both fast skeletal and cardiac muscle. Deletion of either 96-116 or 104-115 residues in fsTnI reduces the inhibition and increases myofilament Ca^{2+} sensitivity of force/ATPase activity (Farah et al. 1994; Szczesna et al. 1999). Individual replacement of each amino acid in the minimal IP (104-115 residues) of rabbit fsTnI with Gly not only causes a decrease in the inhibition of ATPase activity, but also weakens fsTnI-fsTnC interactions in the presence of either Mg^{2+} or Ca^{2+}(Van Eyk and Hodges 1988). These observations, in addition to the Ala scanning of individual amino acids in the IP region (Kobayashi et al. 2009), demonstrate the biological significance of the amino acid sequence, as well as the importance of each amino acid in the IPs of both cTnI and fsTnI.

A comparison of the IP sequences of cTnI and fsTnI reveals high degree of homology except for a significant change—Pro at site 111 of human fsTnI versus Thr at the corresponding site 143 in human cTnI (see Figure 7). The presence of Thr at 143 in cTnI significantly decreases the ability of the peptide to make a turn in the helix, causing the cardiac peptide to become more flexible (Campbell et al. 1992). Different conformational states of the IP in cTnI and fsTnI may have implications for the magnitude of activation. Greater mobility of the IP in cTnI is implicated in attenuating the inhibition of actin-myosin ATPase activity (Campbell et al. 1992). Another study shows that the substitution of Gly for Pro at site 110 of rabbit fsTnI decreases the interaction of the IP with actin-Tm (Van Eyk and Hodges 1988). Interestingly, crystal structures suggest that the manner in which the IP is confined in the vicinity of the central helix of TnC also differs in fast skeletal versus cardiac

Tn (Takeda et al. 2003; Vinogradova et al. 2005). The highly ordered structure of the central helix in fsTnC imparts some rigidity to the IP of fsTnI (Vinogradova et al. 2005). This is not the case in the cardiac Tn complex because the IP of cTnI is flexible and is not observed in the crystal structure (Takeda et al. 2003). This disparity in the structure of IP raises two important questions: 1) is there a cardiac-specific thin filament protein that stabilizes the structure of IP in the Ca^{2+}-activated state?; and 2) is the disordered structure of the IP in cTnI essential for cardiac contractile dynamics? The IP may not work in isolation because complete inhibition by fsTnI is only possible when residues 117-156 are present (Syska et al. 1976; Farah et al. 1994). Likewise, the residues present in the C-terminal region (151-210 residues) of cTnI are also essential for inhibitory activity (Rarick et al. 1997). Collectively, these observations point to the presence of a second IP in both cTnI and fsTnI, but the molecular details of their role in inhibition and their effect on the structure of the primary IP are not fully understood.

4.3.5. Interaction of the Mobile Domain of TnI with Actin and Tm

The significance of the MD was highlighted by a study in which NMR structures of the chicken fsTnI were docked into the high resolution models of the fast skeletal thin filament (Hoffman et al. 2006). In the presence of Ca^{2+}, strong interactions of the SP of fsTnI with the hydrophobic pocket of fsTnC causes the MD to overcome the kinetic barrier for structural transition. This causes the MD to unfold and weaken its interaction with actin-Tm in the presence of Ca^{2+}; however, in the absence of Ca^{2+}, its association with actin-Tm is strengthened by a “fly-casting” mechanism (Hoffman et al. 2006). In lieu of these observations, Hoffman et al. suggest that the MD of TnI is important for both contraction and relaxation kinetics of muscle. The MD of three different isoforms of TnI also exhibit substantial sequence heterogeneity. For instance, a sequence alignment of MD of any two of the three TnI isoforms (residues 164-210 of human cTnI; 133-182 in human fsTnI; and 134-187 in human ssTnI) reveal 30% sequence heterogeneity (Figure 7), suggesting that sequence variation in this region may confer important tissue specific function. The significance of the MD, especially in cTnI, is highlighted by a partial or complete truncation of this structure observed in patients with restrictive cardiomyopathy (Kostareva et al. 2009) and in rat hearts during ischemia-reperfusion (McDonough et al. 1999).

Experimental evidence for the functional significance of MD comes from various biochemical and muscle contractile studies on deletion mutants of cTnI. For example, the deletion of residues 193-210 in cTnI increases Ca^{2+} affinity of cTnC only when the cardiac Tn complex is incorporated into the thin filament (Tachampa et al. 2008). This effect is only possible through a direct interaction of residues 193-210 of cTnI with Tm, actin or both; such interactions may act to stabilize XBs in the strongly-bound state, leading to an increase in myofilament Ca^{2+} sensitivity. Electron microscopy and 3D image reconstruction study shows that the human cTnI lacking residues 193-210 causes preferential shift of Tm to a state that favors the easy binding of myosin to actin (Galinska et al. 2010). The region comprising residues 157-163 of rabbit fsTnI (corresponding to residues 188-194 in human cTnI) interacts with the residue 146 of Tm in a Ca^{2+}-dependent manner (Mudalige et al. 2009). The removal of residues 193-210 weakens the binding of human cTnI to Tm (Zhang et al. 2011). The truncation of 193-210 of human cTnI (Narolska et al. 2006) or its equivalent in mouse/rat cTnI (Foster et al. 2003) not only increasesCa^{2+} sensitivity, but also affects XB cycling kinetics in cardiac myofilaments. Despite its functional importance, the link between sequence heterogeneity in the MD and disparities in activation and relaxation kinetics of thin filaments remains unknown.

4.3.6. The N Terminal Region of TnI

A unique feature of cTnI is its 30-32 amino acid N-terminal extended region, which is absent in both fsTnI and ssTnI isoforms (Syska et al. 1974; Jin et al. 2008). This extended region of cTnI has Ser at sites 23/24 that are known to be phosphorylated by protein kinase A (PKA); phosphorylation of these Ser residues are important for the heart during the β-adrenergic stimulation (Solaro et al. 1976; Swiderek et al. 1988). Previous studies aimed at addressing the functional role of the N-terminal extension of cTnI suggested that either the removal of this extension (Barbato et al. 2005; Biesiadecki et al. 2010) or the PKA-mediated phosphorylation of Ser23/24 sites in cTnI (Chandra et al. 1997; Sadayappan et al. 2008) modulated cardiac myofilament Ca^{2+} sensitivity in a similar manner.

The role of the N-terminal extended region in the Ca^{2+} sensitive switch of cardiac muscle activation is highlighted by our study in which we probed the interaction between fluorescently-labeled cTnI and different fragments of cTnC (Chandra et al. 1997).

Our study shows that the affinity of Ca^{2+} for cTnI-$cTnC_{1-89}$ complex decreases upon PKA-mediated phosphorylation of cTnI, an effect that is comparable with a decrease in the myofilament Ca^{2+} sensitivity of rat cardiac fibers reconstituted with this complex. Therefore, we conclude that the PKA-mediated phosphorylation of Ser23/24 induces conformational changes in the N-domain that are then transmitted to the C-domain of cTnI.

Our observations are in agreement with other recent functional studies (Sadayappan et al. 2008; Akhter et al. 2012), as well as previous NMR studies (Abbott et al. 2001), which have demonstrated that structural alterations in the N-terminal extension of cTnI induce long-range structural changes in the C-domain of cTnI.

Do these long-range effects of cTnI phosphorylation go beyond that of the Tn complex? Experimental evidence have demonstrated that the PKA-mediated phosphorylation of Ser23/24 in cTnI has no impact on either the XB-mediated effect on thin filaments or XB cycling kinetics (Hunlich et al. 2005; Mamidi et al. 2012). These observations point to the idea that the phosphorylation-induced effect of cTnI is a function of cTnC-cTnI interactions modulated by Ca^{2+}.

A recent study has discovered two new PKC phosphorylatable sites, Ser5 and Ser6, within the acidic region of the N-terminal extension of cTnI (Henze et al. 2012). Pseudo-phosphorylation of Ser5 or Ser6 shows contrasting effects on Ca^{2+}-activated maximal tension, myofilament Ca^{2+} sensitivity, and the rate of force redevelopment (Henze et al. 2012). Because this region of cTnI is known to interact with the N-domain of cTnC (Ward et al. 2003; Warren et al. 2009), the significance of these phosphorylatable sites in the cardiac muscle requires further investigation.

4.3.7. Interaction of TnI with TnT

Apart from its interactions with TnC, Tm and actin, there is substantial evidence for extensive interactions between TnI and TnT in both cardiac and skeletal muscle tissue. The following studies have shown interactions between the N-terminus of TnI and TnT: 1) amino acid residues 66-135 of human cTnI interact with human cTnT (Takeda et al. 2003); 2) residues 40-98 of rabbit fsTnI interact with rabbit fsTnT (Hitchcock-De Gregori 1982); 3) residues 1-120 of rabbit fsTnI interact with human fsTnT, the absence of which prevents fsTnI to form a ternary complex with fsTnT and fsTnC (Jha et al. 1996); 4) residues 1-102 residues of chicken fsTnI interact with chicken fsTnT whose presence is important for fsTnI to form a complex with fsTnT and fsTnC (Farah et al. 1994). Similarly, the deletion of residues 54-79 in bovine cTnI not only precludes it from binding to cTnT, but also attenuates Ca^{2+}-activated maximal ATPase activity in bovine cardiac myofibrils (Rarick et al. 1999), suggesting that this specific interaction between cTnT and cTnI is essential for Ca^{2+}-mediated activation of cardiac thin filaments. Similarly, rabbit fsTnI lacking the N-terminal 1-57 residues does not bind to fsTnT (Potter et al. 1995), suggesting that this region of fsTnI contains a binding site for fsTnT. However, since regulated thin filaments containing fsTnI versus fsTnI lacking the first 57 residues demonstrated equivalent ATPase activity, conclusions drawn from Potter et al.'s study suggest that fsTnT-fsTnI interaction is not essential for Ca^{2+} regulation, but this is accomplished via a direct interaction between fsTnT and fsTnC.

4.4. TnT

The third subunit of the Tn complex is termed troponin T because it binds strongly to Tm (Greaser et al. 1972). TnT is the only subunit of the Tn complex that interacts with other thin filament proteins—Tm, TnC, TnI, and nearly half of the actin monomers (reviewed in Gordon et al. 2000; Kobayashi and Solaro 2005). Thus, TnT is pivotally positioned to modulate the activity of Tm, TnI, and TnC; this is how TnT plays a central role in the thin filament activation and force development. TnT acts not only as a bridge between the Ca^{2+} receptor, TnC, and force-bearing XBs, but also as an inhibitor/activator of XBs (Malnic et al. 1998; Maytum et al. 2002; Tobacman et al. 2002). In both cardiac and skeletal muscle, the Ca^{2+}-triggered movement of Tm-Tn on the actin filament regulates force development (reviewed in Gordon et al. 2000; Kobayashi and Solaro 2005). At low intracellular [Ca^{2+}], the function of TnT is to convert TnI-TnC and Tm to a Ca^{2+}-sensitive inhibitor of the actin-myosin ATPase activity (Malnic et al. 1998; Maytum et al. 2002; Tobacman et al. 2002). TnT is thought to accomplish this by reducing the affinity of TnC for TnI and by stabilizing actin-Tm interactions.

There are three known isoforms of striated muscle TnT: cardiac TnT (cTnT), fast skeletal TnT (fsTnT) and slow skeletal TnT (ssTnT), which are encoded by three different genes (Cooper and Ordahl 1984; Breitbart et al. 1985; Smillie et al. 1988). Individual sequence alignment of cTnT, fsTnT, and ssTnT across various species (human, bovine, porcine, rabbit, rat, and mouse) shows high degree of sequence homology: 85-94% (Figure 8A); 86-99% (Figure 8B); and ~94-98% (Figure 8C), respectively. However, sequence heterogeneity becomes apparent when these isoforms are compared between different tissues within the same species. For example, there is a 41% sequence heterogeneity between human cTnT and fsTnT; a 42% heterogeneity between human cTnT and ssTnT; and a 40% heterogeneity between human fsTnT and ssTnT (Figure 9). Interestingly, cardiac-specific TnT contains an extended N-terminus (approximately 30-32 amino acids) that is rich in negative charges; this extension is not present in fsTnT or ssTnT isoforms (Leszyk et al. 1987). Such sequence differences among TnT isoforms are also consistent across species (Jin et al. 2008), suggesting an important heart-specific function for the extended N-terminus in cTnT.

TnT isoforms from vertebrates vary in length ranging from approximately 30-36 kDa, with the cardiac isoform being the largest isoform across all species. In addition to being longer, cTnT may also contain more structured segments within the protein (Snyder et al. 1996). Circular dichroism studies demonstrate significant differences in the helical content of cTnT and fsTnT, with cTnT containing more helical content than fsTnT (Chandra unpublished data). Biochemical properties of a well-studied isoform (rabbit fsTnT) indicate that the N-terminus contains a constellation of acidic amino acids, whereas the C-terminus contains predominantly basic amino acids. This distribution of charges at the N- and C-termini not only appears to be common to all TnT isoforms, but also point to the presence of two domains in TnT. The presence of two domains was predicted earlier based on immuno-electron microscopy studies (Ohtsuki 1975).

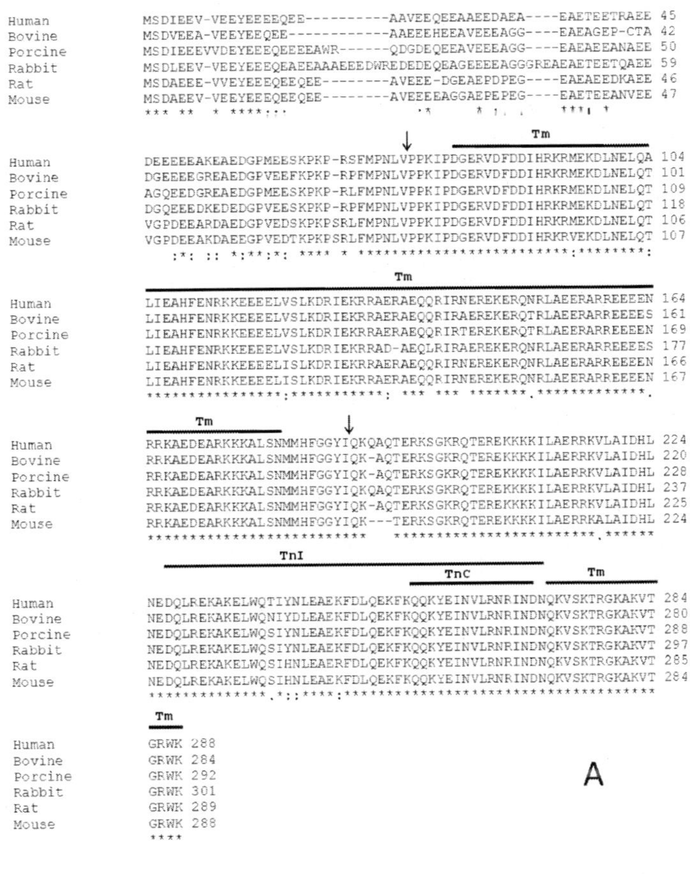

Figure 8. (Continued).

```
                                                                           ↓
Human     MSDEEVEQVEEQYEEEEEAQEEE--------EVQEDTAEEDAEEEKPRPKLTAPKIPEGE 52
Bovine    MSDEEVEHVEEEYEEEEEAQEEAPPPPAEVPEVHEEVHEVHEPEEKPRPRLTAPKIPEGE 60
Porcine   MSDEEVEHVEEEYEEEEEAQEEAPPPPAEVHEVHEEVHEVHEPEEKPRPKLTAPKIPEGE 60
Rabbit    MSDEEVEHVEEQYEEEEEAQEEAP-----SPAEVHEPEEVHE-EEKPRPKLTAPKIPEGE 54
Rat       MSDEETEQVEEQYEEEEEAQEEE------VQEEAPEPEEVQE-EEKPRPKLTAPKIPEGE 53
Mouse     MSDEETEQVEEQYEEEEEAQEEE------VQEEAPEPEEVQE-EEKPRPKLTAPKIPEGE 53
          *****.*:***:**********               :  * .  ******:**********

                                  Tm
                         ____________________________________
Human     KVDFDDIQKKRQNKDLMELQALIDSHFEARKKEEEELVALKERIEKRRAERAEQQRIRAE 112
Bovine    KVDFDDIQKKRQNKDLMELQALIDSHFEARKKEEEELVALKERIEKRRAERAEQQRIRAE 120
Porcine   KVDFDDIQKKRQNKDLMELQALIDSHFEARKKEEEELVALKERIEKRRAERAEQQRIRAE 120
Rabbit    KVDFDDIQKKRQNKDLMELQALIDSHFEARKKEEEELVALKERIEKRRAERAEQQRIRAE 114
Rat       KVDFDDIQKKRQNKDLMELQALIDSHFEARKKEEEELIALKERIEKRRAERAEQQRIRAE 113
Mouse     KVDFDDIQKKRQNKDLMELQALIDSHFEARKKEEEELIALKERIEKRRAERAEQQRIRAE 113
          *************************************:**********************

                     Tm                            ↓      TnC
          ________________________________________  ______________
Human     KERERQNRLAEEKARREEEDAKRRAEDDLKKKKALSSMGANYSSYLAKADQKRGKKQTAR 172
Bovine    KERERQNRLAEEKARREEEDAKRRAEDDLKKKKALSSMGANYSSYLAKADQKRGKKQTAR 180
Porcine   KERERQNRLAEEKARREEEEAKRRAEDDLKKKKALSSMGANYSSYLAKADQKRGKKQTAR 180
Rabbit    KERERQNRLAEEKARREEEDAKRRAEEDLKKKKALSSMGANYSSYLAKADQKRGKKQTAR 174
Rat       KERERQNRLAEEKARREEEDAKRRAEDDLKKKKALSSMGANYSSYLAKADQKRGKKQTAR 173
Mouse     KERERQNRLAEEKARREEEDAKRRAEDDMKKKKALSSMGANYSSYLAKADQKRGKKQTAR 173
          *******************:******:*:*******************************

                                        TnI
                         ____________________________________
                 TnC                                        TnC
          __________________                              _____
Human     EMKKKILAERRKPLNIDHLGEDKLRDKAKELWETLHQLEIDKFEFGEKLKRQKYDITTLR 232
Bovine    EMKKKVLAERRKPLNIDHLSEDKLRDKAKELWDTLYQLETDKFEYGEKLKRQKYDITNLR 240
Porcine   EMKKKVLAERRKPLNIDHLSEDKLRDKAKELWDALYQLEIDKFEYGEKLKRQKYDIINLR 240
Rabbit    EMKKKILAERRKPLNIDHLSDEKLRDKAKELWDTLYQLETDKFEFGEKLKRQKYDIMNVR 234
Rat       EMKKKILAERRKPLNIDHLSDDKLRDKAKELWDTLYQLETDKFEFGEKLKRQKYDIMTVR 233
Mouse     EMKKKILAERRKPLNIDHLSDDKLRDKAKELWDTLYQLETDKFEFGEKLKRQKYDIMTVR 233
          *****:*************.::**********::*:*** ****:***********  .:*

                 TnI
          ________________
                 TnC
          _______________________
                       Tm
                  _______________
Human     SRIDQAQKHSKKAGTPAKGKVGGRWK 258
Bovine    SRIDQAQKHSKKAGTAPKGKVGGRWK 266
Porcine   SRIDQAQKHSKKAGTTPKGKVGGRWK 266              B
Rabbit    ARVEMLAKFSKKAGTTAKGKVGGRWK 260
Rat       ARVEMLAKFSKKAGATAKGKVGGRWK 259
Mouse     ARVEMLAKFSKKAGATAKGKVGGRWK 259
          :*::    *.*****:..*********


Human     MSDTEEQEYEEEQPEEE-AAEEEEEAPEEPEPVAEPEEERPKPSRPVVPPLIPPKIPEGE 59
Bovine    MSDAEEQEYEEEQPEEEEAAEEEEEAPEEPEPAAEPEEERPKPSRPVVPPLIPPKIPEGE 60
Porcine   MSDAEEQEYEEEQPEEEEAAEEEE-APEEPEPVAEREEERPKPSRPVVPPLIPPKIPEGE 59
Rat       MSDTEEQEYEEEQAEDEEAV-EEE-APEEPEPVAEREEERPKPSRPVVPPLIPPKIPEGE 58
Mouse     MSDTEEQEYEEEQAEDEEAVEEEE-APEEPEPVAEREE-RPKPSRPVVPPLIPPKIPEGE 58
          ***:*********.*:* *. *** *******.** ** *********************

Human     RVDFDDIHRKRMEKDLLELQTLIDVHFEQRKKEEEELVALKERIERRRSERAEQQRFRTE 119
Bovine    RVDFDDIHRKRMEKDLLELQTLIDVHFEQRKKEEEELVALKERIERRRAERAEQQRFRTE 120
Porcine   RVDFDDIHRKRMEKDLLELQTLIDVHFEQRKKEEEELVALKERIERRRAERAEQQRFRTE 119
Rat       RVDFDDIHRKRMEKDLLELQTLIDVHFEQRKKEEEELIALKDRIERRRAERAEQQRFRTE 118
Mouse     RVDFDDIHRKRMEKDLLELQTLIDVHFEQRKKEEEELIALKDRIERRRAERAEQQRFRTE 118
          *************************************:***:******:***********

Human     KERERQAKLAEEKMRKEEEEAKKRAEDDAKKKKVLSNMGAHFGGYLVKAEQKRGKRQTGR 179
Bovine    KERERQAKLAEEKMRKEEEEAKKRAEDDAKKKKVLSNMGAHFGGYLVKAEQKRGKRQTGR 180
Porcine   KERERQAKLAEEKMRKEEEEAKKRAEDDAKKKKVLSNMGAHFGGYLVKAEQKRGKRQTGR 179
Rat       KERERQAKLAEEKMRKEEEEAKKRAEDDAKKKKVLSNMGAHFGGYLVKAEQKRGKRQTGR 178
Mouse     KERERQAKLAEEKMRKEEEEAKKRAEDDAKKKKVLSNMGAHFGGYLVKAEQKRGKRQTGR 178
          ************************************************************

Human     EMKVRILSERKKPLDIDYMGEEQLRARSAWLPPSQPSCPAREKAQELSDWIHQLESEKFD 239
Bovine    EMKLRILSERKKPLNIDHMGEEQLR---------------EKAQELSDWIHQLESEKFD 224
Porcine   EMKQRILSERKKPLNIDHMGEDQLR---------------EKAQELSDWIHQLESEKFD 223
Rat       EMKLRILSERKKPLNIDYMGEDQLR---------------EKAQELSEWIHQLESEKFD 222
Mouse     EMKLRILSERKKPLNIDYMGEDQLR---------------EKAQELSEWIHQLESEKFD 222
          *** **********:**:***:***               *******:***********

Human     LMAKLKQQKYEINVLYNRISHAQKFRKGAGKGRVGGRWK 278
Bovine    LMAKLKQQKYEINVLYNRISHAQKFRKGAGKGRVGGRWK 263
Porcine   LMAKLKQQKYEINVLYNRISHAQKFRKGAGKGRVGGRWK 262              C
Rat       LMEKLKQQKYEINVLYNRISHAQKFRKGAGKGRVGGRWK 261
Mouse     LMEKLKQQKYEINVLYNRISHAQKFRKGAGKGRVGGRWK 261
          ** ************************************
```

Figure 8. Sequence alignment of TnT isoforms from various speciesusing ClustalW. (A) cTnT comparison and (B) fsTnT comparison, and (C) ssTnT comparison. "*" indicates fully conserved residues, "." indicate conservation between groups of weakly similar properties, ":" indicates conservation between groups of strongly similar properties, and "-" indicates deletion introduced to maximize sequence similarities. Tm, tropomyosin; TnC, troponin C; TnI, troponin I. Arrows indicate the separation for the N-terminal end region, central region and the T2 region.

```
HcTnT      MSDIEEVVEEYEEEEQEEAAVEEQEEAAEEDAEAEAETEETRAEEDEEEEEAKEAEDGPM 60
HfsTnT     MSDEEV--EQVEEQYEEEEEAQE-EEEVQEDT-AEEDAEEE------------------- 37
HssTnT     MSDTEE--QEYEEEQPEEEAAEE-EEEAPEEPEPVAEPEEE------------------- 38
           *** *   :: **:  **  .:* ** . *:.  .  :.**

HcTnT      EESKPKPR-SFMPNLVPPKIPDGERVDFDDIHRKRMEKDLNELQALIEAHFENRKKEEEE 119
HfsTnT     ---KPRPK------LTAPKIPEGEKVDFDDIQKKRQNKDLMELQALIDSHFEARKKEEEE 88
HssTnT     ---RPKPSRPVVPPLIPPKIPEGERVDFDDIHRKRMEKDLLELQTLIDVHFEQRKKEEEE 95
              :*:*         * .****:**:******::** :*** ***:**: *** *******

HcTnT      LVSLKDRIEKRRAERAEQQRIRNEREKERQNRLAEERARREEEENRRKAEDEARKKKALS 179
HfsTnT     LVALKERIEKRRAERAEQQRIRAEKERERQNRLAEEKARREEEDAKRRAEDDLKKKKALS 148
HssTnT     LVALKERIERRRSERAEQQRFRTEKERERQAKLAEEKMRKEEEEAKKRAEDDAKKKKVLS 155
           **:**:***:**:*******:* *:*:*** :****: *:***: :::***: :***.**

HcTnT      NM-MHFGGYIQKQAQTERKSGKRQTEREKKKKILAERRKVLAIDHLNEDQLR-------- 230
HfsTnT     SMGANYSSYLAKADQ---KRGKKQTAREMKKKILAERRKPLNIDHLGEDKLR-------- 197
HssTnT     NMGAHFGGYLVKAEQ---KRGKRQTGREMKVRILSERKKPLDIDYMGEEQLRARSAWLPP 212
           .*  ::..*: *  *   * **:** ** * :**:**:* * **::.*::**

HcTnT      --------EKAKELWQTIYNLEAEKFDLQEKFKQQKYEINVLRNRINDNQKVSKTRG--- 279
HfsTnT     --------DKAKELWETLHQLEIDKFEFGEKLKRQKYDITTLRSRIDQAQKHSKKAGTPA 249
HssTnT     SQPSCPAREKAQELSDWIHQLESEKFDLMAKLKQQKYEINVLYNRISHAQKFRKGAG--- 269
                   :**:** : :::** :**::  *:*:***:*..* .**.. **   *  *

HcTnT      KAKVTGRWK 288
HfsTnT     KGKVGGRWK 258
HssTnT     KGRVGGRWK 278
           *.:* ****
```

Figure 9. Comparison of human cTnI, fsTnI, and ssTnI isoforms using ClustalW. "*" indicates fully conserved residues, "." indicate conservation between groups of weakly similar properties, ":" indicates conservation between groups of strongly similar properties, and "-" indicates deletion introduced to maximize sequence similarities.

Rabbit fsTnT can be separated into fragments (T1 and T2) by using mild chymotrypsin treatment; hence the name T1 and T2 (Ohtsuki 1979). In rabbit fsTnT, the T1 region (fsT1) corresponds to residues 1-158 and the T2 region (fsT2) corresponds to residues 159-259. Based on extensive biochemical and *in vitro* studies (Schaertl et al. 1995), it is suggested that these two domains of TnT are associated with the following functions: 1) the T1 region is important for establishing the size of the functional unit (number of actins turned on) and for modulating kinetic rates of transition between *on* and *off* states of Tm-Tn; and 2) the T2 region contains the site for Ca^{2+} regulation of thin filaments and is considered to be important for regulating the rate of XB binding to actin. TnT is thought to accomplish these functions because of its ability to interact with several thin filament proteins as described below.

In fsTnT, there are two known binding sites for Tm; the first binding site is in the fsT1 region and the second site is in the fsT2 region (Ohtsuki 1979). The first binding site for Tm in rabbit fsT1 is mapped to residues 70-160 (Jackson et al. 1975) and the second one to residues 243-259 in the rabbit fsT2 (Tanokura et al. 1983). With regard to its interaction with fsTnC, two different sites are suggested to be involved; residues 160-193 and 228-259 of rabbit fsT2 (Blumenschein et al. 2001). The T2 region of chicken fsTnT also contains binding sites for actin and TnI which correspond to residues158-191 (Malnic et al. 1998; Oliveira et al. 2000) and 217-263(Malnic et al. 1998), respectively.

Chymotryptic digestion of cTnT also generates two distinct fragments called cT1 and cT2, but in this case the cT1 fragment is approximately 45-50Å longer due to 30-32 amino acid N-terminal extension. Although both cTnT and fsTnT share some similar features at the gross level, substantial sequence heterogeneity is evident throughout the molecule. For example, sequence comparisons of cT1 and fsT1 domains across species show a sequence homology of 56-62%, while those of cT2 and fsT2 domains show a sequence homology of

56-61% (Figures 8 and 9), suggesting that sequence heterogeneity between different domains of TnT confer tissue-specific function.

4.4.1. Biological Significance of the T1 Domain in cTnT and fsTnT

Recent studies from our laboratory have shown that two distinct regions within the T1 domain enable TnT to execute specialized functions that may relate to its modulatory role in the Ca^{2+}-mediated regulation of the thin filament (Chandra et al. 1999; Mamidi et al. 2013; Gollapudi et al. 2013). Accordingly, the T1 domain may be separated into two parts: (1) the extreme N-terminal end region (residues 1-45 in rabbit fsTnT and residues 1-75 in human cTnT; NTE), which does not interact with any other sarcomeric protein; and (2) the central region (residues 46-158 in rabbit fsTnT and residues 76-188 in human cTnT; CR), which interacts strongly with Tm.

4.4.1.1. The Extreme N-Terminal End Region of the T1 Domain

The deletion of first 45 residues in rabbit fsTnT attenuates Ca^{2+}-mediated activation of actin-myosin ATPase activity, suggesting that the NTE of fsTnT is important for the activation of thin filaments (Pan et al. 1991). Evidence for the functional significance of the cardiac-specific NTE comes from a study which demonstrates that the deletion of the first 76 amino acids in rat cTnT (RcTnT) enhances the binding of RcTnT to Tm and attenuates Ca^{2+}-activated force and ATPase activity in rat cardiac muscle fibers (Chandra et al. 1999). Conclusions drawn from both of these studies suggest that actin-myosin interactions are inhibited due to a stronger interaction between the mutant TnT and Tm, which stabilizes cardiac thin filaments in a submaximally-activated state. Supporting evidence for this observation comes from two other independent studies, which demonstrate that the deletion of either the first 91 residues or caspase-3 activated deletion of the first 96 residues in cTnT impairsCa^{2+}-activated force and ATPase activity (Communal et al. 2002; Sumandea et al. 2009). A recent study from our lab also demonstrates that residues 1-44 of cTnT is sufficient for attenuating maximal activation because the deletion of residues 45-74 has no effect on maximal activation (Mamidi et al. 2013).

Insofar as the effect of the NTE on thin filaments in both cardiac and fast skeletal muscle, the unique nature of the cardiac-specific NTE raises important questions because it is not only extended by ~30-32 amino acids, but is also rich in acidic amino acids. We addressed this question by using a transgenic mouse line that expressed a recombinant chimeric cTnT in which the cardiac-specific NTE was removed by replacing amino acid residues 1-73 of mouse cTnT with the corresponding 1-41 residues of mouse fsTnT (Gollapudi et al. 2012). The absence of NTE in the transgenic cTnT construct increases myofilament Ca^{2+} sensitivity at both short and long sarcomere lengths: inferences drawn from our contractile dynamic measurements suggest that the NTE constrains cardiac thin filament activation such that the transition of the thin filament from the blocked-to-closed state becomes less responsive to Ca^{2+}(Gollapudi et al. 2012). To determine whether the effect of NTE observed in Gollapudi et al.'s work was not affected by the fsTnT sequence, a subsequent study tested the effect of a cTnT mutant in which residues 45-74 of mouse cTnT were deleted (Mamidi et al. 2013). This cTnT deletion mutant behaves similar to the transgenic cTnT construct used in Gollapudi et al.'s (2012) work, clearly demonstrating that the sequence (45-74) of cardiac-specific NTE reduces cardiac thin filament activation by making it less responsive to Ca^{2+}. Because the twitch kinetics of the cardiac muscle are influenced by the activating effects of Ca^{2+} on thin filaments, fine-tuning of the thin filament responsiveness to Ca^{2+} may have significant implications for the heart function. It is well appreciated that both tissue- and disease-related changes in the N-terminus of cTnT affects twitch dynamics and function.

The NTE of cTnT is not known to interact with cTnC, cTnI, Tm, and actin: how does it modulate the function of cTnT? To address this question, we proposed a refined model of the cTn-mediated regulation of cardiac thin filaments in which the NTE was shown to exert its influence on the thin filament activation by modulating the activity of the CR (Gollapudi et al. 2013). Because the CR participates in strong interactions with Tm, our model proposes that the NTE of cTnT may interact with the CR to modulate its interaction with Tm (Figure 10). The concept of intramolecular interactions is not alien to other Tn subunits: as pointed out earlier, such intramolecular interactions are known to modulate the function of both TnC and TnI (see sections 4. 2 and 4. 3).

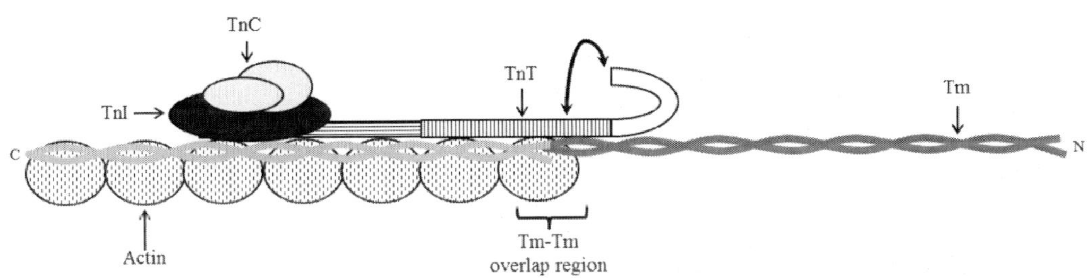

Figure 10. Schematic representation of the refined model of Tn-mediated regulation of cardiac thin filaments. The region with horizontal stripes in cTnT represents the T2 domain (residues 189-288 of human cTnT), which interacts with TnC and TnI. The region with vertical stripes in cTnT represents the CR (residues 76-188 of human cTnT). The CR interacts with Tm near the head-to-tail overlap region of two contiguous Tm molecules. The unshaded portion of the cTnT molecule corresponds to the N-terminal end region (residues 1-75of human cTnT). The N-terminal end region of cTnT (horse shoe-shaped ends of cTnT) may synergistically modulate the interaction of CR with Tm.

4.4.1.2. The Central Region of TnT

Biochemical and *in vitro* studies provide important clues regarding the role of the CR of both cTnT and fsTnT in the thin filament activation. For example, a region comprising 1-70 residues of rabbit fsTnT or 1-76 residues of chicken fsTnT does not interact with Tm (Pearlstone and Smillie 1977; Oliveira et al. 2000). Likewise, the deletion of 70-150 residues in rabbit fsTnT reduces the binding of TnT to Tm (Fisher et al. 1995), suggesting that this region is essential for strong Tm-binding and the activation of actin-myosin ATPase activity. Consistent with this notion, chicken fsTnT fragments comprising residues 1-191 or1-216 activate the ATPase activity in the presence of Tm to a level similar to that of the whole Tn complex (Oliveira et al. 2000; Malnic et al. 1998). The CR of rabbit fsTnT contains nearly 80-90% α-helix and accounts for most of the helical content in the intact fsTnT (Jackson et al. 1975; Pearlstone and Smillie 1977). This highly-structured segment of rabbit fsTnT participates in strong interactions with Tm near the head-to-tail overlap region of two contiguous Tm molecules. Previous biochemical solution studies have suggested the following with regard to functional significance of the CR in both fsTnT and cTnT: (1) it promotes the binding of Tm to actin (Heeley et al. 1987); (2) it aids in the assembly of Tm on the actin filament by promoting Tm-Tm polymerization (Lehrer and Geeves 1998); (3) it is important for the folding stability and flexibility of TnT (Palm et al. 2001; Hinkle and Tobacman 2003); (4) it regulates the myosin binding to actin and cooperative activation of the thin filament (Maytum et al. 2002; Regnier et al. 2002; Tobacman et al. 2002); and (5) it modulates the length-mediated XB recruitment dynamics via cooperative feedback of strong XBs on the thin filament (Gollapudi et al. 2013).

While it is understood that contractile activation in both fast skeletal and cardiac muscle is highly cooperative, they differ significantly in the way they respond to the activating effects of Ca^{2+} and strong XBs (Regnier et al. 2002; Smith et al. 2009). Because the spread of strong XB-mediated effects to neighboring structural units depends on CR-Tm interactions, the stability and flexibility of the CR will affect cooperative mechanisms operating within the thin filament (Palm et al. 2001; Hinkle and Tobacman 2003). It is likely that differences in the allosteric properties of the CR in cTnT versus fsTnT are tightly-dependent on the isoform-specific sequence of amino acids. For example, biochemical and three-dimensional reconstruction studies of bovine cT1 peptide suggest that the blocked state—in which substantial proportion of myosin heads are prevented to interact with actin—is promoted by the cT1 domain (Tobacman et al. 2002). Because the CR of T1 is the main segment that modulates Tm activity on actin, inferences drawn from their study suggest that portions of bovine cardiac CR help to position Tm on the actin outer domain, thereby inhibiting muscle contraction in the absence of Ca^{2+}. Another *in vitro* biochemical study suggests that the rabbit fsT1 fragment inhibits myofilament activation, not by the formation of the blocked state, but rather by stabilizing the thin filaments in the closed state (which can bind myosin weakly). Collectively, these studies demonstrate that the T1 domain (which includes CR) plays a significant role in myofilament activation and that the CRs of cardiac and skeletal TnT isoforms have different effects on thin filament activation.

Because the α-Tm isoform from cardiac and fast skeletal muscle is identical, different effects of cardiac and fast skeletal muscle CR is strongly suggestive of a dominant role for the CR in modulating thin filament activation. In this regard, it is important to note that the CR of both cTnT and fsTnT share sequence homology of only ~66-69% across species (Figures 8 and 9). Given how small perturbations within the CR affect its stability and flexibility, a 31-34% heterogeneity between fast skeletal CR and cardiac CR must have tissue-specific functional relevance. We recently showed that the substitution of fsTnT CR in cTnT not only decreased the magnitude of the muscle length-mediated XB recruitment, but also increased the rate of length-mediated XB recruitment in rat cardiac muscle fibers (Gollapudi et al. 2013). Our data suggest that the modification of the CR in cTnT causes a rigid CR-Tm interaction which attenuates XB recruitment.

4.4.2. Biological Significance of the T2 Domain in TnT

Studies show that the C-domain of TnT has two important roles. First, the T2 region anchors the TnC-TnI complex onto the thin filament via its interactions with TnI-TnC-Tm, and possibly actin. In support of this notion, fragment 1-216 of chicken fsTnT reduces its interaction with chicken fsTnC and fsTnI (Malnic et al. 1998), while fragment 1-201 of human fast TnT reduces its interaction with rabbit fsTnC and fsTnI (Jha et al. 1996), suggesting that the presence of the rest of the C-domain in these specie fsTnT isoforms is important for the formation of the fsTnC-fsTnI-fsTnT ternary complex. Second, the T2 region of fsTnT interacts directly with fsTnI, fsTnC, Tm, and possibly actin to regulate the inhibition of the actin-myosin ATPase activity in a Ca^{2+}-dependent manner.

4.4.2.1. Interactions between T2 and TnI

The first experimental evidence for the binding region for fsTnI on fsTnT comes from biochemical studies, which show that rabbit fsTnT fragments containing amino acid residues 159-227 and 159-242, but not 159-222, bind strongly to fsTnI (Tanokura and Ohtsuki 1982; Tanokura et al. 1983). These observations suggest that the rabbit fsT2 region comprising residues 223-242 contains an important binding interface for interaction with fsTnI. Studies on chicken fsTnT peptide fragment have suggested that a region comprising residues 216-263 may contain the regulatory binding site for fsTnI (Malnic et al. 1998). Another study has shown that a region within the human fsT2 domain (residues 198-251) contains a conserved hydrophobic heptad repeat motif that is similar to a region comprising residues 57-106 in rabbit fsTnI; these two regions of fsTnT and fsTnI interact to form a α-helical coiled-coil (Stefancsik et al. 1998). This coiled-coil interaction is also revealed in the crystal structure of the core domain of human cardiac Tn complex (Takeda et al. 2003). The significance of this region is highlighted by the fact that structural alterations in and around the heptad repeat motifs of both fsTnT and fsTnI are shown to weaken their coiled-coil interactions, which attenuates ATPase activity (Stefancsik et al. 1998).

4.4.2.2. Interactions between T2 and TnC

Biochemical study of rabbit fsTnT peptides have revealed two binding sites for the interaction fsTnT and fsTnC: residues 176-230 and 206-258 in the fsT2 domain, both of which strengthen their interaction with fsTnC in a Ca^{2+}-dependent manner (Pearlstone and Smillie 1978). This is in agreement with another study which demonstrates that the smallest fragment of rabbit fsTnT that is capable of forming a Ca^{2+}-stabilized complex with fsTnC contains residues 159-222 (Tanokura et al. 1983). Subsequent peptide studies have also confirmed that residues 160-193 and 228-260 of rabbit fsTnT interact with fsTnC (Blumenschein et al. 2001). In contrast, the crystal structure of human cardiac Tn complex reveals only the regulatory binding site for cTnC, which corresponds to region 256-270 in human cTnT (Takeda et al. 2003).

4.4.2.3. Interaction of T2 Domain with Tm and Actin

Biochemical studies of TnT peptide fragments show that residues 243-259 of rabbit fsTnT contain the second binding site for Tm (Pearlstone and Smillie 1981; Tanokura et al. 1983). In human cTnT, this binding region has been shown to be located in the region comprising residues 272-288 (Takeda et al. 2003). However, the sites of actin binding in both cTnT and fsTnT are poorly defined. While the initial studies have suggested that only the fsT2 domain is important for anchoring the Tn complex on the actin filament (Heeley et al. 1987; Oliveira et al. 2000), subsequent studies have shown that residues 95-153 in the CR of bovine cTnT also participate in anchoring Tm on the actinfilament (Hinkle et al 1999; Hinkle and Tobacman 2003).

4.4.3. Functional Studies Related to the T2-Domain

Crystallography studies of both cardiac and skeletal Tn complex have shown that the T2 domain binds to TnC and that the T2-TnC-TnI complex acts as a Ca^{2+} sensor (Takeda et al. 2003; Vinogradova et al. 2005). Previous studies have shown that Ca^{2+} regulation of the actin-myosin ATPase activity is achieved through direct fsTnT-fsTnC interactions and that fsTnT-fsTnI interactions are not essential for Ca^{2+} regulation (Potter et al. 1995). Thus, the isoform-related sequence heterogeneity in the T2 region leads to the expectation that its interaction with TnC—and possibly TnI—is altered in a tissue-specific manner.

Supporting this notion, we have recently shown that the substitution of rat fsT2 into RcTnT significantly attenuates both the myofilament Ca^{2+} sensitivity and the length-dependent increase in the recruitment of strong XBs (Gollapudi et al. 2013). Deletion of 14-31 residues in the extreme C-terminus of cTnT increases myofilamentCa^{2+} sensitivity and decreases maximal activation (Redwood et al. 2000; Montgomery et al. 2001a; Stelzer et al. 2004).

4.4.4. Functional Differences between cTnT, fsTnT and ssTnT

All three TnT isoforms differ substantially in many functional aspects including, but not limited to, magnitude of force development, myofilament Ca^{2+} sensitivity, thin filament cooperativity, XB cycling kinetics, and length-mediated activation of thin filaments. For example, human ssTnT isoform, which is most commonly expressed in adult slow-twitch skeletal muscle, decreases myofilament Ca^{2+} sensitivity and augments force recovery in porcine cardiac myofibers (Pinto et al. 2012). Co-expression of mouse cTnT and chicken fsTnT isoforms increases cooperativity of force development (Huang et al. 1999), modulates Ca^{2+}-dependence of force redevelopment in transgenic mouse cardiac myofibers (MacFarland et al. 2002), and decreases contractility in transgenic mouse hearts (Huang et al. 2008; Yu et al. 2012).

In addition, co-expression of mouse cTnT with chicken fsTnT abolishes the PKC phosphorylation-induced decrease in tension of transgenic mouse cardiac fibers (Montgomery et al. 2001b). Partial replacement of cTnT by chicken fsTnT in the mouse heart attenuates force and Ca^{2+} sensitivity to a greater extent at acidic pH (Nosek et al. 2004). Similarly, another study demonstrates that Ca^{2+}-binding to pyrene-labeled cTnC is reduced at acidic pH when chicken fsTnT replaces porcine cTnT (Liou and Chang 2004).

Interestingly, the substitution of chicken fsTnT-fsTnI for porcine cTnT-cTnI attenuatesthis effect, suggesting that isoform-specific TnT-TnI interactions confer Ca^{2+} sensitivity to changes in pH. We have demonstrated that the substitution of rat fsTnT for RcTnT decreases the magnitude of length-mediated increase in XB recruitment and increases the speed of XB recruitment in rat cardiac myofibers (Chandra et al. 2006). Collectively, these observations suggest that isoform specific TnT-TnI and TnT-TnC interactions vary significantly in different muscle types.

References

Abbott, M. B. V. Gaponenko, et al. (2000). "Regulatory domain conformational exchange and linker region flexibility in cardiac troponin C bound to cardiac troponin I." *J. Biol. Chem.* 275(27): 20610-20617.

Akhter, S. Z. Zhang, et al. (2012). "The heart-specific NH2-terminal extension regulates the molecular conformation and function of cardiac troponin I." *Am J Physiol Heart Circ Physiol* 302(4): H923-33.

Babu, A., W. Lehman, et al. (1989). "Characterization of the Ca2+-switch in skeletal and cardiac muscles." *FEBS Lett* 251(1-2): 177-82.

Ball, K. L., M. D. Johnson, et al. (1994). "Isoform specific interactions of troponin I and troponin C determine pH sensitivity of myofibrillar Ca2+ activation." *Biochemistry* 33(28): 8464-8471.

Barbato, J. C., Q. Q. Huang, et al. (2005). "Proteolytic N-terminal truncation of cardiac troponin I enhances ventricular diastolic function." *J Biol Chem* 280(8): 6602-9.

Biesiadecki, B. J., K. Tachampa, et al. (2010). "Removal of the cardiac troponin I N-terminal extension improves cardiac function in aged mice." *J Biol Chem* 285(25): 19688-98.

Blumenschein, T. M. A., B. P. Tripet, et al. (2001). "Mapping the interacting regions between troponins T and C - Binding of TnT and TnI peptides to TnC and NMR mapping of the TnT-binding site on TnC." *J. Biol. Chem.* 276(39): 36606-36612.

Breitbart, R. E., H. T. Nguyen, et al. (1985). "Intricate combinatorial patterns of exon splicing generate multiple regulated troponin T isoforms from a single gene." *Cell* 41(1): 67-82.

Brito, R. M., G. A. Krudy, et al. (1993). "Calcium plays distinctive structural roles in the N- and C-terminal domains of cardiac troponin C." *J Biol Chem* 268(28): 20966-73.

Burkart, E. M., G. M. Arteaga, et al. (2003). "Altered signaling surrounding the C-lobe of cardiac troponin C in myofilaments containing an alpha-tropomyosin mutation linked to familial hypertrophic cardiomyopathy." *J Mol Cell Cardiol* 35(10): 1285-93.

Calvert, M. J., D. G. Ward, et al. (2000). "The importance of the carboxyl-terminal domain of cardiac troponin C in Ca^{2+}-sensitive muscle regulation." *J. Biol. Chem.* 275(42): 32508-32515.

Campbell, A. P., J. E. Van Eyk, et al. (1992). "Interaction of troponin I and troponin C: use of the two-dimensional transferred nuclear Over hauser effect to determine the structure of a Gly-110 inhibitory troponin I peptide analog when bound to cardiac troponin C." *Biochim Biophys Acta* 1160(1): 35-54.

Campbell, K. (1997). "Rate constant of muscle force redevelopment reflects cooperative activation as well as cross-bridge kinetics." *Biophys J* 72(1): 254-62.

Chandra, M., E. F. da Silva, et al. (1994). "The effects of N helix deletion and mutant F29W on the Ca2+ binding and functional properties of chicken skeletal muscle troponin." *J Biol Chem* 269(21): 14988-94.

Chandra, M., W. J. Dong, et al. (1997). "Effects of protein kinase A phosphorylation on signaling between cardiac troponin I and the N-terminal domain of cardiac troponin C." *Biochemistry* 36(43): 13305-11.

Chandra, M., D. E. Montgomery, et al. (1999). "The N-terminal region of troponin T is essential for the maximal activation of rat cardiac myofilaments." *J. Mol. Cell. Cardiol.* 31(4): 867-880.

Chandra, M., M. L. Tschirgi, et al. (2006). "Troponin T modulates sarcomere length-dependent recruitment of cross-bridges in cardiac muscle." *Biophys. J.* 90(8): 2867-2876.

Chiu, Y. L., E. W. Ballou, et al. (1982). "Velocity transients and viscoelastic resistance to active shortening in cat papillary muscle." *Biophys J* 40(2): 121-8.

Communal, C., M. Sumandea, et al. (2002). "Functional consequences of caspase activation in cardiac myocytes." *Proc. Natl. Sci. Acad. U. S. A.* 99(9): 6252-6256.

Cooper, T. A. and C. P. Ordahl (1984). "A single troponin T gene regulated by different programs in cardiac and skeletal muscle development." *Science* 226(4677): 979-82.

de Tombe, P. P., R. D. Mateja, et al. (2010). "Myofilament length dependent activation." *J Mol Cell Cardiol* 48(5): 851-8.

Dobesh, D. P., J. P. Konhilas, et al. (2002). "Cooperative activation in cardiac muscle: impact of sarcomere length." *Am J Physiol Heart Circ Physiol* 282(3): H1055-62.

Dotson, D. G. and J. A. Putkey (1993). "Differential recovery of Ca2+ binding activity in mutated EF-hands of cardiac troponin C." *J Biol Chem* 268(32): 24067-73.

Ebashi, S. (1963). "Third Component Participating in the Super precipitation of 'Natural Actomyosin'." *Nature* 200: 1010.

Farah, C. S., C. A. Miyamoto, et al. (1994). "Structural and regulatory functions of the NH2- and COOH-terminal regions of skeletal muscle troponin I." *J Biol Chem* 269(7): 5230-40.

Farman, G. P., E. J. Allen, et al. (2010). "The role of thin filament cooperativity in cardiac length-dependent calcium activation." *Biophys J* 99(9): 2978-2986.

Fisher, D., G. Wang, et al. (1995). "NH2-terminal truncation of skeletal muscle troponin T does not alter the Ca2+ sensitivity of thin filament assembly." *J Biol Chem* 270(43): 25455-60.

Fitzsimons, D. P., J. R. Patel, et al. (2001a). "Cooperative mechanisms in the activation dependence of the rate of force development in rabbit skinned skeletal muscle fibers." *J Gen Physiol* 117(2): 133-48.

Fitzsimons, D. P., J. R. Patel, et al. (2001b). "Cross-bridge interaction kinetics in rat myocardium are accelerated by strong binding of myosin to the thin filament." *J Physiol* 530(Pt 2): 263-72.

Foster, D. B., T. Noguchi, et al. (2003). "C-terminal truncation of cardiac troponin I causes divergent effects on ATPase and force: implications for the pathophysiology of myocardial stunning." *Circ Res* 93(10): 917-24.

Fuchs, F. and Z. Grabarek (2011). "The Ca2+/Mg2+ sites of troponin C modulate crossbridge-mediated thin filament activation in cardiac myofibrils." *Biochem Biophys Res Commun* 408(4): 697-700.

Galinska, A., V. Hatch, et al. (2010). "The C terminus of cardiac troponin I stabilizes the Ca2+-activated state of tropomyosin on actin filaments." *Circ Res* 106(4): 705-11.

Gergely, J., Z. Grabarek, et al. (1993). "The molecular switch in troponin C." *Adv. Exp. Med. Biol.* 332: 117-23.

Gillis, T. E., D. A. Martyn, A. J. Rivera, and M. Regnier. (2007). "Investigation of thinfilamentnear-neighbourregulatoryunitinteractionsduringforcedevelopment in skinned cardiac and skeletal muscle." *J Physiol* 580(Pt. 2):561-576.

Gollapudi, S. K., C. E. Gallon, et al. (2013). "The Tropomyosin Binding Region of Cardiac Troponin T Modulates Crossbridge Recruitment Dynamics in Rat Cardiac Muscle Fibers." *J Mol Biol.*

Gollapudi, S. K., R. Mamidi, et al. (2012). "The N-terminal extension of cardiac troponin T stabilizes the blocked state of cardiac thin filament." *Biophys J* 103(5): 940-8.

Gordon, A. M., E. Homsher, et al. (2000). "Regulation of contraction in striated muscle." *Physiol. Rev.* 80(2): 853-924.

Grabarek, Z., J. Grabarek, et al. (1983). "Cooperative binding to the Ca2+-specific sites of troponin C in regulated actin and actomyosin." *J Biol Chem* 258(23): 14098-102.

Grabarek, Z. and J. Gergely (1987). "Troponin I binds to the N-terminal twelve residue segment of actin." *Biophys J* 51 (331a).

Greaser, M. L., M. Yamaguchi, et al. (1972). "Troponin subunits and their interactions." *Cold Spring Harbor Symp.* Quant Biol. 37: 235-49.

Gulati, J., A. B. Akella, et al. (1995). "Functional role of arginine-11 in the N-terminal helix of skeletal troponin C: combined mutagenesis and molecular dynamics investigation." *Biochemistry* 34(22): 7348-55.

Gulati, J., A. Babu, et al. (1992). "Functional delineation of the Ca(2+)-deficient EF-hand in cardiac muscle, with genetically engineered cardiac-skeletal chimeric troponin C." *J Biol Chem* 267(35): 25073-7.

Gulati, J., E. Sonnenblick, and A. Babu. (1991). "The role of troponin C in thelengthdependence of Ca(2+)-sensitive force of mammalian skeletal and cardiac muscles." J Physiol 441:305-324.

Guth, K., K. J. Poole, et al. (1987). "The apparent rates of crossbridge attachment and detachment estimated from ATPase activity in insect flight muscle." *Biophys J* 52(6): 1039-45.

Haselgrove, J. C. (1972). "X-ray evidence for a conformational change in the actin containing filaments of vertebrate striated muscle." *Cold Spring Harbor Symp. Quant. Biol.* 37: 341-52.

Heeley, D. H., K. Golosinska, et al. (1987). "The effects of troponin T fragments T1 and T2 on the binding of nonpolymerizable tropomyosin to F-actin in the presence and absence of troponin I and troponin C." *J. Biol. Chem.* 262(21): 9971-8.

Henze, M., S. E. Patrick, et al. (2012). "New insights into the functional significance of the acidic region of the unique N-terminal extension of cardiac troponin I." *Biochim Biophys Acta.*

Herzberg, O. and M. N. James (1985). "Structure of the calcium regulatory muscle protein troponin-C at 2. 8 A resolution." *Nature* 313(6004): 653-9.

Herzberg, O. and M. N. James (1988). "Refined crystal structure of troponin C from turkey skeletal muscle at 2. 0 A resolution." *J Mol Biol* 203(3): 761-79.

Herzberg, O., J. Moult, et al. (1986). "A model for the Ca2+-induced conformational transition of troponin C. A trigger for muscle contraction." *J Biol Chem* 261(6): 2638-44.

Hinkle, A., A. Goranson, et al. (1999). "Roles for the troponin tail domain in thin filament assembly and regulation. A deletional study of cardiac troponin T." *J Biol Chem* 274(11): 7157-64.

Hinkle, A. and L. S. Tobacman (2003). "Folding and function of the troponin tail domain. Effects of cardiomyopathic troponin T mutations." *J. Biol. Chem.* 278(1): 506-513.

Hitchcock-De Gregori, S. E. (1982). "Study of the structure of troponin-I by measuring the relative reactivities of lysines with acetic anhydride." *J Biol Chem* 257(13): 7372-80.

Hoffman, R. M., T. M. Blumenschein, et al. (2006). "An interplay between protein disorder and structure confers the Ca2+ regulation of striated muscle." *J Mol Biol* 361(4): 625-33.

Houdusse, A., M. L. Love, et al. (1997). "Structures of four Ca2+-bound troponin C at 2. 0 A resolution: further insights into the Ca2+-switch in the calmodulin superfamily." *Structure* 5(12): 1695-1711.

Huang, Q. Q., F. V. Brozovich, et al. (1999). "Fast skeletal muscle troponin T increases the cooperativity of transgenic mouse cardiac muscle contraction." *J. Physiol.* 520 Pt 1: 231-42.

Huang, Q. Q., H. Z. Feng, et al. (2008). "Co-expression of skeletal and cardiac troponin T decreases mouse cardiac function." *Am J Physiol Cell Physiol* 294(1): C213-22.

Huang, X., and W. Miller. 1991. "A time-efficient, linear-space local similarity algorithm." *Adv. Appl. Math* 12:337–357.

Hunlich, M., K. J. Begin, et al. (2005). "Protein kinase A mediated modulation of acto-myosin kinetics." *J Mol Cell Cardiol* 38(1): 119-25.

Huxley, A. F. (1957). "Muscle structure and theories of contraction." *Prog. Biophys. Biophys. Chem.* 7: 255-318.

Jackson, P., G. W. Amphlett, et al. (1975). "The primary structure of troponin T and the interaction with tropomyosin." *Biochem. J.* 151(1): 85-97.

Jha, P. K., P. C. Leavis, et al. (1996). "Interaction of deletion mutants of troponins I and T: COOH-terminal truncation of troponin T abolishes troponin I binding and reduces Ca^{2+} sensitivity of the reconstituted regulatory system." *Biochemistry* 35(51): 16573-80.

Jin, J. P., Z. Zhang, et al. (2008). "Isoform diversity, regulation, and functional adaptation of troponin and calponin." *Crit Rev Eukaryot Gene Expr* 18(2): 93-124.

Johnson, J. D. and J. D. Potter (1978). "Detection of two classes of Ca2+ binding sites in troponin C with circular dichroism and tyrosine fluorescence." *J Biol Chem* 253(11): 3775-7.

Kobayashi, T. and R. J. Solaro (2005). "Calcium, thin filaments, and the integrative biology of cardiac contractility." *Annu. Rev. Physiol.* 67: 39-67.

Kobayashi, T., S. E. Patrick, et al. (2009). "Ala scanning of the inhibitory region of cardiac troponin I." *J Biol Chem* 284(30): 20052-20060.

Konhilas, J. P., T. C. Irving, et al. (2002). "Length-dependent activation in three striated muscle types of the rat." *J Physiol* 544(Pt 1): 225-36.

Kostareva, A., A. Gudkova, et al. (2009). "Deletion in TNNI3 gene is associated with restrictive cardiomyopathy." *Int J Cardiol* 131(3): 410-412.

Kreutziger, K. L., T. E. Gillis, et al. (2007). "Influence of enhanced troponin C Ca2+-binding affinity on cooperative thin filament activation in rabbit skeletal muscle." *J Physiol* 583(Pt 1): 337-50.

Kreutziger, K. L., N. Piroddi, et al. (2011). "Calcium binding kinetics of troponin C strongly modulate cooperative activation and tension kinetics in cardiac muscle." *J Mol Cell Cardiol* 50(1): 165-74.

Leavis, P. C. and E. L. Kraft (1978). "Calcium binding to cardiac troponin C." *Arch Biochem Biophys* 186(2): 411-5.

Lehrer, S. S. and M. A. Geeves (1998). "The muscle thin filament as a classical cooperative/allosteric regulatory system." *J Mol Biol* 277(5): 1081-9.

Leszyk, J., R. Dumaswala, et al. (1987). "Bovine cardiac troponin T: amino acid sequences of the two isoforms." *Biochemistry* 26(22): 7035-42.

Li, M. X., L. Spyracopoulos, et al. (1999). "Binding of cardiac troponin-I147-163 induces a structural opening in human cardiac troponin-C." *Biochemistry* 38(26): 8289-98.

Liao, R., C. K. Wang, et al. (1994). "Coupling of calcium to the interaction of troponin I with troponin C from cardiac muscle." *Biochemistry* 33(42): 12729-12734.

Liou, Y. M. and J. C. Chang (2004). "Differential pH effect on calcium-induced conformational changes of cardiac troponin C complexed with cardiac and fast skeletal isoforms of troponin I and troponin T." *J Biochem* 136(5): 683-92.

Little, S. C., B. J. Biesiadecki, et al. (2012). "The rates of Ca2+ dissociation and cross-bridge detachment from ventricular myofibrils as reported by a fluorescent cardiac troponin C." *J Biol Chem* 287(33): 27930-40.

Little, S. C., S. B. Tikunova, et al. (2011). "Measurement of calcium dissociation rates from troponin C in rigor skeletal myofibrils." *Front Physiol* 2: 70.

MacFarland, S. M., J. P. Jin, et al. (2002). "Troponin T isoforms modulate calcium dependence of the kinetics of the cross-bridge cycle: studies using a transgenic mouse line." *Arch Biochem Biophys* 405(2): 241-6.

Malnic, B., C. S. Farah, et al. (1998). "Regulatory properties of the NH_2- and COOH-terminal domains of Troponin T. ATPase activation and binding to troponin I and troponin C." *J. Biol. Chem.* 273(17): 10594-10601.

Mamidi, R., S. K. Gollapudi, et al. (2012). "Alanine or aspartic acid substitutions at serine23/24 of cardiac troponin I decrease thin filament activation, with no effect on crossbridge detachment kinetics." *Arch Biochem Biophys* 525(1): 1-8.

Mamidi, R., S. L. Mallampalli, et al. (2013). "Identification of two new regions in the N-Terminal region of cardiac troponin T that have divergent effects on cardiac contractile function." *J Physiol.* 591(Pt 5):1217-34

Martyn, D. A., and A. M. Gordon. (2001). "Influence of length on force and activation-dependent changes in troponin C structure in skinned cardiac and fast skeletal muscle." *Biophys J* 80(6):2798-2808.

Maytum, R., M. A. Geeves, et al. (2002). "A modulatory role for the troponin T tail domain in thin filament regulation." *J. Biol. Chem.* 277(33): 29774-29780.

McDonald, K. S., L. J. Field, et al. (1995). "Length dependence of Ca2+ sensitivity of tension in mouse cardiac myocytes expressing skeletal troponin C." *J Physiol* 483 (Pt 1): 131-9.

McDonough, J. L., D. K. Arrell, et al. (1999). "Troponin I degradation and covalent complex formation accompanies myocardial ischemia/reperfusion injury." *Circ Res* 84(1): 9-20.

McKay, R. T., L. F. Saltibus, et al. (2000). "Energetics of the induced structural change in a Ca^{2+} regulatory protein: Ca^{2+} and troponin I peptide binding to the E41A mutant of the N-domain of skeletal troponin C." *Biochemistry* 39(41): 12731-12738.

Mercier, P., M. X. Li, et al. (2000). "Role of the structural domain of troponin C in muscle regulation: NMR studies of Ca^{2+} binding and subsequent interactions with regions 1-40 and 96-115 of troponin I." *Biochemistry* 39(11): 2902-2911.

Montgomery, D. E., J. C. Tardiff, et al. (2001a). "Cardiac troponin T mutations: correlation between the type of mutation and the nature of myofilament dysfunction in transgenic mice." *J. Physiol. (Lond.)* 536(2): 583-592.

Montgomery, D. E., M. Chandra, et al. (2001b). "Transgenic incorporation of skeletal TnT into cardiac myofilaments blunts PKC-mediated depression of force." *Am J Physiol Heart Circ Physiol* 280(3): H1011-8.

Morimoto, S. and I. Ohtsuki (1987). "Ca2+- and Sr2+-sensitivity of the ATPase activity of rabbit skeletal myofibrils: effect of the complete substitution of troponin C with cardiac troponin C, calmodulin, and parvalbumins." *J Biochem* 101(2): 291-301.

Morimoto, S. and I. Ohtsuki (1988). "Effect of substitution of troponin C in cardiac myofibrils with skeletal troponin C or calmodulin on the Ca2+- and Sr2+-sensitive ATPase activity." *J Biochem* 104(1): 149-154.

Moss, R. L., M. R. Lauer, et al. (1986). "Altered Ca2+ dependence of tension development in skinned skeletal muscle fibers following modification of troponin by partial substitution with cardiac troponin C." *J Biol Chem* 261(13): 6096-9.

Moss, R. L., L. O. Nwoye, et al. (1991). "Substitution of cardiac troponin C into rabbit muscle does not alter the length dependence of Ca2+ sensitivity of tension." *J Physiol* 440: 273-89.

Moss, R. L., M. Razumova, et al. (2004). "Myosin crossbridge activation of cardiac thin filaments: implications for myocardial function in health and disease." *Circ Res* 94(10): 1290-300.

Mudalige, W. A., T. C. Tao, et al. (2009). "Ca2+-dependent photo crosslinking of tropomyosin residue 146 to residues 157-163 in the C-terminal domain of troponin I in reconstituted skeletal muscle thin filaments." *J Mol Biol* 389(3): 575-83.

Narolska, N. A., N. Piroddi, et al. (2006). "Impaired diastolic function after exchange of endogenous troponin I with C-terminal truncated troponin I in human cardiac muscle." *Circ Res* 99(9): 1012-20.

Negele, J. C., D. G. Dotson, et al. (1992). "Mutation of the high affinity calcium binding sites in cardiac troponin C." *J Biol Chem* 267(2): 825-831.

Nosek, T. M., M. A. Brotto, et al. (2004). "Troponin T isoforms alter the tolerance of transgenic mouse cardiac muscle to acidosis." *Arch. Biochem. Biophys.* 430(2): 178-184.

Ohtsuki, I. (1975). "Distribution of troponin components in the thin filament studied by immunoelectron microscopy." *J Biochem* 77(3): 633-9.

Ohtsuki, I. (1979). "Molecular arrangement of troponin T in the thin filament." *J. Biochem. (Tokyo)* 86(2): 491–497.

Ohtsuki, I. and F. Shiraishi (2002). "Periodic binding of troponin C. I and troponin I to tropomyosin-actin filaments." J *Biochem* 131(5): 739-43.

Oliveira, D. M., C. R. Nakaie, et al. (2000). "Mapping the domain of troponin T responsible for the activation of actomyosin ATPase activity. Identification of residues involved in binding to actin." *J. Biol. Chem.* 275(36): 27513-27519.

Palm, T., S. Graboski, et al. (2001). "Disease-causing mutations in cardiac troponin T: Identification of a critical tropomyosin-binding region." *Biophys. J.* 81(5): 2827-2837.

Pan, B. S., A. M. Gordon, et al. (1991). "Deletion of the first 45 NH2-terminal residues of rabbit skeletal troponin T strengthens binding of troponin to immobilized tropomyosin." *J. Biol. Chem.* 266(19): 12432-8.

Pearlstone, J. R. and L. B. Smillie (1977). "The binding site of skeletal alpha-tropomyosin on troponin-T." *Can. J. Biochem.* 55(10): 1032-1038.

Pearlstone, J. R. and L. B. Smillie (1978). "Troponin T fragments: physical properties and binding to troponin C." *Can J Biochem* 56(6): 521-7.

Pearlstone, J. R. and L. B. Smillie (1981). "Identification of a second binding region on rabbit skeletal troponin-T for a-tropomyosin." *FEBS Lett.* 128(1): 119-122.

Pinto, J. R., A. V. Gomes, et al. (2012). "The functional properties of human slow skeletal troponin T isoforms in cardiac muscle regulation." *J Biol Chem* 287(44): 37362-70.

Pinto, J. R., T. Veltri, et al. (2008). "Modulation of troponin C affinity for the thin filament by different cross-bridge states in skinned skeletal muscle fibers." *Pflugers Arch* 456(6): 1177-87.

Potter, J. D. and J. Gergely (1975). "The calcium and magnesium binding sites on troponin and their role in the regulation of myofibrillar adenosine triphosphatase." *J Biol Chem* 250(12): 4628-33.

Potter, J. D., S. P. Robertson, et al. (1981). "Magnesium and the regulation of muscle contraction." *Fed Proc* 40(12): 2653-6.

Potter, J. D., Z. Sheng, et al. (1995). "A direct regulatory role for troponin T and a dual role for troponin C in the Ca^{2+} regulation of muscle contraction." *J. Biol. Chem.* 270(6): 2557-2562.

Rao, V. G., A. B. Akella, et al. (1995). "Molecular mobility of the Ca(2+)-deficient EF-hand of cardiac troponin C as revealed by fluorescence polarization of genetically inserted tryptophan." *Biochemistry* 34(2): 562-8.

Rarick, H. M., X. H. Tu, et al. (1997). "The C terminus of cardiac troponin I is essential for full inhibitory activity and Ca2+ sensitivity of rat myofibrils." *J Biol Chem* 272(43): 26887-26892.

Rarick, H. M., H. P. Tang, et al. (1999). "Interactions at the NH2-terminal interface of cardiac troponin I modulate myofilament activation." *J Mol Cell Cardiol* 31(2): 363-375.

Redwood, C., K. Lohmann, et al. (2000). "Investigation of a truncated cardiac troponin T that causes familial hypertrophic cardiomyopathy: Ca(2+) regulatory properties of reconstituted thin filaments depend on the ratio of mutant to wild-type protein." *Circ Res* 86(11): 1146-52.

Regnier, M., A. J. Rivera, et al. (2002). "Thin filament near-neighbour regulatory unit interactions affect rabbit skeletal muscle steady-state force-Ca(2+) relations." *J Physiol* 540(Pt 2): 485-97.

Sadayappan, S., N. Finley, et al. (2008). "Role of the acidic N' region of cardiac troponin I in regulating myocardial function." *Faseb J* 22(4): 1246-1257.

Satyshur, K. A., S. T. Rao, et al. (1988). "Refined structure of chicken skeletal muscle troponin C in the two-calcium state at 2-A resolution." *J Biol Chem* 263(4): 1628-47.

Schaertl, S., S. S. Lehrer, et al. (1995). "Separation and characterization of the two functional regions of troponin involved in muscle thin filament regulation." *Biochemistry* 34(49): 15890-4.

Schiaffino, S. and C. Reggiani (1996). "Molecular diversity of myofibrillar proteins: gene regulation and functional significance." *Physiol Rev* 76(2): 371-423.

Sia, S. K., M. X. Li, et al. (1997). "Structure of cardiac muscle troponin C unexpectedly reveals a closed regulatory domain." *J Biol Chem* 272(29): 18216-21.

Smillie, L. B., K. Golosinska, et al. (1988). "Sequences of complete cDNAs encoding four variants of chicken skeletal muscle troponin T." *J Biol Chem* 263(35): 18816-20.

Smith, L., N. J. Greenfield, et al. (1994). "The effects of deletion of the amino-terminal helix on troponin C function and stability." *J Biol Chem* 269(13): 9857-63.

Smith, L., N. J. Greenfield, et al. (1999). "Mutations in the N- and D-helices of the N-domain of troponin C affect the C-domain and regulatory function." *Biophys J* 76(1 Pt 1): 400-8.

Smith, L., C. Tainter, et al. (2009). "Cooperative cross-bridge activation of thin filaments contributes to the Frank-Starling mechanism in cardiac muscle." *Biophys J* 96(9): 3692-702.

Snyder, M. R., J. D. Potter, et al. (1996). "Secondary and tertiary structural analysis of troponin." *Biophys J* 70: A381.

Solaro, R. J., A. J. Moir, et al. (1976). "Phosphorylation of troponin I and the inotropic effect of adrenaline in the perfused rabbit heart." *Nature* 262(5569): 615-7.

Sorenson, M. M., A. C. da Silva, et al. (1995). "Concerted action of the high affinity calcium binding sites in skeletal muscle troponin C." *J Biol Chem* 270(17): 9770-7.

Spyracopoulos, L., M. X. Li, et al. (1997). "Calcium-induced structural transition in the regulatory domain of human cardiac troponin C." *Biochemistry* 36(40): 12138-46.

Stefancsik, R., P. K. Jha, et al. (1998). "Identification and mutagenesis of a highly conserved domain in troponin T responsible for troponin I binding: Potential role for coiled coil interaction." *Proc. Natl. Acad. Sci. U. S. A.* 95(3): 957-962.

Stelzer, J. E., J. R. Patel, et al. (2004). "Expression of cardiac troponin T with COOH-terminal truncation accelerates cross-bridge interaction kinetics in mouse myocardium." *Am. J. Physiol. Heart Circul. Physiol.* 287(4): H1756-H1761.

Strauss, J. D., C. Zeugner, et al. (1992). "The positive inotropic calcium sensitizer EMD 53998 antagonizes phosphate action on cross-bridges in cardiac skinned fibers." *Eur J Pharmacol*227(4): 437-41.

Strynadka, N. C., M. Cherney, et al. (1997). "Structural details of a calcium-induced molecular switch: X-ray crystallographic analysis of the calcium-saturated N-terminal domain of troponin C at 1. 75 A resolution." *J Mol Biol* 273(1): 238-55.

Sumandea, M. P., S. Vahebi, et al. (2009). "Impact of cardiac troponin T N-terminal deletion and phosphorylation on myofilament function." *Biochemistry* 48(32): 7722-31.

Sun, Y. B., B. Brandmeier, et al. (2006). "Structural changes in troponin in response to Ca2+ and myosin binding to thin filaments during activation of skeletal muscle." *Proc Natl Acad Sci U S A*103(47): 17771-6.

Sundaralingam, M., R. Bergstrom, et al. (1985). "Molecular structure of troponin C from chicken skeletal muscle at 3-angstrom resolution." *Science* 227(4689): 945-8.

Swiderek, K., K. Jaquet, et al. (1988). "Cardiac troponin I, isolated from bovine heart, contains two adjacent phosphoserines. A first example of phosphoserine determination by derivatization to S-ethylcysteine." *Eur J Biochem* 176(2): 335-42.

Syska, H., S. V. Perry, et al. (1974). "A new method of preparation of troponin I (inhibitory protein) using affinity chromatography. Evidence for three different forms of troponin I in striated muscle." *FEBS Lett* 40(2): 253-7.

Syska, H., J. M. Wilkinson, et al. (1976). "The relationship between biological activity and primary structure of troponin I from white skeletal muscle of the rabbit." *Biochem J* 153(2): 375-387.

Szczesna, D., G. Guzman, et al. (1996). "The role of the four Ca2+ binding sites of troponin C in the regulation of skeletal muscle contraction." *J Biol Chem* 271(14): 8381-6.

Szczesna, D., R. Zhang, et al. (1999). "The role of the NH(2)- and COOH-terminal domains of the inhibitory region of troponin I in the regulation of skeletal muscle contraction." *J Biol Chem* 274(41): 29536-42.

Tachampa, K., T. Kobayashi, et al. (2008). "Increased cross-bridge cycling kinetics after exchange of C-terminal truncated troponin I in skinned rat cardiac muscle." *J Biol Chem* 283(22): 15114-15121.

Takeda, S., A. Yamashita, et al. (2003). "Structure of the core domain of human cardiac troponin in the Ca^{2+}-saturated form." *Nature* 424(6944): 35-41.

Talbot, J. A. and R. S. Hodges (1981). "Synthetic studies on the inhibitory region of rabbit skeletal troponin I. Relationship of amino acid sequence to biological activity." *J Biol Chem* 256(6): 2798-802.

Tanokura, M. and I. Ohtsuki (1982). "Location of troponin I-binding on troponin T sequence." *FEBS Lett* 145(1): 147-9.

Tanokura, M., Y. Tawada, et al. (1983). "Chymotryptic subfragments of troponin T from rabbit skeletal muscle. Interaction with tropomyosin, troponin I and troponin C." *J. Biochem. (Tokyo)* 93(2): 331-7.

Tobacman, L. S., M. Nihli, et al. (2002). "The troponin tail domain promotes a conformational state of the thin filament that suppresses myosin activity." *J. Biol. Chem.* 277(31): 27636-42.

Tobacman, L. S. and D. Sawyer (1990). "Calcium binds cooperatively to the regulatory sites of the cardiac thin filament." *J Biol Chem* 265(2): 931-9.

Tripet, B., J. E. Van Eyk, et al. (1997). "Mapping of a second actin-tropomyosin and a second troponin C binding site within the C terminus of troponin I, and their importance in the Ca2+-dependent regulation of muscle contraction." *J Mol Biol* 271(5): 728-750.

Vahebi, S., T. Kobayashi, et al. (2005). "Functional effects of rho-kinase-dependent phosphorylation of specific sites on cardiac troponin." *Circ. Res.* 96(7): 740-747.

Van Eyk, J. E. and R. S. Hodges (1988). "The biological importance of each amino acid residue of the troponin I inhibitory sequence 104-115 in the interaction with troponin C and tropomyosin-actin." *J Biol Chem* 263(4): 1726-32.

Vinogradova, M. V., D. B. Stone, et al. (2005). "Ca^{2+}-regulated structural changes in troponin." *Proc. Natl. Acad. Sci. U. S. A.* 102(14): 5038-5043.

Wang, Y. P., F. Fuchs, et al. (1994). "Length, force, and Ca(2+)-troponin C affinity in cardiac and slow skeletal muscle." *Am J Physiol*266(4 Pt 1): C1077-82.

Ward, D. G., S. M. Brewer, et al. (2003). "A cross-linking study of the N-terminal extension of human cardiac troponin I." *Biochemistry* 42(34): 10324-32.

Warren, C. M., T. Kobayashi, et al. (2009). "Sites of intra- and intermolecular cross-linking of the N-terminal extension of troponin I in human cardiac whole troponin complex." *J Biol Chem* 284(21): 14258-66.

Yu, Z. B., H. Wei, et al. (2012). "Chronic coexistence of two troponin T isoforms in adult transgenic mouse cardiomyocytes decreased contractile kinetics and caused dilatative remodeling." *Am J Physiol Cell Physiol* 303(1): C24-32.

Zhang, Z., S. Akhter, et al. (2011). "Calcium-regulated conformational change in the C-terminal end segment of troponin I and its binding to tropomyosin." *Febs J* 278(18): 3348-59.

Zot, H. and J. Potter (1982). "A structural role for the Ca^{2+}-Mg^{2+} sites on troponin C in the regulation of muscle contraction. Preparation and properties of troponin C depleted myofibrils." *J. Biol. Chem.* 257(13): 7678-7683.

In: Troponin
Editor: J.-P. Jin

ISBN: 978-1-62948-320-7

Chapter V

Conformational States and Behavior of the Heterotrimeric Troponin Complex

***Daniel C. Rieck, MSc*[1] *and Wen-Ji Dong, PhD*[1,2*]**

[1] Gene and Linda Voiland School of Chemical Engineering and Bioengineering, Washington State University, Pullman, Washington, US

[2] Department of Integrative Physiology and Neuroscience, Washington State University, Pullman, Washington, US

Abstract

Complete understanding of the functions of the heterotrimeric troponin complex rests upon knowledge of its structure. Insights into the physiological structure and conformational behavior of troponin are thus reviewed in this chapter. The primary experimental settings and biophysical methods that have been used to investigate troponin structure are covered first, wherein a critical understanding of the differences between approaches is emphasized. General aspects of the structure and molecular environment of troponin are treated next, starting at the subunit level and ending with a description of how troponin fits into the ultrastructure of the myofilament lattice. Building off of this foundation, the changes in troponin conformational state that lead to its complex regulatory behaviors are then discussed. Overall, it has become clear that the conformational states of troponin and the nature of conversions between them collectively play a central role in determining the physiological contractile behavior of the myocyte sarcomere.

Introduction

The physiological structure and conformational behavior of the heterotrimeric troponin (TN) complex has long been a topic of considerable scientific importance. It is as easy to see

* Correspondence: wdong@vetmed.wsu.edu.

why as the act of seeing itself. Right now, your eyes are moving along this text because of the Ca^{2+}-dependent conformational behavior of TN. TN structure also plays an essential role in maintaining your body posture as you read. When you manipulate the remainder of this text, whether it is in print or digital form, Ca^{2+}-induced changes in the conformational state of TN will enable you to turn your thoughts into muscular action. While reading, your heart will faithfully and involuntarily pump blood throughout your body thanks also to Ca^{2+}-dependent structural changes in TN. The Ca^{2+} responsiveness of TN conformational state not only provides these types of control over muscular force generation in *Homo sapiens*, but also in an evolutionary panoply of vertebrates and invertebrates. It is this impressive biological milieu in which the science of TN structure finds both its context and aim.

This science has advanced considerably in recent decades as the conformational behavior of TN has been increasingly illuminated by numerous lines of investigation. The field necessarily started with studying TN and its subunits in relative biochemical isolation. Invaluably detailed crystal structure models of the TN core domain have emerged from X-ray diffraction (XRD) studies on *H. sapiens* cardiac TN (cTNC) [1] and *Gallus gallus* fast skeletal TN (sTN) [2]. As important as this structural information is, our field has built substantially on the relatively static, artificial, and isolated picture of TN given by X-ray crystallography. Through a variety of other techniques, the dynamic and contextual nature of the relationship between TN conformation and function has progressively come to light under increasing levels of physiological approximation. Methods for studying TN structure *in situ* have even been developed. Overall, it is now much clearer how the Ca^{2+}-dependent conformational behavior of TN is intricately complex, intensely allosteric, isoform specific, and continually modulated by the presence and activity of other proteins found in the sarcomere.

To explore the structural behavior of TN is to immerse oneself in the mystery and wonder of what protein chemistry achieves in muscle. The primary goal of this chapter is to fortify within the reader a critical familiarity with what has been learned about TN structure and its functional role in muscle physiology. An important secondary goal is to help the reader grasp the key scientific problems that remain in the field of TN structural study. The chapter is broadly structured to first review the basics of TN structure from a cellular physiology perspective and then discuss the conformational behavior of TNin detail while focusing on the more advanced developments of our field that have been made in the last decade. Accordingly, after introducing the complexities inherent in the study of TN conformational behavior, we will begin with a critical overview of the key experimental methods that have been employed. Next, we will explore the structure of TN starting at the subunit level and proceeding to its arrangement in the architecture of the sarcomeric thin filament (TF). This will serve as a foundation for discussing the effects of Ca^{2+}binding and dissociation and strong cross-bridge (sXB) formation on the conformational state of TN. We will thus conclude with a mechanistic treatment of how these changes in TN conformation serve as the structural basis for the ability of TN to regulate sarcomeric force generation.

The Importance of Conformational State to Troponin Function

The concept that a protein's structure dictates its function is the axiomatic premise behind the quintessential hypothesis of structural biology investigation. When engrossing oneself in the study of TN structure, it is therefore important to remain cognizant of the physiological context and evolutionary purpose of TN. TN serves a regulatory role in the TFs found in the highly organized and force-generating myocyte sarcomere, one of Nature's most complex macromolecular assemblies [3]. Accordingly, the *prima facie* identity of TN is that ofa complex of binding proteins. TN binds divalent cations like Ca^{2+} and other TF proteins such as actin and tropomyosin, but seems to perform no other discernible chemistry. It is only after the structural behavior of TN is closely examined that the subtlety of its characteristics becomes more evident. Though TN is not strictly an enzyme, the Michaelis Menten model nevertheless accurately describes the kinetics of Ca^{2+}-induced TN conformational changes[4]. This suggests that TN serves as both "catalyst" and "substrate" for producing reversible changes in its own structure. The useful "product" that TN affords the myocyte is the Ca^{2+} dependent conversion between its conformations because those conformations either inhibit or promote sarcomeric force generation. The integrity of not only the conformational conversion process but also the conformations themselves is therefore physiologically significant [5]. Therein lays the importance of studying TN structure.

The conformational conversion function of TN has been referred to in the past as Ca^{2+}-regulated switching (CRS) [6,7] and TN itself has been called a Ca^{2+}-sensitive molecular switch [8]. The mechanism underlying CRS is deceptively simple. TN possesses Ca^{2+} sensing functionality akin to calmodulin (CaM) such that its conformation changes in response to Ca^{2+} binding or dissociation. ItsCa^{2+}-dependent structural changes have evolved to function as a binary switch in turning sarcomeric force generation "on" or "off." Thus CRS represents a highly specialized form of second messenger regulation that has been integrated into the sarcomeric thin filament (TF). TN transduces second messenger Ca^{2+} signals into TF protein conformational changes and thereby provides the regulatory signaling mechanism necessary for controlling muscular force. Though some details of the mechanism involved are isoform specific, the binary switching concept of TN function appears universal [9]. Nevertheless, the more that has been learned about the conformations of TN and how they change, the clearer it has become that CRS is far more complex than a simple matter of "on or off."

CRS involves ensemble behavior, energetic balance, dynamics, and kinetics. Ca^{2+}-triggered conformational changes in TN and other TF regulatory proteins take time, and it also takes time for the actomyosin cross-bridge cycle to respond accordingly. This adds a layer of additional complexity to TN structural study that is compounded by the significant differences between the *in vivo* molecular environment and what is present during experimentation. Thus, the physiological kinetics of CRS are still largely unknown, as are any changes to normal kinetics that are caused by mutations and post translational modifications (PTMs). There are many other fundamental uncertainties surrounding the structural mechanism underlying *in vivo* CRS that need to be addressed. Particularly, the structural basis for the role of TN in effecting the cooperativity and sarcomere length dependence of Ca^{2+}-dependent TF activation remains largely uncharacterized. Hence, in spite of years of considerable progress made in the study of TN structure, there remains much to learn through additional scientific inquiry about the physiological conformational behavior of TN.

Methods for Studying Troponin Structure

The scientific goal of our field is to gain a complete understanding of the how *the in vivo* conformational behavior of TN relates to its physiological function. We aim to evolve this understanding from accumulating experimental observations made under a particular setting using some combination of methods. Every overall approach has both benefits and drawbacks, and it is vital to be cognizant of such pros and cons. A brief critical survey of the settings and methods that have been used in structural studies of TN reveals the difficulty of our goal: the elucidation of the physiological relationship between TN structure and function.

Experimental Settings

In the context of this review, "setting" represents which isoform of TN is being studied and what kind of molecular environment is being used in an experiment. Each type of environment aims to achieve some form of balance between interpretational simplicity and reproduction of native complexity. One must nevertheless recognize that protein folding and binding are inexorably a function of molecular environment. It is not only the primary sequence of TN subunits that will dictate their stable conformations and binding affinities. Other important determinants include solvent properties and the influence of the other biomolecules present in the sarcomere. Keeping that caveat in mind, the three settings wherein significant advances have been made in our knowledge of TN structure are *in vitro*, *in situ*, and *in silico*.

The In Vitro Setting

TN structural study began *in vitro* and enjoys a long and fruitful history with this setting. Typically, recombinant isoforms of one or more TN subunits are first bacterially expressed and purified. They are combined with each other and other purified sarcomeric proteins if desired. They are then structurally characterized under some biochemical condition. The major advantages of *in vitro* investigation are isolation and controlled integration. A TN subunit may first be studied alone, next when integrated into the whole complex, and finally when integrated into reconstituted TFs. This allows for the relatively unambiguous interpretation of changes in structural behavior and the potential determinants of physiological conformational behavior can thus be identified more easily. Higher resolution structural methods such as XRD and nuclear magnetic resonance (NMR) are still almost solely compatible with the *in vitro* experimental format due to its simpler preparations.

The major disadvantage of the *in vitro* setting is that the level of physiological approximation is very slight even under the best case. The most physiologically representative *in vitro* setting has been the reconstituted TF, wherein whole TN is integrated into F-actin along with TM. Additionally, myosin subfragment-1 (S1) may be added to simulate the presence of XBs. However, a familiarity with the field of sarcomerogenesis makes one realize just how different reconstituted TFs are from what occurs naturally in myocytes [3, 10]. Molecular rulers are missing *in vitro*, structural and elastic elements are not present, and components are "assembled" *via* unguided binding equilibria. Force cannot be generated and sarcomere length dependence cannot be studied because the sarcomere lattice is absent. Since the cooperativity of Ca^{2+}-dependent TF activation is greatly attenuated *in vitro* relative to *in situ*, significant allosteric pathways for modulating TN conformational state must also be missing. Nevertheless, the *in vitro* setting has produced many useful insights into TN conformational states and the Ca^{2+} and XB dependent conversions between them.

The In Situ Setting

Our field has recently developed abilities to investigate TN structure when found in the relatively intact sarcomeres of isolated muscle tissue. Though the *in situ* setting is not technically *in vivo*, it has allowed structural investigation to begin approaching the physiological environments in which TN is naturally found. Generally, tissue is dissected from a particular organism and chemically skinned to remove myocyte membranes and their associated proteins. Demembranation allows direct access to sarcomeric proteins and experimental control over the solution that bathes myofilaments. Also, TN constructs with any mutations and fluorescent labels may be verifiably swapped with endogenous counterparts. The stability of chemically skinned *in situ* preparations is fortunately quite good even at room temperature when reducing agents and protease inhibitors are present [11]. Thus, the advantage of the *in situ* environment is that the experimentalist works with preparations that are the product of actual sarcomerogenesis. Components absent *in vitro* are now present, and one can investigate how TN conformational behavior relates to force generation, sarcomere length, and a more realistic level of cooperativity.

The disadvantage of the *in situ* setting is that control over experimental variables comes at a compromise. Demembranation changes sarcomere lattice spacing [12], and skinned fibers have poor stability at physiological temperatures such that room temperature *in situ* studies are preferred[13]. Muscles can come from many different organisms, leaving one to wonder about their *in vivo* differences and consequent interpretational subtleties. Most *in situ* structural techniques to date have been based on fluorescence spectroscopy and therefore require that endogenous TN subunits are substituted with fluorescently labeled recombinant preparations. Labeling often requires site directed mutagenesis in addition to the covalent addition of bulky fluorophores to protein structure. However, since protein structures frequently mutate during evolution with little functional consequence, there is good reason to take seriously the data from labeled TN constructs that have been carefully experimentally controlled. In spite of its limitations, the *in situ* setting is much closer to *in vivo* than *in vitro* alternatives and can add much new information to the science of TN structure.

The In Silico Setting

It may seem that use of the *in silico* setting is moving even further away from the physiological environment of TN than the *in vitro* setting. Nevertheless, computational simulation has an important and useful place in the study of TN structure. The principal approach used *in silico* is the Monte Carlo method. Simulation proceeds from a solvated starting structure of the proteins under investigation. As simulated time passes, the protein is allowed to stochastically search through its conformational space according to the same thermodynamic constraints that determine real world conformational behavior. Thus the major advantage of the *in silico* setting is that it can provide powerful answers about the likelihood of conformational possibilities that cannot readily be studied. It is useful for determining how TN mutation likely changes intra-protein residue contacts known from XRD or NMR, or the probable behavior of TN regions too dynamic to be confidently resolved from XRD data.

The obvious disadvantage of *in silico* methods is that modeling the reality of protein chemistry is computationally intensive and fundamentally challenging. This is why the virtual environment of probabilistic, stochastic simulation, the quintessential *in silico* approach, resembles the *in vitro* rather than the *in situ* setting. Another consequence of computational limitations is that conformational behavior may currently be simulated up to nanosecond time scales at most. Many physiologically important events that occur on slower millisecond time scales cannot yet be simulated. Finally, *in silico* methods are geared toward equilibrium and finding the stable conformations of a molecule. Important but short lived intermediates that may be involved in conversions between conformations will likely be missed. Though one should be careful when interpreting *in silico* predictions for these reasons, the probabilistic information provided about TN structure is still very useful.

Structural Measurement Techniques

Quantum physics is the fundamental science that makes the biophysical measurement of TN structure possible. Even a basic understanding of molecules as quantum systems can prove illuminating when thinking critically about TN structural studies. Like any molecule, TN is an entity with a dynamic structure wherein spinning electrons in constant motion are localized to spinning nuclei in a manner describable by a three dimensional wave function. This behavior stems from how underlying fundamental forces driven by charges are kept in a continual balance. Electron orbitals may be thought of accordingly as energetically optimized "clouds" of dynamic nuclear localization. If an electron were localized any more "tightly" than the cloud it is stably occupying, its electric potential energy would decrease less than its kinetic energy would increase; hence the balance achieved by the orbital. Of course, the dynamic balance of forces in a molecule may be changed to a higher energy state if the right amount of energy is somehow supplied. All structural techniques depend on observing how the balance of forces in a molecule interacts with some form of externally applied force.

Fluorescence Spectroscopy

Many studies of TN structure have relied on fluorescence spectroscopy, wherein one observes the complex interaction between an externally applied electromagnetic force and a fluorescent molecule [14]. Photons are energy in the form of a harmonically oscillating electromagnetic field propagated through space. If an incident photon is oscillating at a certain frequency, it can supply the exact amount of applied electromagnetic force needed to change the localization of a valence electron within a fluorophore. Within femtoseconds, photonic energy will "excite" the electron away from its ground state orbital to occupy an orbital of higher electric potential energy. The distribution of Coulombic charge within the fluorophore, represented by a dipole moment, consequently changes. Excitation can also cause the fluorophore to vibrate more energetically. Accordingly, it is common for a photon absorbing fluorophore to become both vibrationally and electronically, or vibronically, excited. Compared to the energy required to change a fluorophore's vibrational mode, the energy required to change valence electron localization is two orders of magnitude greater. Thus, many vibrational modes might be accessible from the supplied excitation energy.

The lowest energy excited state is only semi-stable such that fluorophores typically exhibit ~10 nanosecond fluorescence lifetimes. An excited fluorophore will spontaneously relax to its ground state through one of several possible pathways. Emission occurs when the excited electron spontaneously returns to its ground state orbital and in the process emits self-propagating electromagnetic waves. Relaxation can also occur non-radiatively through external conversion wherein electronic excitation energy is dissipated through collision with neighboring solvent molecules. Collisional quenching is similar but involves solute instead of solvent molecules. A fluorophore can even resonantly transfer its excitation energy to a neighboring fluorophore if certain conditions are met. Ultimately, the quantum yield, or probability that emission will be the electronic relaxation pathway that follows an excitation event, depends both on a fluorophore's structure and the nature of its molecular environment.

Though the energy of an emitted photon can be equal to that of the exciting photon, it is generally less. This happens because some of the exciting photon's absorbed energy can be lost in several ways prior to emission. For instance, through picosecond internal conversion an excited electron will non-radiatively relax to an orbital at one electronic energy level above the ground state regardless of how many energy levels it initially gained. Similarly, a vibrationally excited fluorophore will thermally transfer its excess vibrational energy to its environment within picoseconds. Some energy may also be lost when surrounding solvent dipoles are induced into alignment with the excited state dipole of the fluorophore. Additionally, though emission returns the excited electron back to its ground electronic energy level, it can leave the fluorophore in one of generally many possible excited vibrational modes. These types of energy losses are responsible for the Stokes shift, or energetic difference between excitation and emission maxima, that makes fluorophores so useful as a molecular label.

Environmentally Sensitive Fluorescent Probes

The high sensitivity of some fluorophores to environmental conditions can be employed for structural investigation. A conversion between a protein's conformational states can involve changes in the polarity of the local environment at certain locations within the protein. When a fluorophore is attached to such locations, its maximal emission intensity will change as a function of conformational state if its quantum yield is significantly affected by polarity. This is the idea behind using environmentally sensitive fluorophores as an indicator of TN structural transitions. The method is well suited to TN structural study because CRS involves the binding of positively charged Ca^{2+} and changes in the solvent exposure of hydrophobic regions within TN [15]. Environmentally sensitive fluorophores have been used extensively to study the binding of Ca^{2+} to TN *in vitro* and have recently also been employed *in situ* [16, 17]. One major advantage of environmentally sensitive fluorophores is that they can be used to measure the kinetics of changes in local polarity associated with changes in TN structure. The main disadvantage of the method is that it provides little structural detail other than indicating that a conformational change has occurred. Furthermore, if a TN structural transition occurs without producing a change in the polarity of the probe's local environment, the transition will not be detected [18].

Fluorescence Anisotropy

The experimentalist can use anisotropy to exploit the polarized nature of fluorescence in measuring regional protein flexibility and orientation [14]. Photons are absorbed and radiated by a fluorophore along its excitation and emission transition dipole moments, respectively. If vertically polarized excitation light is focused on a fluorescently labeled sample, those fluorophores whose excitation transition dipole moments happen to be vertically aligned will be preferentially excited. The orientation of these selectively excited fluorophores will then begin to randomize due to rotational diffusion associated with stochastic molecular motion. The vertically and horizontally polarized emission intensity from the sample can thus be measured to determine the extent of rotational diffusion that occurs over the duration of the excited state. The more the vertical and horizontal emission intensities overlap, the more the orientations of the selectively excited fluorophore population must have randomized from rotational diffusion. Hence, anisotropy gives information about the inertial susceptibility of labeled species to thermal forces.

Anisotropy can be measured under both steady-state and time-resolved conditions. When time-dependent anisotropy measurements are conducted, fluorescence anisotropy will decay with time according to the kinetics of the underlying rotational diffusion processes that contribute to changes in excited probe orientation. Processes for TN can include the stochastic motions of the thin filament, the core domain, regional subunit structure, and the fluorophore itself [19]. Fluorophore tumbling is generally much faster than the fluorescence lifetime, whereas thin filament and core domain motions are practically negligible due to the large molecular sizes involved. Regional protein structural motions occur on the order of the fluorescence lifetime. Anisotropy can therefore provide information about regional protein dynamics [19] and spatial orientations [20].

The main advantage of anisotropy is its ability to provide information about protein dynamics in very large molecules such as TN when integrated into the thin filament. It has worked not only *in vitro* [19], but also in *situ*. Two bifunctional rhodamines can been attached to an α-helix in TN and crosslinked. A fiber containing labeled TN can then be illuminated with excitation light aligned either in parallel with or perpendicular to the fiber axis [20-22]. This provides a quantitative measure of changes in regional α-helical orientation [21] or even TN core domain orientation [20] during CRS. The main disadvantage of anisotropy is that time window for observation of rotational diffusion is limited to the fluorescence lifetime of the probe. Also, like any fluorescence technique, anisotropy requires careful fluorescence labeling that may even involve site directed mutagenesis to generate single or double cysteine mutants amenable to the desired placement of probes.

Förster Resonance Energy Transfer

Förster resonance energy transfer (FRET) is a powerful and versatile structural technique often described as a spectroscopic ruler [14]. To understand how FRET works, note that an excited fluorophore can be accurately described as an oscillating dipole, which is analogous to a tuning fork. Accordingly, every fluorophore has a resonant frequency. If exposed to an electromagnetic field oscillating at that frequency, a fluorophore will convert the energy of that field into its own oscillation. Oscillating dipoles generate "near" and "far" field components. The near field is dominated by the strength of the dipole's own electromagnetic field that, which has time variant electric and magnetic field components. Since time variant electric fields generate magnetic fields and *vice versa*, dipole oscillation produces secondary radiative field components whose strength dominates the far field. The dipole and radiative field strengths decay according to inverse cube and square power laws, respectively. FRET takes place within the near field, where absorption of electromagnetic field energy by the acceptor causes the excited state donor dipole to cease oscillating. Thus the excited donor valance electron becomes relaxed while the acceptor valance electron becomes excited.

FRET can occur between a donor and an accepter fluorophore pair where the donor emission spectrum overlaps with the acceptor excitation spectrum. This translates to overlap between the energetics of the vibronic emission and excitation transitions of the donor and acceptor, respectively. Another requirement is that the transition dipole moments of the donor and acceptor are parallel. If these conditions are met, then FRET will occur with an efficiency governed by an inverse 6th power law and involving a unique Förster critical distance commonly denoted R_0. Note that R_0 depends on the orientational freedom of the fluorophores. When the donor and acceptor are spaced apart by R_0, FRET is 50% efficient. FRET efficiency can be determined by measuring donor and acceptor fluorescence intensities or lifetimes, the latter of which is preferred because it is independent of fluorophore concentration.

Depending on the pair of fluorophores chosen by the experimentalist, FRET is useful for measuring molecular distances ranging from approximately 10 to 100 Å. FRET is fortuitously independent of molecular environment. Other advantages of FRET are its high sensitivity to alterations in inter-probe distance, suitability for measuring a range of distances significant to protein conformational changes, compatibility with the *in situ* setting, and ability to measure the kinetics of changes in molecular distance. Multi-acceptor FRET can even be used to determine the orientation of TN in the TF, but the experiment involved is complex and challenging to interpret correctly [23]. Its two main disadvantages are that it requires labeling with multiple fluorophores, and that R_0 must be determined indirectly in a manner that includes assumptions about relative dipole orientations and freedom of movement. Furthermore, FRET measurements must be carefully controlled. Measurements of the donor emission in the absence of acceptors are a must, and conformational changes must not end up limiting the freedom of probe orientation.

Protein Crystallization and X-Ray Diffraction

XRD is another structural technique wherein the interaction between electrons and electromagnetic radiation is observed [24]. Compared to fluorescence, one works with higher energy X-ray photons that are 0.5–2.5 Å in wavelength; and instead of depending on the excitatory absorption of photons by electrons, XRD instead relies on electrons to scatter photons. Scattering is caused by electromagnetic field oscillations associated with an incident X-ray as it is propagated into matter. Electrons are acted upon by the electromagnetic field to oscillate at the same frequency as the X-ray, and the associated charge motions induce a secondary electromagnetic field oscillation also of the same frequency. This secondary electromagnetic field oscillation is propagated in all directions. If scattering occurs from an array of atoms that are appropriately spaced and basically immobile, the X-ray waves produced by that scattering can undergo interference to produce diffracted rays arranged in a pattern representative of underlying atomic structure. Crystal lattices contain the ordered and immobilized atomic arrays needed to diffract X-rays in this manner.

Crystallization is the process whereby a folded protein is converted from its dynamic state in solution into repeating units of ordered and immobilized atomic arrangement suitable for XRD. Protein crystallization is still something of an art form but generally involves the achievement of a supersaturated protein-precipitant solution, prepared from highly purified protein, whose conditions are ideal for the stability and crystallization of the protein of interest [25]. The crystals produced are principally held together by hydrogen bonding between hydrated protein surfaces and must be of adequate size for handling. If a crystal can obtained and imaged, then its XRD patterns can be interpreted into an electron density map. Since atoms of greater electron density will produce diffracted rays of greater intensity, XRD gives information about atomic position and identity.

The advantage of a good XRD study is the detailed Å resolution structural information it provides. Because crystals are formed from folded proteins, it is reasonable to expect that crystal structures will be representative of *in vivo* conformational states. Though XRD is an incredibly useful structural tool, it does have several disadvantages. Because a crystal lattice is required to produce interpretable XRD patterns, TN must first be rendered into crystalline form. This makes high resolution XRD visualization of TN when incorporated into the reconstituted TF impossible, much less an *in situ* XRD study of Å level resolution. Another issue is when TN constructs have successfully been crystallized in the past, their numerous dynamic regions have stayed disordered and proved unresolvable [1,2]. Finally, the conformational state that TN adopts in a crystal may differ from *in vivo* wherein the molecular environment is different and all binding partners are present. This is an important caveat applicable to any high resolution XRD study.

Nuclear Magnetic Resonance

Whereas fluorescence spectroscopy and XRD are focused primarily on electrons, NMR spectroscopy is concerned with an interaction that can occur between certain atomic nuclei and radio frequency electromagnetic radiation [26]. When nuclei possessing net magnetic spin are immersed in a static magnetic field, their magnetic moment will spontaneously align with the external field. The amount of applied electromagnetic force required to invert the moment against the external field is within the range of energy stored in radio frequency photons. The frequency of incident photon required to invert the spin of a particular nucleus depends on the effective strength of the external field felt by the nucleus and a gyromagnetic ratio that is determined by nuclear structure. The effective field strength is generally less than the external field strength because the electrons localized to the nucleus will create secondary magnetic fields that generally oppose the external field. Because electron localization depends on the type of nuclei and bonds present in a molecule, photon absorbance indicates not only the presence of particular nuclei but also the nature of their local chemical environment.

The selective spin inversion concept underlying NMR can be cleverly exploited to gain a wealth of structural information and makes NMR an amazingly versatile tool for structural study. NMR experiments can characterize the energetic coupling between spin nuclei mediated through covalent bonding and thereby map out molecular connectivity and associated bond distances and angles. Measurement of the nuclear Overhauser effect can provide inter-nuclear distances between localized nuclei that are not connected by bonding. By performing multidomain heteronuclear measurements on the resonant frequencies of multiple types of spin nuclei (e.g. ^{1}H, ^{13}C, ^{15}N), three dimensional structures of entire proteins can be resolved from NMR data. Heteronuclear experiments do require that isotopically substituted recombinant constructs of the protein of interest are generated. Finally, NMR can be conducted in a time resolved format to obtain information about the dynamics and kinetics of protein conformational changes.

The major advantages of NMR are its versatility and ability to provide dynamic structural information of atomic resolution about protein structure in solution. Unfortunately, for the time being NMR is effectively an *in vitro* technique as far as high resolution is concerned.

Resonance congestion becomes problematic at large molecular sizes, and conventional multidimensional heteronuclear NMR spectroscopy is limited to ~35 kDa proteins. Even whole fast skeletal TN from *H. sapiens* is 70.1 kDa and is therefore too large, so past NMR studies have been on TN subunits either in isolation or in the presence of peptidic reductions of binding partners. The recent development of protein deuteration techniques to reduce the complicating spin inversion relaxation pathways associated with ^{1}H has allowed NMR to be useful for studying ~100 kDa sized proteins [27]. Nevertheless, just one *H. sapiens* cardiac TF structural regulatory unit (SRU) is a staggering ~731 kDa. It may be possible to study isotopically substituted recombinant TN subunits in cultured myocytes, but a truly *in situ* high resolution NMR experiment is currently beyond the reach of the technique. However, molecular-scale magnetic resonance imaging may one day be possible [28].

Electron Paramagnetic Resonance

Electron paramagnetic resonance (EPR) is a powerful technique that is very similar to NMR, but like FRET it can be used to measure long range atomic distances [29]. EPR involves the energetics of inverting electron spin in an externally applied magnetic field. This typically requires that a protein is labeled with a probe containing an unpaired electron whose spin can be freely manipulated by electromagnetic field changes. Electron spin inversion is triggered by microwave photons under the typical approach. As in NMR, spin inversion is energetically coupled to the molecular environment of the spin label. Spins that are sufficiently localized near each other can become energetically coupled through a long-range dipole–dipole interaction, analogous to the concept that underlies FRET. Pulsed electron-electron double resonance (PELDOR) is the experimental approach that takes advantage of this phenomenon and it can be used to measure atomic distances over 8 nm in length.

The advantage of PELDOR over FRET is that spin labels are much smaller and less structurally perturbing than fluorophores. Furthermore, the same spin label is attached to both locations of interest in the protein, whereas FRET requires a more complex labeling scheme involving distinct donor and an acceptor fluorophores. Finally, PELDOR distances are determined directly and therefore do not depend on the uncertain orientation parameter required for converting FRET efficiency to molecular distance. However, FRET is technically more sensitive on a per-probe basis and can be conducted *in situ* and at more physiological temperatures. The advantage of general EPR methods is the incredible specificity of spin labeling because the native electrons found in proteins and typical solvents are generally paired. Much less sample is consequently needed for EPR than NMR or XRD and molecular size is no longer a problem; EPR has been used to study TN *in situ*[30], as has PELDOR [31]. The disadvantages of EPR are otherwise the same as NMR, except that spin labeling is required in EPR whereas in NMR it is not.

Electron Microscopy, Hydrogen/Deuterium Exchange and Other Methods

A variety of other structural measurement techniques have proven very useful in the study of TN structure. Low dose transmission electron microscopy (EM) can visualize the structure and orientation of TN on reconstituted thin filaments *in vitro* [32]. EM takes advantage of the wave-particle duality of electrons to image specimens with subnanometer resolution using a magnetically and electrostatically manipulated electron beam. The Compton wavelength of an electron is ~2.42 pm, which is >10^5 times shorter than highest energy visible light. Amide hydrogen/deuterium (H/D) exchange can be combined with mass spectrometry in what represents a powerful and sensitive new approach for determining Ca^{2+}-dependent changes in regional TN dynamics and solvent exposure [33]. Small angle neutron scattering can be used to probe the size and shape of TN [34]. Circular dichroismis another fluorescence spectroscopy technique, and it can be used to determine protein secondary structure because chiral excitation light interacts uniquely with α-helical chirality [35]. Linear dichroism can measure the degree of ordered orientation in TN regional structure with respect to the thin filament axis [36]. Single molecule methods have the potential to identify subpopulations of conformational states normally obscured by ensemble averaging [37]. Finally, biochemical methods can identify structural regions within TN and clarify their functional importance and interaction sites, paving the way for further biophysical investigation [38].

Many useful techniques are clearly available to the experimentalist interested in studying TN structure. However, if one thing can be learned from the above review of the methods used in TN conformational studies, it's that no technique is without its drawbacks. Every technique also has its strong points and these strengths are often unique. Overall, the field of TN structural study is thus at its best when a variety of techniques are being employed and structural information is steadily accumulating from numerous lines of investigation. As will be demonstrated below, it is this kind of multifaceted investigative environment that has led to our field's recent advancements in a contextual understanding of TN structure.

The Structure and Conformational Environment of Troponin

The physiological conformational behavior of TN finds its context in the ultrastructure of the sarcomeres found in striated muscle. To properly interpret TN structural information, it is therefore essential to have at least a basic awareness of the complex structural arrangement of proteins from which the sarcomere is built. This section will review what is known about how TN fits into sarcomeric ultrastructure. The most straightforward way to work with TN has been when it is out of context, and the field of TN structural study began in this manner. However, as our science has progressed, TN conformational behavior has progressively been investigated at increasing levels of physiological approximation. This section will begin accordingly by concentrating on the structural details of each subunit within the TN complex. The view will then broaden to inspect how TN subunits are combined and integrated together into the architecture of the TF. Finally, the ultra structural relationship between the TF and other sarcomeric components, the thick filament especially, will then be discussed. The information reviewed in this section is not only from TN structural studies [8, 39] of interest here but also a considerable history of biochemical investigation [40, 41].

Subunit Level Structure

Regardless of isoform or tissue type, the physiological TN complex is assembled from three subunits: troponin C (TNC), troponin I (TNI), and troponin T (TNT). The letters in this terse naming scheme hint at what is traditionally perceived to be the primary function of each subunit. TNC is the ~18 kDa Ca^{2+} sensor of the TN complex that is ultimately required for transducing second messenger Ca^{2+} signals into regulatory conformational changes. TNI is a ~24 kDa extended binding protein that inhibits actomyosin ATPase activity, and thereby grants TN its regulatory power over muscular force generation. TNT is a larger 30-35 kDa extended binding protein that not only binds to TNC and TNI, but also binds strongly to tropomyosin to play a scaffolding role in anchoring the TN complex to the thin filament. However, structural information has emerged that suggests that TNC, TNI, and TNT are much more than a Ca^{2+} sensor, inhibitor and anchor, respectively. It is now known that each of these subunits plays a complex, modulatable role in the physiological changes in TN conformation that underlie TF regulation of muscle contraction.

Troponin C

TNC is the most evolutionarily conserved subunit of TN. Its structure closely resembles that of CaM [42], the Ca^{2+} binding messenger protein ubiquitously found in eukaryotes [43], and may be crudely described as "dumbbell shaped." Like CaM, it possesses two globular domains (e.g., *H.sapiens*cTNC$_{1-81}$ and cTNC$_{96-161}$) tethered to one another through a central flexible linker (e.g., *H. sapiens* cTNC$_{82-95}$). Both the N- and C-domains of TNC consist primarily of two Ca^{2+}-binding EF-hand motifs. Each EF-hand is essentially a helix-turn-helix motif consisting of two α-helices linked by a 12 residue Ca^{2+}chelation loop. This makes for a total of eight EF-hand α-helices (sequentially labeled A-H from N- to C-terminus) and four potential EF-hand Ca^{2+}-binding sites (labeled I-IV) within every TNC isoform. The chelating ability of TNC E-F hand loops stems from electronegative residues positioned ideally for coordinating cations in a pentagonal-bipyramidal geometry [44]. The α-helices of each TNC E-F hand motif are amphiphilic, containing both polar and nonpolar residues. This allows both the N- and C-domains of TNC to achieve their globular fold which exhibits a buried hydrophobic core and polar solvent exposed surface. Thus the ligand free form of TNC bears a striking resemblance to ligand free CaM.

Unsurprisingly, the Ca^{2+}-saturated conformation of TNC is also very similar to that of CaM. When Ca^{2+} binds to an active TNC Ca^{2+}-binding site, the E-F hand motif reorients itself around Ca^{2+} to facilitate cation coordination [45]. Generally, when Ca^{2+} ions are bound to both binding loops within a TNC globular domain, the reorientation of that domain's E-F hands "opens" the domain and reorganizes its hydrophobic core into a solvent-exposed patch. As is the case with CaM, the exposed hydrophobic patch represents a previously concealed surface for protein–protein interaction that can bind an amphiphilic α-helix which bears complimentary hydrophobic regions. This functionality allows TNC to be integrated into the TN core domain and interact with TNI as part of CRS. Also, it is the change in polarity that accompanies Ca^{2+} binding and hydrophobic patch exposure that has made environmentally sensitive fluorophores such successful indicators of changes in TNC conformational state.

The structure of each globular domain andTNCCa^{2+}-binding site is tuned by evolution to either achieve a particular Ca^{2+}-binding affinity and specificity or even deactivate divalent cation binding altogether [46]. Site activity is very important for Ca^{2+}-induced domain opening. It is determined by the integrity of the Ca^{2+}-chelation loop and its highly conserved coordinating residues [47]. Site affinity and specificity depend on the primary sequences of the Ca^{2+}-chelation loop [48], its flanking helices [47], and other nearby helices [49]. Affinity is also influenced by interaction between neighboring loops within a domain [50]. Sites III and IV in the C-domain of TNC (C-TNC) have a relatively high affinities for Ca^{2+} ($K_d \approx 10^{-7}$

M) and Mg^{2+} ($K_d \approx 10^{-3}$ M) if active [35]. Sites I and II of the N-domain of TNC (N-TNC) have a lower Ca^{2+} affinity ($K_d \approx 10^{-5.3}$ M), but bind Ca^{2+} with high specificity.

Structural diversity in TNC especially concerns Ca^{2+}-binding site activity. Sites II-IV are active throughout vertebrate striated muscle TNC isoforms. Site I is active in STNC, but disabled in CTNC due to chelation loop mutations [35]. This structural difference is a contributor to the unique conformational behaviors of N-CTNC and N-STNC that are important for achieving tissue-specific CRS mechanisms. Particularly, Ca^{2+}binding to N-CTNC is essential but by itself inadequate for completely exposing the hydrophobic patch [51-54]. Invertebrate muscle contains additional TNC variants in indirect flight muscle (IFM) with surprising differences from vertebrate homologues [9]. Sites I-III are disabled in F1-TNC and the remaining site IV binds Ca^{2+} with K_d = 0.16 μM. F1-TNC thus lacks regulatory Ca^{2+}-sensing ability. Furthermore, unlike vertebrate TNC, Ca^{2+}binding to F1-TNC site IV will completely expose the C-domain hydrophobic patch [55]. In F2-TNC, sites I and III are disabled, and the Ca^{2+} affinities of sites II and IV are described by K_d = 180 and 2.6 μM, respectively. The apparent evolutionary purpose behind CRS in IFM is correspondingly different in IFM. Force generation is regulated by Ca^{2+} and modulated by sarcomere length in vertebrate muscle, whereas it is regulated by both Ca^{2+} and sarcomere length in IFM [56].

Owing to its high Ca^{2+}sensitivity, it is commonly found that divalent cations are constitutively bound to the active Ca^{2+} binding sites of C-TNC *in vivo*[44]. The C-domain hydrophobic core is accordingly constitutively exposed, affording to C-TNC a scaffolding role wherein it spontaneously integrates itself with the regions of the other TN subunits that form the core domain [57]. In contrast, the low Ca^{2+}sensitivity of active Ca^{2+}-binding sites of N-TNC ensures that they are ligand free at basal Ca^{2+} levels. In addition, the Ca^{2+}-binding affinity of these sites is also tuned such that physiological increases in Ca^{2+} will result in the binding of Ca^{2+} that is necessary for complete opening of N-TNC and concomitant exposure of the hydrophobic patch. As will be explained further, N-TNC hydrophobic patch interacts with the C-terminus of TNI (C-TNI) as part of CRS. Thus N-TNC exhibits the regulatory conformational behavior that would be expected for a CaM-like protein that has evolved to be the sarcomere's directly-integrated, dedicated Ca^{2+}-sensor.

Troponin I

TNI is the subunit whose switchable protein–protein interactions are key to the regulatory power of TN. TNI exhibits an elongated structure [58] possessing the conformational plasticity required for TNI to exchange its binding partners as part of CRS [8]. Six consecutive regions within an N-domain and a C-domain may be identified within vertebrate TNI according to their diverse functions. Starting from the N-terminus, the structure of the N-domain of TNI(N-TNI) begins with a 30 residue cardiac specific region termed the N-terminal extension (N-EXT)[8]. N-EXT-CTNI contains PKA-mediated phosphorylation sites at Ser-22 and Ser-23 (i.e., residue locations in the *H. sapiens* isoform) and may be subdivided into an acidic N-terminus ($CTNI_{1-10}$), a rigid linker ($CTNI_{11-19}$), and a C-terminal α-helix [59].The N-EXT-CTNI is followed by a C-TNC binding region ($CTNI_{34-71}$ and $STNI_{1-40}$) and a TNT2 binding region ($CTNI_{80-136}$ and $STNI_{50-106}$), which completes N-CTNI [8]. The C-domain of CTNI (C-CTNI) includes an inhibitory region (IR) that strongly binds actin ($CTNI_{128-147}$ and $STNI_{96-115}$), a switch region (SR) that strongly binds N-CTNC ($CTNI_{147-163}$ and $STNI_{115-131}$), and a mobile domain (MD) that interacts either strongly or weakly with actin and TM

($cTNI_{164-210}$ and $sTNI_{132-180}$). Hence, in terms of structure, the "I" in TNI could just as well stand for "interactive" instead of "inhibitory."

The conformational plasticity of TNI makes its structural properties impossible to understand unless the influence of its binding partners is considered. Consistent with this propensity of TNI for protein–protein interaction, the conformational behavior of TNI puts the "switching" in CRS. The only persistently rigid regions within TNI are the C-TNC and TNT2 binding regions of TNI [60], which participate in forming the IT arm of the TN core domain. However, this scaffolding role is important to CRS structure transitions, since *in vitro* H/D exchange and *in situ* anisotropy evidence suggests that the IT arm stabilizes TN and localizes TNI to the TF in a way that optimally positions the other TNI regions for their switchable protein interactions [20, 33, 60]. TheN-EXT-CTNI is directly attached to the C-TNC binding region of CTNI and it extends up CTNC to interact with N-CTNC when unphosphorylated [59]. This interaction may be switched "off" by PKA phosphorylation. FRET evidence suggests that the TNI-IR adopts a collapsed β-hairpin structure when bound to actin and a more extended conformation once switched by Ca^{2+}binding to N-TNC into interacting with TNC [61-63]. Ca^{2+}binding also switches CTNI-MD from a collapsed state [32] that strongly interacts with actin and TM [7] into an unstructured, weakly interacting state [19, 64]. At the center of all this switching activity lies TNI-SR, which is also called the switch peptide and regulatory region, and its ability to interact with N-TNC.

It was previously mentioned that the exposed hydrophobic patch of N-TNC facilitates interaction with a complimentary amphiphilic α-helix. Interestingly, the TNI-SR exhibits a basic amphiphilic α-helix structural motif. This motif is the quintessential binding target of Ca^{2+}-bound CaM [43]. NMR verified that TNI-SR interacts with the exposed hydrophobic patch of Ca^{2+}-saturated N-TNC [52, 65], and XRD demonstrated that the TNI-SR takes on α-helical character when so bound [1, 2]. NMR [64, 66], FRET [67], and anisotropy [19] suggest that the TNI-SR is less structured when dissociated from N-TNC. Thus, Ca^{2+} binding to N-TNC essentially switches "on" the TNI-SR–N-TNC interaction, whereas Ca^{2+}-dissociation switches it "off." Like a master switch within a Boolean network, the conformational state of TNI-SR thereby determines the ultimate conformations and protein–protein interactions of C-TNI's switchable protein regions. The TNI-SR–N-TNC interaction is also the primary means by which the Ca^{2+}sensitivity of CRS can be modulated due to its energetic coupling with regulatory Ca^{2+} binding and N-TNC opening [68].

The ability of TNI-SR to modulate N-TNC Ca^{2+}-binding affinity is just one example of how TNI can go beyond regulating force generation to influence the CRS process itself. This is made possible by TNI's higher degree of structural diversity than TNC, which is traced to the existence of N-EXT-CTNI in cardiac isoforms. NMR showed that N-EXT-CTNI can interact weakly with N-CTNC [69, 70]. This N-EXT-CTNI–N-CTNC interaction enhances the strength of the CTNI-SR–N-CTNC interaction and thereby increases the Ca^{2+}-binding affinity of N-CTNC [71, 72]. An NMR study also suggested that the conformational state of N-EXT-CTNI can be changed to position its acidic N-terminus for interaction with CTNI-IR [59]. These modulatory actions of N-EXT-CTNI are regulated by the effect of PKA-mediated PTMs made to Ser-22 and Ser-23 on the structure of TNI [59, 73]. Furthermore, PKC mediated PTMs of Ser-43, Ser-45, and Thr-144 (i.e., in *Mus musculus*) [74] were demonstrated through NMR and FRET [75] to also impact the N-EXT-CTNI conformation dominantly over PKA signaling [76]. CTNI also possesses a site at Thr-31 (i.e., in *H. sapiens*) that can be phosphorylated by the apoptosis inducing Mst1 [77], and a site at Ser-151 (i.e., in *Rattus norvegicus*) that can be phosphorylated by p21-activated kinase [78] and AMP-activated protein kinase [79].

Lastly, when considering the basics of TNI structure, it is important to understand that its inhibitory action alone is inadequate for proper regulation of XB cycling. TNI is partnered with TM in achieving the regulatory inhibition of muscular force generation over an entire TF structural regulatory unit. Though TNI directly contacts TM *via* the TNI-MD [7], an equally important indirect contact between TM and TNI is mediated through TNT, the remaining TN subunit yet to be discussed.

Troponin T

Originally envisioned as the scaffolding anchor of the TN complex, structural studies have increasingly uncovered the important role of in TNT in CRS. TNT structure is remarkably similar to TNI structure and epitopic analysis has even detected evolutionarily suppressed TNI-like epitopes in TNT [80]. Like TNI, TNT is an elongated molecule that participates in numerous protein–protein interactions and whose N-terminus exhibits a high degree of structural diversity [81]. Unlike TNI, TNT contains a proteolytic cleavage site that divides the protein into two fragments upon mild chymotryptic digestion. The resulting N-terminal and C-terminal fragments are simply named TNT1 and TNT2, respectively. From the N-terminus, TNT1 structure begins with a hypervariable region (N-VAR-TNT) that can be highly acidic and whose length can reach 70 residues depending on the vertebrate isoform [82]. Though N-VAR-TNT currently has no structurally verified binding partner [81], its important functional roles in maximal TF activation, TF cooperativity, and TM flexibility were recently biochemically identified [83, 84]. N-VAR-TNT is followed by a ~81 residue conserved central region that interacts with the head-to-tail junction of axially joined TM dimers through an N-terminal 39 residue TM-binding region [38]. Thus, TNT not only anchors TN to a single TM dimer, but it also involves TN in the allosteric coupling that occurs between axially adjacent regulatory units.

TNT1 is followed by the conserved ~100 residue TNT2 chymotryptic fragment that contains scaffolding functionality. The N-terminus of TNT2 contains a 25 residue TM binding region that anchors TN onto TM [38]. This is followed by an extended region that binds N-TNI and C-TNC along its length and at its C-terminus, respectively, and thus integrates TNT into the core domain [81]. These structural properties of TNT2 indicate its important role in conformational signal transduction wherein it serves as the allosteric connection between CRS and TM movement. Overall, TNT structure conceptually resembles an inverted, non-switching version of TNI with the extension region maintained at the N-terminus. The N-TNI and C-TNC binding behavior of the C-terminus of TNT2 is reminiscent of N-TNI, and the TM binding sites in TNT1 and TNT2are reminiscent of the actin binding CTNI-IR and CTNI-MD. These similarities are in keeping with the evolutionary divergence of TNT and TNI from a common ancestral protein to occupy unique functional niches in TF regulation [80].

Like CTNI, CTNT contains PKC phosphorylation sites. Four locations have been identified: Thr-197, Ser-201, Thr-206, and Thr287 (i.e., in adult *H. sapiens*). The most functionally important site is Thr-206 [85], which is located close to the N-terminal cap of the α-helix that makes up the TNT2 outer prong of the IT arm W-scaffold. Computational simulation suggested that PKC-mediated phosphorylation of Thr-206 would extend the α-helix [85], which may somehow affect the conformational behavior of TNT; however, the mechanism remains unknown [86]. In contrast to TNI, TNT is not known to be phosphorylated by PKA. Though TNT is highly regulated both developmentally and by restrictive proteolysis [81], the structural consequences of the associated modifications has not been characterized.

Interfacing between Troponin Subunits

As indicated above, TNC, TNI, and TNT are highly interactive proteins that act as an interdependent, integrated whole to regulate force generation *in vivo*. The integration of TN subunits produces the TN core domain (Fig. 1), which has been well visualized by XRD [1,2] and consists of the regulatory head anchored into the IT arm. The regulatory head (Fig. 1A) consists of N-TNC and the TNC central linker. The IT arm is built from the C-TNC- and N-TNI-binding regions of TNT2; the C-TNC- and TNT2-binding regions of N-TNI; and C-TNC, whose sites III and IV are constitutively bound to either Mg^{2+} or Ca^{2+} and whose hydrophobic patch is consequently exposed. The N-TNI-binding region of TNT2 and the TNT2-binding region of N-TNI are integrated as follows. TNT2 and N-TNI each fold into a helix-turn-helix motif wherein the C-terminal helix of the motif contains the binding domain. Thus, the C-terminal helices of each motif integrate into a coiled-coil. This results in a "W"-like scaffolding substructure (Fig. 1B) [2] somewhat resembling the head of a three-pronged kitchen fork, which for the convenience of further discussion will be named the "W-scaffold."

The base of the W-scaffold contains the turns of the TNT2 and N-TNI helix-turn-helix motifs. Further TN integration is therefore accomplished by the "prongs" of the W-scaffold. One of the W-scaffold's outer prongs consists of the N-terminal portion of TNT2, and it thus binds to TM [38]to directly anchor the W-scaffold onto the TF (Fig. 1B). The central prong of the W-scaffold consists of the TNT2–N-TNI coiled-coil, and as Fig. 1B shows it is at the upper tip of this central prong where TNT2 interacts with C-TNC. The tip is also the point from where C-TNI projects out of the TN core domain, which makes C-TNI well positioned for its switchable interactions with actin, TM, and the TN regulatory head. The W-scaffold's remaining outer prong consists of the N-terminal portion of N-TNI and projects toward C-TNC. In all vertebrate TN isoforms, this remaining prong contains an amphiphilic α-helix that interacts with the exposed hydrophobic patch of C-TNC. C-TNC is thus "gripped" between the N-TNI outer prong and the central, coiled-coil prong of the W-scaffold (Fig. 1C) [2]. In cTN, N-EXT-cTNI also projects out from the tip of the N-TNI outer prong of the W-scaffold to interact with N-cTNC and the IR in C-cTNI (Fig. 1B).

It is evident from this description ofthe W-scaffold that it couples TN to the TF and is important for organizing TN subunits into a configuration optimal for actomyosin regulation. The spatial arrangement achieved by scaffolding facilitates additional interfacing between TN subunits that involves the regional protein binding and dissociation events associated with CRS [87]. XRD has shown that when N-TNC is saturated with Ca^{2+}, TNI-SR interacts with the exposed hydrophobic patch of open N-cTNC (Fig. 1D) [1, 2]. XRD of Ca^{2+}-saturated sTN has also shown that TNI-IR interacts with a structured TNC central linker in the regulatory head [1, 2], a conclusion supported by neutron-scattering data showing that sTNC is extended when Ca^{2+} saturated [88-90]. However, it is notable that the TNC central linker and the TNI-IR could not be visualized in the Ca^{2+} saturated cTN crystal structure. When N-sTNC was free of Ca^{2+}, it was in its closed conformation, the central linker of sTNC was melted, the interactions between N-sTnC and C-sTnI were absent. The clear implication was that TnI-Ir would have switched to interacting with actin if it were present.

One remarkable facet of core domain structure is that in spite of the structural diversity found in vertebrate Tn subunits, the organization of Tn and the conformational behavior associated with CRS is very similar between sTn and cTn. In fact, some differences between the cTnC [1] and sTnC [2] crystal structures can be reconciled by considering the results of other studies. NMR [87, 91, 92], FRET [93], and H/D exchange [33] have showed that the cTnC central linker becomes more rigid when N-cTnC has Ca^{2+} bound in the presence of C-cTnI. NMR has also indicated that cTnI-Ir interacts with the TnC central linker under these conditions [92]. Finally, fluorescence anisotropy and amide H/D exchange have indicated that

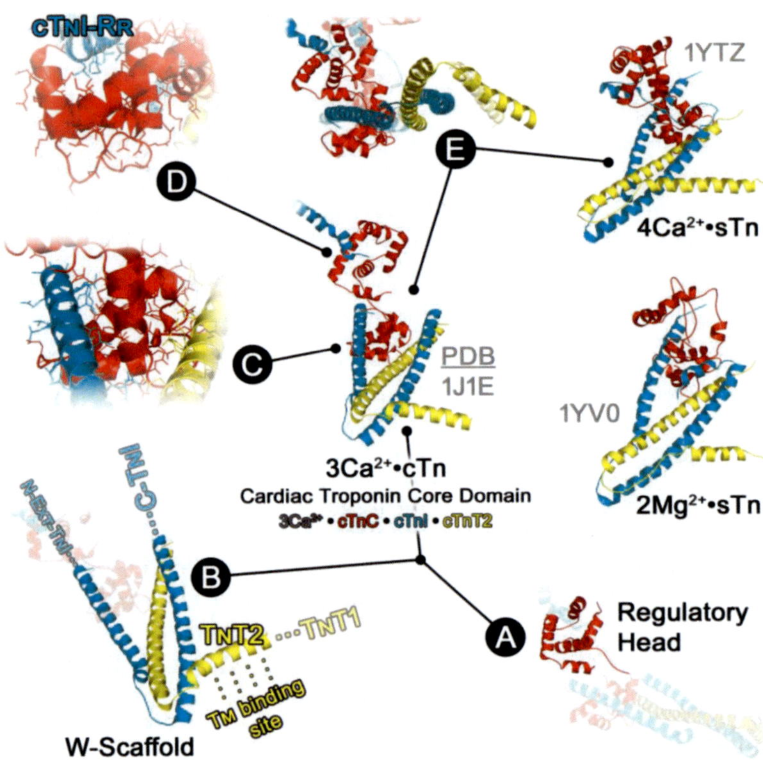

Figure 1. Shown here are various structural aspects of the Tn core domain based on the crystal structures produced by Takeda et al.[1] and Vinogradova et al. [2]. Crystal structures were rendered using PyMOL [94]. The entire Ca^{2+}-saturated cTn crystal structure is shown in the center of the diagram. Ca^{2+}-saturated sTn appears to the right of cTn, and Ca^{2+}-free sTn to the upper right, for comparison. A-E highlight important aspects of Tn subunit integration that are evident in the structures. **A)** The regulatory head is shown highlighted with the W-scaffold faded out. **B)** The W-scaffold of cTn is now highlighted with the regulatory head faded out. The relative locations of structural regions that were missing from the crystal structure are also indicated. **C)** This zoomed-in view depicts how the W-scaffold "grips" C-TnC *via* protein-protein interactions with N-TnI and TnT2. The exposed hydrophobic patch of C-TnC can be seen interacting with N-TnI on the left. **D)** Here, one can see how TnI-Sr interacts with the exposed hydrophobic patch of N-TnC under Ca^{2+} saturated conditions. **E)** Ca^{2+}-saturated cTn and sTn have been aligned here along the W-scaffold. Though the overall organization of the core domain is similar in both crystal structures, one can see minor differences in W-scaffold structure and how the significant structural differences between cTnC and sTnC affect the regulatory head and its interaction with TnI-Sr.

cTnI-IR is rigid under Ca^{2+} saturated conditions [19, 33], which further implies that it interacts with the cTnC central linker. When this additional evidence is considered, the most reasonable conclusion is that the unstructured nature of cTnI-IR and the cTnC central linker in the crystallized cTn core domain was due to the effects of crystal packing and missing binding partners on Tn structure [2].

In spite of their general structural similarities, there nevertheless are important, subtle differences between the organization of sTn and cTn core domains (Fig. 1E). The first and most obvious difference is the presence of the N-Ext in cTnI. NMR implicates residues 22-34 of this region as weakly interacting with the N-cTnC when unphosphorylated, which stabilizes the open conformation of N-cTnC [95, 96]. This interestingly suggests that the stabilizing effect of Ca^{2+} binding to site I [35], now missing due to evolutionary mutations, is being substituted to some extent with a switchable protein–protein interaction. Neutron scattering showed that compared to sTnC, cTnC is more compact yet less intertwined with other subunits when integrated into the core domain [97]. Though isoform specific structural differences do lead to subtle differences in the CRS mechanism that will be discussed later, the structural mechanism by which Tn subunits are integrated with each other is largely preserved.

Integration of Troponin into Thin Filament Architecture

We will now expand our scope to focus on the Tf, which is the macromolecular assembly in which Tn finds its most immediate *in vivo* molecular context. The Tf itself consists of F-actin with each strand decorated with coiled-coil Tm dimers and Tn complexes at regular intervals. The geometry involved gives rise to the sRU, which consists of seven actins bound to one Tm dimer and one Tn complex. EM has nicely shown how Tn is found at ~37 nm intervals along the Tf [32], and this periodicity matches the span of a seven-actin monomer stretch. A Tm dimer is ~42 nm [40], which allows axially adjacent Tm dimers to overlap, interact, and form a continuous strand that runs along F-actin [32]. Tm–Tm overlap interactions structurally consist of a flexible, interlocking head-to-tail junction that has been visualized by NMR [98]. Overlap junctions help maintain an alignment between Tm and actin optimal for Tm localization through its seven quasi-repeating motifs that weakly interact with the actin surface [98,99]. Therefore, Tn localizes to Tm through TnT1 and TnT2, and Tm is localizes to actin. Each strand of the Tf is thus built from a long chain of sRUs into which many Tn complexes are organized.

The structure of Tn when so incorporated into the *in vitro* Tf is consistent with its structure when in isolation and the structural effects expected from the presence of Tm and actin. Neutron scattering measurements of deuterated TnC incorporated into *Bos taurus* native myocardial Tfs suggested that TnC retains an extended dumbbell structure [100]. C-TnI structural regions that are dynamic in solution become structured in the presence of Tm and actin. FRET has indicated that the TnI-IR is collapsed into a helix-loop-helix motif when bound to actin in the relaxed state [63], and EM has shown that the TnI-Md is collapsed onto Tm and actin [101]. Anisotropy has provided supplementary observations that the TnI-IR and TnI-Md have low flexibility under relaxation, while the TnI-SR remains more flexible [19]. FRET has determined that the Ca^{2+}-dependent molecular distances between C-TnI functional regions, N-cTnC, and actin are consistent with the protein conformations expected during

CRS [6, 67, 78, 102, 103]. Interestingly, FRET has shown one perhaps unexpected structural effect of incorporating TN into the TF. In the absence of Ca^{2+}, the presence of actin causes the closed conformation of N-TNC to become more collapsed and increases the separation between TNI-SR and N-TNC [6]. PELDOR measurements have provided consistent observations [104]. This implies that a partial interaction between TNI-SR and N-TNC occurs when the TN complex is not incorporated into the TF, and that this partial interaction is prevented by the TNI-IR–actin and TNI-MD–actin interactions.

To further understand the effects of TF incorporation on the conformational behavior of TN, it is important to consider how TN becomes part of a greater ultrastructure. Lengths of randomly sampled TFs from papillary murine myocardium were determined by electron tomography and found to average 1.04 ± 0.03 μm (Fig. 2A) [105]. Each strand of these TFS would therefore contain ~27 sRUs, for a total of ~54 TN complexes organized into each murine TF. Rather than being in isolation, every TN complex thus is generally located next to three nearest neighbors. A neighboring TN is ~40 nm away on either axial side of aTN being considered, and a third neighbor is located ~8 nm radially away on the other lateral side of the TF (Fig. 2D). Conformational changes inTF-incorporated TN thus occur under the possibility of structural influence from neighboring TNs. The orientation of TN with respect to the actin surface also becomes structurally relevant because TN is anchored to TM and TM position changes during CRS. Cryo-EM suggested that TN–TM is located on the outer actin domain under relaxing conditions, and moves azimuthally around the lateral actin surface to the inner actin domain when Ca^{2+}-saturated [106]. Since the switching state and azimuthal position of each of the ~54 TN complexes in a TF is determined by a combination of sarcomeric Ca^{2+} level, TF architecture, and stochastic molecular motion, even an *in vitro* reconstituted TF becomes quite complicated.

The potential the TF ultrastructure creates for allosteric communication between TF-incorporated TN complexes imbues the conformational behavior of Tn with a high degree of cooperativity. This is clearly seen under the *in situ* setting, wherein the steady state relationship between force generation and sarcomeric Ca^{2+}levels is described by a Hill coefficient that is consistently reported to be3–4[107]. This is surprisingly steep, since TN contains two regulatory Ca^{2+} binding sites at most. Such steepness suggests that CRS is cooperatively triggered by Ca^{2+} in a manner requiring allosteric communication between neighboring TN complexes. Axially directed allosteric communication (Fig. 2D) occurs because each TN is allosterically coupled with its axial neighbors through its TNT subunit and TM head-to-tail junctions. Furthermore, the persistence length of TM is sufficient for facilitating axial allostery between neighboring sRUs[108], and saturation transfer EPR has shown that actin and TN exert a rigidifying effect on TM such that TM backbone dynamics are not a major factor in TM movement during CRS and sXB binding [109]. Similarly, it is now understood that TM is azimuthally braced between radially neighboring TNs under relaxing conditions [32,106], which may represent another pathway for allosteric communication (Fig. 2D). TF CRS thus involves a high degree of allostery mediated by numerous protein–protein interactions which occur over impressively long molecular distances.

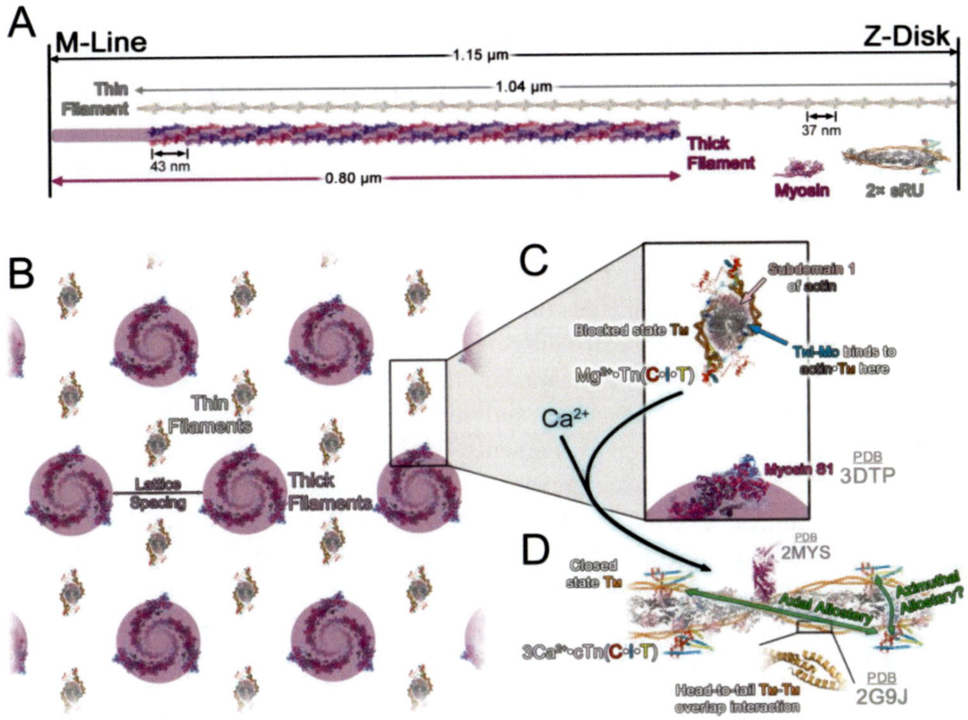

Figure 2. This figure presents the ultrastructural context that TN experiences once it is integrated into the TF. The protein structures seen in this figure are from the indicated PDB coordinate files rendered using PyMOL [94]. The exception to this is that all TF protein structures were rendered from supplementary data provided by Hoffman et al. [64]. **A)** Shown here is a model of a single TF overlapping with a single thick filament at a sarcomere length of 2.3 μm. The indicated lengths of the thin [105] and thick filaments [110] are based on measurements that have been made by Burgoyne *et al.* and Kensler, respectively, using EM. One can see from the diagram how the TF consists of a coiled-coil built from strands of sequential repeats of ~37 nm long sRUs. The rendered myosin heads shown in the thick filament cartoon depict how thick filament architecture results in coils of myosin molecules wherein myosin heads are oriented for interaction with neighboring TFs. The structural arrangement shown is based on the measurements made by AL-Khayat et al. using EM and single particle analysis [111]. The axial spacing between crowns of myosin heads is ~14.3 nm [112], and the azimuthal orientation with which a myosin head projects out of the thick filament is repeated every ~43 nm [110, 111]. **B)** The model in panel A is flipped to an end-on view looking down the central axes of the filaments and expanded to depict a sarcomere lattice. As in panel A, the myosin head structures shown in the thick filaments depict head projection points based on EM [110, 111] and a stereospecific model by Suzuki et al. [112]. **C)** A thin and thick filament from panel B are shown in a close-up view looking down their filament axes. One can see how TN–TM sterically blocks the myosin binding sites found on the outer actin domain [113]. TM position is controlled by a pair of TN complexes working together across the lateral surface of F-actin such that TM is sandwiched in place between TNT and TNI [32]. **D)** The TF from panel C is now viewed laterally under Ca^{2+}-saturated conditions with a XB about to strongly bind actin. This diagram thus emphasizes the potential for allosteric communication between neighboring TN complexes during activation. Axial allostery between the TNs of neighboring sRUs is mediated through TM–TM overlap interactions [98, 108] known to interact with TNT1 [38, 83, 84]. Azimuthal allostery may also be possible due to the counterpunching mechanism of TN-mediated TM positioning. Since S1 also controls the position TM on actin [114, 115], there is the potential for both TF intrinsic and sXB mediated allosteric effects on CRS [107, 116].

Structural Basis of Role in Interactions between Thick and Thin Filaments

Until now, structural details of the process regulated by TN-mediated CRS have yet to be considered in this chapter. That process is the interaction between thin and thick filaments that culminates in force generation (Fig. 2). In order to understand the structural relationship between TN conformational behavior and sXB formation, the ultrastructure of the sarcomere must be considered once again. TFs are assembled from their anchoring point in the Z-disk of the half-sarcomere, which is a scaffolding macromolecular assembly built from numerous proteins including α-actinin, CapZ, desmin, etc. [39]. TFs extend toward the M-line of the half-sarcomere, which is another scaffolding macromolecular assembly built from myomesin, M-protein, etc., into which thick filaments are anchored. The giant, spring-like ruler protein titin connects the Z-disk to the M-line, and myosin-binding Protein-C provides a radial connection between thick and thin filaments [111]. Thus, the TF possesses a spatial relationship with surrounding thick filaments at all times in the physiological myocyte. It is the structural consequences of this spatial relationship on TF regulation and force generation that give sarcomere length (Fig. 2A) and lattice spacing (Fig. 2B) their physiological significance [117, 118].

Thick filaments are primed for interaction with the TFs. Vertebrate thick filaments are built from three-headed crowns of myosin molecules [111, 112]. In each crown, myosin heads project out of the thick filament lateral surface at 0°, 120°, and 240° degrees around the thick filament axis in an internal frame of reference (Fig. 2B). Crowns are stacked with a ~14.3 nm periodicity such that each successive crown is rotated ~40° around the thick filament axis; however, it should be noted that the thick filament does not conform to an exact, unperturbed helical arrangement according to EM evidence [111]. Since a sarcomere unit lattice has a hexagonal arrangement (Fig. 2B), a TF will "see" a myosin head whose lateral orientation is optimal for actin binding every ~43 nm. However, every ~14.3 nm there is a myosin head whose lateral orientation permits some chance of binding actin. The result as far as TN is concerned is that many myosin heads are readily available for binding actin along a significant length of the TF (Fig.2A). TN is able to block the onrush of myosin binding and force generation through its regulatory partnership with TM (Fig. 2C).Under relaxing conditions, TN–TM occupies a position on the outer domain of actin wherein myosin binding sites are sterically blocked [113]. Optical laser trap measurements have suggested that this blocking position of TN–TM reduces the rate of sXB attachment by ~100× [37]. Interestingly, EM evidence has also suggested that the blocked state TM position is locked in place by azimuthally paired TN complexes working across the actin surface in what has been described as a counterpunching mechanism [32, 101, 106].

There are some notable idiosyncrasies in the structural relationship between thin and thick filaments. For example, their mismatch in periodicity is particularly interesting. An optical laser trap study demonstrated that myosin heads bind native *in vitro* TFs with distances of either 36, 72, or 108 nm between bound heads [112]. Furthermore, EM measurements in IFM have shown that sXB attachment predominantly occurs toward the center of a sRU [119].This indicates that myosin binding to actin occurs according to TN–TM and not thick filament periodicity and implies that only one myosin head can be bound to an sRU at a time. It also suggests that the fully activated TF may not necessarily be saturated with XBs *in vivo*. In fact, due to the nature of myofilament lengths and arrangements, approximately one third of the TF does not even overlap with the XB-bearing zone of the merely disabled but was hampering N-cTnC opening by site II on some level. This evolutionary energetic tuning significantly affects Tn conformational behavior.

One of the more fascinating isoform-specific differences between sTnC and cTnC is how the Ca^{2+}-dependent conformational conversion process involved in the CaM-like opening of the N-domain occurs. sTnC behaves as would be intuitively expected. Ca^{2+} binding to sites I and II of sTnC opens up N-sTnC, which stably exposes an interaction site that triggers CRS through TnI-Sr. It turns out that site I in N-cTnC is not only disabled, but also mutated to introduce an additional entropic cost for opening N-cTnC [35]. This ensures that N-cTnC will not stably open upon Ca^{2+} binding to site II [50, 51, 53]. However, the Rosevear lab demonstrated through NMR that N-cTnC undergoes a partial opening [124]. This was later elaborated on by the Sykes lab, who showed that this partially open conformation involves a mere ~36 Å^2 increase in the solvent exposed surface area of the N-cTnC hydrophobic core[125]. This is compared to the ~554 Å^2 increase in surface area experienced by N-sTnC [15]. FRET measurements also indicated that no major intra domain distance increases occur in N-cTnC unless cTnI-Sr is present [53]. Hence, unlike N-sTnC, both Ca^{2+} and cTnI-Sr are needed for N-cTnC to stably enter the open conformation.

The fact that simultaneous Ca^{2+} binding and cTnI-Sr interaction are needed for N-cTnC opening may seem like a mechanistic paradox. If Ca^{2+} represents the signal that is supposed to trigger the cTnI-Sr–N-cTnC interaction and subsequent CRS, what happens when Ca^{2+} can't open N-cTnC on its own? Strong evidence toward an answer was recently provided by a paramagnetic relaxation enhancement NMR study by Cordina et al. conducted on isolated, spin labeled cTnC [126]. N-cTnC was shown to exist in an equilibrium between its opened and closed conformations. This finding is in agreement with interpretations made in past NMR studies [69, 127, 128] and FRET studies from our lab [6, 129, 130]. Cordina et al. observed that when Ca^{2+} was absent, the N-cTnC conformational equilibrium was dominated by the closed conformation. Ca^{2+} binding partially stabilized the open conformation, which shifted the equilibrium enough to create a probabilistic opportunity for the cTnI-Sr to access the N-cTnI hydrophobic core. Once Ca^{2+} and cTnI-Sr were both bound to N-cTnC, the N-cTnC conformational equilibrium became dominated by the open conformation. The evolutionary disabling of N-cTnC site I thus leads to a conformational selection mechanism for the exposure of the N-cTnC hydrophobic patch. This implies that sTn and cTn rely on the same fundamental TnI-Sr dependent CRS mechanism to achieve Tf regulation.

Switching of the C-domain of Troponin I

Once the CRS is triggered by the N-TnC Ca^{2+}binding or dissociation, the switching of the protein–protein interactions of C-TnI represents the next step in the conformational conversion mechanism. Ca^{2+}-binding kinetics are extremely rapid, with an apparent bimolecular rate constant of 5.1×10^{7} $M^{-1}\cdot s^{-1}$ for binding to the whole complex at 4 °C. However, the conformational changes associated with CRS progress at much slower rates and follow a defined series of events in which C-TnI is heavily involved. The biophysical principle underlying the role of C-TnI in CRS is that of energetic balance. Ca^{2+}-binding to N-TnC and TnI-Sr's interaction with the exposed hydrophobic patch of N-TnC work together against an energetic "counterweight" in the form of the TnI-Ir–actin and TnI-Md–actin

thick filament (Fig. 2A). Another interesting facet of the structural relationship bet
and myosin is the ability of sXB attachment to cause Ca^{2+}-sensitizing change
conformation in what is termed the "positive feedback mechanism"[107,120,121
myosin itself moves TM, and TN is so intimately associated with TM, myosin-ind
movement affects TN conformational behavior. Numerous *in vitro* reconstituted TF ar
studies have in fact observed S1-induced effects on changes to TF allostery (Fig.
changes in N-TNC structure consistent with Ca^{2+}-sensitization [6, 19, 21, 36, 67, 1
122].

Changes in Troponin Conformational State

In the preceding sections, we have discussed how various structural meth
elucidated the structural arrangement and stable conformations of TN when incorpo
the TF. We have thus laid the groundwork for discussing the advances in our conte
mechanistic understanding of how TN conformational changes occur during CRS. T
most advanced aspect of TN structural study. The highly allosteric regulatory confo
behavior of TN involves conversions between stable conformations, and such co
entail energetics, dynamics, and kinetics. This makes CRS an impressively
mechanism for regulating muscular force generation. However, it also means that s
innocuous residue substitutions can have serious pathophysiological consequence
molecular mechanism underlying CRS. Therefore, characterizing physiological C
important scientific goal.

Though *in vivo* CRS behavior is incredibly complex, our field has made co
advancements in building a more accurate and dynamic structural picture of this
evolutionary adaptation of Ca^{2+} regulation for controlling muscle contractili
biophysical insights into the molecular mechanisms underlying CRS have been ga
will begin by exploring what has been learned about how CRS itself starts: with th
of Ca^{2+} to N-cTNC.

Ca^{2+} Binding and Opening of the N-domain of Troponin C

The binding of Ca^{2+} to N-TNC supplies the energy needed for TN to begin c
from its stable force-inhibitory conformation into an activating conformation th
force generation [35, 68]. This energy input is required for exposure of th
hydrophobic patch to solvent and concomitant access by TNI-SR. The conf
behavior of TNC is thus energetically tuned for converting a Ca^{2+}-signal into a c
conformational changes within TN. This energetic tuning is complex. For exam
binding to sites I and II are unequal energetic contributors to the N-sTNC openin
NMR has shown that Ca^{2+} binding occurs in a stepwise manner first to site II follow
I [123]. Furthermore, circular dichroism has indicated that Ca^{2+}-binding to site II
dominant role in opening N-sTNC, whereas site I stabiles the open conforma
Interestingly, the ability of N-cTNC site II to stimulate similar changes in Ca^{2+}-
ellipticity was attenuated relative to N-sTNC site II, suggesting that N-cTNC site

interactions to control the overall conformational behavior of TN in effecting TF activation and deactivation.

Thin filament Activation

Structural evidence strongly indicates that a sRU is activated by CRS through what is known as the "drag and release" mechanism [6,19]. The sRU begins in the relaxed state; TN is anchored *via* TNT2 and TNT1 onto the rigid TM dimer of the sRU [109]. The rigid IT arm is holding the regulatory head close to the actin surface in a position optimal for interaction with TNI-SR and therefore the triggering of CRS [20]. The TNC central linker is melted. TNI-IR and TNI-MD are bound to actin [67,102]and thus inflexible [5, 19, 60], whereas the TNI-SR is unbound, disordered, and rapidly sampling the space near N-cTNC [64]. TNI-MD is also interacting with the TM dimer of the cross-filament sRU [32, 131]. Ca^{2+} first rapidly binds to N-TNC. This causes stable hydrophobic patch exposure in N-sTNC, whereas in N-cTNC the conformational selection mechanism is initiated [126]. The flexible TNI-SR stochastically encounters and recognizes the exposed hydrophobic patch and collapses into a strong interaction with N-TNC that stabilizes N-TNCin the open conformation. The interaction between Ca^{2+}-saturated N-TNC and TNI-SR "drags" TNI-IR and TNI-MD out of interaction with actin. This "releases" TNI-IR and TNI-MD from actin, which causes the TN-mediated confinement of TM in the blocked-state position to cease. TM spontaneously moves toward the inner domain of actin to occupy an energetically minimized closed-state position [115], which releases the inhibition of force generation. TNI-IR switches to interacting with the central linker of TNC, which stabilizes the linker into a more rigid conformation. Finally, the TNI-MD becomes flexible, but maintains cross-filament transient contacts with its TM binding partner [64].

FRET measurements of reconstituted thin filaments mixed with Ca^{2+}*via* stopped flow suggested that a three-step kinetic mechanism underlies the N-cTNC opening transition required for triggering the drag and release mechanism [132, 133]. The first step was too fast to be resolved, but the other two steps were slower. These observations suggested that N-TNC experiences an initial rapid and dramatic opening, followed by more modest and slower structural changes. Based on the TF deactivation mechanism to be discussed next, it is likely that the fast step relates to the folding of previously unstructured cTNI-SR into an interaction with the exposed hydrophobic patch of N-TNC. The slow steps then likely represent conformational changes associated with the drag-and-release mechanism triggered by the TNI-SR–N-TNC interaction. Whether the kinetics of the drag-and-release mechanism can rate-limit force generation is a point still being debated. While some studies seem to have ruled out the possibility that the kinetics of TF-activating TN conformational changes can be rate limiting [134], a recent study by Kreutziger et al. has suggested that the possibility exists when certain mutations are present within TN subunits [135].

Thin filament Deactivation

TF deactivation does not occur as a symmetric reversal of activation. The best structural evidence indicates that a sRU is deactivated by CRS through what is called the "fly-casting mechanism" [64]. Fly-casting refers to a conformational behavior of TNI-MD wherein it maintains periodic contacts with TM on the actin surface during its rapid conformational sampling under Ca^{2+}-activated conditions. When Ca^{2+} dissociates from TNC, weakening of the N-TNC–TNI-SR interaction allows TNI-MD to rapidly refold into its binding interaction with actin–TM. This rapid collapsing of TNI-MD back down onto the actin surface helps to further destabilize the TNI-SR–N-cTNC interaction, and TNI-IRis eventually pulled out of interaction with the TNC central linker and back down onto actin. TM is moved to the blocked position concomitantly with the return of C-TNI to the surface of actin. The TNI-MD is thus thought to serve as an accelerator of relaxation.

There is strong empirical support for the fly-casting mechanism from the kinetic studies of *in vitro* TF deactivation. Time-resolved anisotropy measurements have demonstrated differences in the rates of changes in C-cTNI regional protein dynamics that match the predictions of the fly-casting mechanism[19]. Changes in the regional dynamics of cTNI-MD as induced by Ca^{2+} dissociation were the most rapid, followed by changes in the cTNI-SR and then cTNI-IR. FRET has also indicated that the kinetics of changes in the distance between each of these structural regions and cTNC, as induced by Ca^{2+} dissociation, follow the same pattern [103]. A recent H/D exchange study has implicated the dynamic regions of TN as prominently involved in CRS [5], which is a key implication of the fly-casting mechanism. Finally, an EM study showed that truncation of the C-terminus of cTNI-MD destabilized closed-state TM, suggesting that TNI-MD normally makes stabilizing contacts with closed-state TM as predicted by the fly-casting mechanism [131].

The debate over whether the kinetics of TF conformational changes can rate limit the kinetics of contractility also extends to relaxation [134]. An *in situ* kinetics study weighed in on this issue using an environmentally sensitive fluorophore to measure the N-sTNC Ca^{2+} dissociation rate in psoas muscle from *Oryctolagus cuniculus* [16]. The study demonstrated that the Ca^{2+}-dissociation rate matched the rate of force relaxation. Interestingly, another *in situ* environmentally sensitive fluorescence study on ventricular myofibrils from *O. cuniculus* showed evidence that at physiological temperatures, CRS and XB cycling might be kinetically similar[17].

Movement of Tropomyosin

The ultimate functional purpose of CRS is to control the position of TM on the actin surface. However, we have already alluded to the interdependent structural relationship between TM movement and CRS. The three-state model of TM positioning provides the basis for understanding how CRS and TM movement are structurally intertwined [113]. As mentioned previously, blocked-state TM covers myosin binding sites on the outer domain of actin and sterically blocks the attachment of sXBs. Closed-state TM is moved toward the actin inner domain and leaves myosin binding sites partially uncovered. As far as CRS is concerned, TN regulates TM occupancies of the blocked-and closed-state positions. As far as TM is concerned, the closed state is its "natural" location on actin [115]. Open-state TM is moved even further into the act in inner domain due to the formation of a sXB that pushes TM out of the way and thereby completely uncovers the myosin binding sites of the sRU. Such XB-induced movement of TM inevitably impacts TN structure. Labeled TN exchange in *O. cuniculus* myofibrils suggested that the three different states of TM produced three states of TN [136]. Along different lines, EM has shown that increasing the affinity of TM for actin will actually reduce the affinity of myosin for actin. Thus, the mutual interaction of TN, TM, and myosin with the actin surface involves an important energetic balance. Along similar lines, troponin counterpunching [32] recognizes that a mutual balance exists between cross-filament neighboring sRUs that work together to control the position of their respective TM dimers. That a sXB induced change in the position of TM can control the conformation of TN is the idea that led to the proposal of a reversible fourth state of TF regulation [116].

Fourth State of Thin Filament Regulation

When Ca^{2+} dissociates from an activated sRU, the fly-casting mechanism is expected to pull TNI-SR and TNI-IR off of TNC and down onto the actin surface. Ideally, TM will be moved and confined to the blocked state during this process. However, what if myosin is already bound to the sRU and blocking interaction between CTNI and actin? The answer to that question is the reversible fourth state of TF regulation as proposed by Sherwin Lehrer [116]. The concept underlying the fourth state is that there are times during physiological muscle function wherein a sXBs will activate an sRU even though N-TNChas no Ca^{2+}bound to it. The associated removal of conformational constraints on TNI-SR should indirectly promote the Ca^{2+}-sensitizing TNI-SR–N-TNC interaction[19]. The consequence is that the affinity of N-TNC for Ca^{2+}will be significantly enhanced in the fourth state because the unconstrained TNI-SR stabilizes the open N-TNC conformation optimal for Ca^{2+}chelation [116, 130].

Troponin Counterpunching

The sRU is typically thought to regulate its own capacity for myosin binding. EM measurements have turned this notion on its head by showing that sRUs work as cross-filament pairs to regulate TM position across the lateral surface of actin [32, 101, 106]. One cross filament neighbor binds TM with its TNT2 region on one side, whereas the TNI-MD of the other cross-filament neighbor is collapsed onto actin and TM on the other side (Fig. 2C). TM is therefore laterally sandwiched between two TN complexes in the blocked-state position. This raises several interesting mechanistic structural questions that warrant further investigation. For example, say that Ca^{2+} binds to the N-TNC of one sRU in a cross-filament pair, which we will call $(TN–TM)_1$. $(TN–TM)_1$ will then undergo CRS and its TNI-MD will unbind actin and therefore no longer constrain the TM of the other, Ca^{2+}-free sRU of the pair, which we will call $(TN–TM)_2$. Interestingly, the azimuthal direction that newly Ca^{2+}-saturated $(TM–TN)_1$ must move in traversing from the blocked-state position to the closed-state position is in the direction of the still-collapsed and actin-bound TNI-MDof Ca^{2+}-free $(TM–TN)_2$. How these events would thus impact the conformational state of $(TM–TN)_2$ is unknown, but it is reasonable to hypothesize that TN counterpunching represents an azimuthal pathway for allosteric communication that is likely Ca^{2+}-sensitizing.

Structural Effects of Strong Cross-bridge Attachment

One of the more surprising properties of CRS is that the very process it regulates, the actomyosin XB cycle, has the ability to regulate CRS through the positive feedback mechanism. Positive feedback complicates our understanding of how CRS and TM movement occur *in vivo*. The fact that Ca^{2+}-saturated TN undergoes conformational changes as TM moves to the open position suggests that sXBs exert important structural effects on TN that are additional to the effects of Ca^{2+}-binding. Numerous lines of *in vitro* and *in situ* biophysical evidence have borne this out.

Ca^{2+}-Sensitizing Structural Feedback

It has long been known that the attachment of sXBs to the TF enhances the Ca^{2+}sensitivity of N-TNC through the positive feedback mechanism [120]. As structural evidence has accumulated, the molecular mechanism that underlies positive feedback has become clearer. *In vitro* FRET studies have shown three key pieces of evidence. First, the attachment of sXBs to the TF causes Ca^{2+}-bound N-cTNC to adopt a more open conformation and increasesCa^{2+}-sensitivity [133]. Second, the distance between TNI-SR and N-TNC decreases [6]. Third, the distance between TNI-SR and actin increases [67]. Taken together, these results suggest that sXB attachment exerts a Ca^{2+}-independent stabilizing effect on the TNI-SR–N-TNC interaction by releasing the conformational constraints that are placed on the TNI-SR when the TNI-IR and the TNI-MD are interacting with actin. That this mechanism occurs *in situ* was supported by the work of Sun *et al.* that showed how E-F hand helix orientation in cTNC has both Ca^{2+}- and myosin-dependent components, and how the presence of sXBs was Ca^{2+}sensitizing [21]. It was later confirmed by Rieck et al. [130] using *in situ* FRET in skinned myocardial fibers that the structural basis for the positive feedback is sXB-mediated stabilization of the Ca^{2+}-sensitizing TNI-SR–N-TNC interaction. This stabilization increased the ensemble-averaged FRET distance associated with the open conformation of N-cTNCby ~1.4-fold. Due to the nature of ensemble FRET measurements, it is still unclear whether sXB-mediated stabilization causes the open conformation of N-TNC to open up further, or whether Ca^{2+}-saturated N-TNC is in a conformational equilibrium that is shifted further toward the open conformation.

Role of Cross-Bridges in Thin Filament Cooperativity

The role of sXBs in cooperative TF activation has recently fallen under renewed scientific scrutiny. It was long thought that the high degree of steepness observed in the force-Ca^{2+} relationship of myocardial and fast-skeletal vertebrate muscle was due to sXB positive feedback [107]. This is a reasonable hypothesis, since the energy of the XB-induced movement of TM in one sRU can be expected to transmit itself into neighboring sRUs *via* TM–TM overlap interactions. Seminal biochemical studies had shown that exposure of skinned muscle to a strong-binding derivative of myosin subfragment-1 (NEM-S1) attenuated the cooperativity of Ca^{2+}-dependent activation [120]. Since NEM-S1 competitively inhibited sXB formation, sXBs were hypothesized as the missing component whose presence would normally drive allostery. However, more recent structural studies have presented evidence that requires this hypothesis to be reevaluated.

In situ anisotropy of bifunctional rhodamines attached to N-cTNC [20] and FRET measurements of N-cTNC opening [130] have both shown that the Ca^{2+}-dependent opening of N-cTNC occurs steeply when sXB attachments are absent. When blebbistatin or vanadate were used block sXBs, N-cTNC opening in response to Ca^{2+}binding occurred as steeply as myocardial force generation under conditions of cycling XBs, albeit with a reduced Ca^{2+}sensitivity. Furthermore, while the cooperativity of N-cTNC opening was blunted by non-cycling sXBs in sRUs located in the XB-bearing zone of the half-sarcomere, N-cTNC opening still occurred steeply in the XB-free zone where sXBs could exert no influence. Another *in situ* structural study used linear dichroism to observe that the majority of the structural changes that occur in cTNC are triggered by Ca^{2+} alone [36]. Importantly, vanadate had no significant effect on the steepness of cTNC structural changes either in study in question, or in another study employing *in situ* anisotropy [137].

These observations imply that the capacity for cooperativity in TF activation is a property intrinsic to the TF itself [107,130] and fits well with current knowledge of CRS. The energy of TM movement in one sRU, whether induced by Ca^{2+} or myosin, should be transmitted through TM–TM overlap interactions into any its neighboring sRUs that are switched off [108]. Given the possible allosteric effects of counterpunching during CRS, cross-filament allostery may also play a role. Whatever allosteric communication pathways are involved, energy transmitted into neighboring sRUs should promote TM movement and thereby produce Ca^{2+}-sensitizing structural feedback on TN. Since cTNC is especially sensitive to energetic modulation [126], this may explain why the cardiac CRS mechanism is more dependent on allosteric communication than the skeletal mechanism [138, 139].

Possibility of Structural Effects from Weakly Bound Cross-Bridges

Since myosin can weakly bind to the TF, whether weak XBs can promote TM movement or changes in TN structure is an important scientific question that is still under investigation. Since myosin binding strength is a key source of energy required for moving TM to the open position [37,115,140], it seems unlikely that weak XBs would be a major contributor to TM movement. Structural evidence has recently showed that weak XBs are at least unable to move TM to the open position. *In situ* anisotropy and blebbistatin have showed that cycling XBs exert effects on N-cTNC α-helix orientation additional to changes induced by Ca^{2+}[21]. Rigor XBs produced further structural feedback effects on cTN. Such XB-dependent structural changes would be very difficult to explain if weak XBs had already promoted TM fully to the open position. Evidence from XRD in *Lethocerus indicus* IFM was more direct. While Ca^{2+}-binding to TN was shown to move TM from the blocked to closed states, vanadate-induced weak XBs could not detectably impact the position of closed-state TM on the actin surface [141]. However, it has not yet been conclusively shown that weak XBs have no impact whatsoever on the energy landscape associated with CRS and TM movement.

Switching Concept of Post Translational Modifications

It was mentioned earlier in the chapter how N-Ext-cTNI interacts with N-cTNC and boosts its Ca^{2+}sensitivity. It has been suggested in a paramagnetic relaxation enhancement NMR study that N-EXT-TNI exerts its Ca^{2+}-sensitizing effect on N-cTNC by shifting its conformational equilibrium toward the open state and thereby enhancing the conformational selection mechanism [69, 126]. NMR data supports this notion by having shown that N-EXT-TNI enhances N-cTNC affinity for cTNI-SR and not Ca^{2+} directly [72]. This view is also supported by NMR measurement of the effect of phosphorylation on the binding affinity of N-cTNC for cTNI-SR [71]. PKA-mediated phosphorylation of sites Ser-23 and Ser-24 introduces a bend into the N-EXT-TNI that switches it from interacting with N-cTNI to interacting with TNI-IR via its acidic N-terminus [59, 73, 142]. What these structural observation simply is that the N-EXT-cTNI is an additional switching system that evolution has added onto the TN molecular switch to provide the cardiomyocyte even more control over force generation.

Conclusion

The TN-mediated CRS mechanism for regulating muscular force generation is truly a complex and amazing evolutionary achievement. As a field, we have progressed from trying to understand whether N-cTNC even opens as part of regulation to exploring its structural modulation by a variety of pathways. We are now characterizing TN structure *in situ*, measuring the kinetics of conversions between conformations, determining TN dynamics, and elucidating previously obscured molecular mechanisms. Nevertheless, the careful experimentalist should always be at least a little haunted by the reality that the observation of TN structure paradoxically requires its perturbation and decontextualization to some degree. The key to continued advancement in our field is that no particular approach is inordinately trusted. Rather, the field of TN structure is at its best when it is both self-critical and, like TN itself, collaborative. The intricate complexity of *in vivo* TN conformational behavior defies easy exploration. We may thus gain confidence in our observations and interpretations only as knowledge is steadily accumulated from numerous lines of investigation. Many exciting advances remain on the horizon for us to pursue in our scientific quest to understand the physiological conformational behavior of TN.

Acknowledgments

The authors would like to thank Murali Chandra for helpful discussions regarding the content and scope of the chapter narrative. We are also grateful to J.P. Jin for his assistance with the narrative. The authors were supported by NIH grant number R01 HL80186-5S1 and Award Number T32GM008336 from the NIGMS during the preparation of this book chapter. The content is solely the responsibility of the authors and does not necessarily represent the official views of the National Institute of General Medical Sciences or the National Institutes of Health.

References

[1] S. Takeda, A. Yamashita, K. Maeda, Y. Maeda, *Nature* 424 (2003) 35-41.
[2] M.V. Vinogradova, D.B. Stone, G.G. Malanina, C. Karatzaferi, R. Cooke, R.A. Mendelson, R.J. Fletterick, *Proc Natl Acad Sci U S A* 102 (2005) 5038-5043.
[3] S.Y. Boateng, P.H. Goldspink, *Cardiovasc Res* 77 (2008) 667-675.
[4] W. Dong, S.S. Rosenfeld, C.K. Wang, A.M. Gordon, H.C. Cheung, *J Biol Chem* 271 (1996) 688-694.
[5] D. Kowlessur, L.S. Tobacman, *J Biol Chem* 287 (2012) 42299-42311.
[6] J.M. Robinson, W.J. Dong, J. Xing, H.C. Cheung, *J Mol Biol* 340 (2004) 295-305.
[7] Z. Zhang, S. Akhter, S. Mottl, J.P. Jin, *FEBS J* 278 (2011) 3348-3359.
[8] M.X. Li, X. Wang, B.D. Sykes, *J Muscle Res Cell Motil* 25 (2004) 559-579.
[9] B. Bullard, A. Pastore, *J Muscle Res Cell Motil* 32 (2011) 303-313.
[10] E. Ehler, M. Gautel, *Adv Exp Med Biol* 642 (2008) 1-14.

[11] M.F. Patterson, D.G. Stephenson, *Pflugers Arch* 440 (2000) 745-750.
[12] T.C. Irving, J. Konhilas, D. Perry, R. Fischetti, P.P. de Tombe, *Am J Physiol Heart Circ Physiol* 279 (2000) H2568-2573.
[13] S.M. Harrison, D.M. Bers, *J Gen Physiol* 93 (1989) 411-428.
[14] J.R. Lakowicz, Principles of fluorescence spectroscopy, *Springer*, New York, 2006.
[15] N.C. Strynadka, M. Cherney, A.R. Sielecki, N.C. Strynadka, L.B. Smillie, M.N. James, *J Mol Biol* 273 (1997) 238-255.
[16] S.C. Little, S.B. Tikunova, C. Norman, D.R. Swartz, J.P. Davis, *Front Physiol* 2 (2011) 70.
[17] S.C. Little, B.J. Biesiadecki, A. Kilic, R.S. Higgins, P.M. Janssen, J.P. Davis, *J Biol Chem* 287 (2012) 27930-27940.
[18] W.J. Dong, C.K. Wang, A.M. Gordon, H.C. Cheung, *Biophys J* 72 (1997) 850-857.
[19] Z. Zhou, K.L. Li, D. Rieck, Y. Ouyang, M. Chandra, W.J. Dong, *J Biol Chem* 287 (2012) 7661-7674.
[20] A.C. Knowles, M. Irving, Y.B. Sun, *J Mol Biol* 421 (2012) 125-137.
[21] Y.B. Sun, F. Lou, M. Irving, *J Physiol* 587 (2009) 155-163.
[22] R.E. Ferguson, Y.B. Sun, P. Mercier, A.S. Brack, B.D. Sykes, J.E. Corrie, D.R. Trentham, M. Irving, *Mol Cell* 11 (2003) 865-874.
[23] J.M. Robinson, W.J. Dong, H.C. Cheung, *J Mol Biol* 329 (2003) 371-380.
[24] D. Liang, Fundamentals of X-ray crystallography, *Science Press/Alpha Science International Ltd.*, Beijing ; Oxford, 2011.
[25] G. Rhodes, Crystallography made crystal clear : a guide for users of macromolecular models, *Elsevier/Academic Press*, Amsterdam ; Boston, 2006.
[26] J. Cavanagh, Protein NMR spectroscopy : principles and practice, *Academic Press*, Amsterdam ; Boston, 2007.
[27] V. Tugarinov, W.Y. Choy, V.Y. Orekhov, L.E. Kay, *Proc Natl Acad Sci U S A* 102 (2005) 622-627.
[28] P. Hemmer, *Science* 339 (2013) 529-530.
[29] S.K. Misra, J. Freed, Multifrequency Electron Paramagnetic Resonance, *Wiley-VCH*, Berlin, 2011.
[30] H.C. Li, P.G. Fajer, *Biochemistry* 37 (1998) 6628-6635.
[31] T. Arata, T. Aihara, K. Ueda, M. Nakamura, S. Ueki, *Adv Exp Med Biol* 592 (2007) 125-135.
[32] W. Lehman, A. Galinska-Rakoczy, V. Hatch, L.S. Tobacman, R. Craig, *J Mol Biol* 388 (2009) 673-681.
[33] D. Kowlessur, L.S. Tobacman, *J Biol Chem* 285 (2010) 2686-2694.
[34] W.T. Heller, N.L. Finley, W.J. Dong, P. Timmins, H.C. Cheung, P.R. Rosevear, J. Trewhella, *Biochemistry* 42 (2003) 7790-7800.
[35] J.R. Pearlstone, M. Chandra, M.M. Sorenson, L.B. Smillie, *J Biol Chem* 275 (2000) 35106-35115.
[36] L. Smith, C. Tainter, M. Regnier, D.A. Martyn, *Biophys J* 96 (2009) 3692-3702.
[37] N.M. Kad, S. Kim, D.M. Warshaw, P. VanBuren, J.E. Baker, *Proc Natl Acad Sci U S A* 102 (2005) 16990-16995.
[38] J.P. Jin, S.M. Chong, *Arch Biochem Biophys* 500 (2010) 144-150.
[39] Y. Au, *Cell Mol Life Sci* 61 (2004) 3016-3033.
[40] A.M. Gordon, E. Homsher, M. Regnier, *Physiol Rev* 80 (2000) 853-924.

[41] W. Lehman, R. Craig, *J Muscle Res Cell Motil* 25 (2004) 455-466.
[42] H.L. Sweeney, R.M. Brito, P.R. Rosevear, J.A. Putkey, *Proc Natl Acad Sci U S A* 87 (1990) 9538-9542.
[43] D. Chin, A.R. Means, *Trends Cell Biol* 10 (2000) 322-328.
[44] N. Finley, A. Dvoretsky, P.R. Rosevear, *J Mol Cell Cardiol* 32 (2000) 1439-1446.
[45] G.A. Krudy, R.M. Brito, J.A. Putkey, P.R. Rosevear, *Biochemistry* 31 (1992) 1595-1602.
[46] V.L. Filatov, A.G. Katrukha, T.V. Bulargina, N.B. Gusev, *Biochemistry (Mosc)* 64 (1999) 969-985.
[47] G. Trigo-Gonzalez, G. Awang, K. Racher, K. Neden, T. Borgford, *Biochemistry* 32 (1993) 9826-9831.
[48] H. Kawasaki, R.H. Kretsinger, *Protein Profile* 1 (1994) 343-517.
[49] W. Liu, D.G. Dotson, X. Lin, J.J. Mullen, 3rd, M.L. Gonzalez-Garay, Q. Lu, J.A. Putkey, *FEBS Lett* 347 (1994) 152-156.
[50] M.X. Li, S.M. Gagne, L. Spyracopoulos, C.P. Kloks, G. Audette, M. Chandra, R.J. Solaro, L.B. Smillie, B.D. Sykes, *Biochemistry* 36 (1997) 12519-12525.
[51] S.K. Sia, M.X. Li, L. Spyracopoulos, S.M. Gagne, W. Liu, J.A. Putkey, B.D. Sykes, *J Biol Chem* 272 (1997) 18216-18221.
[52] M.X. Li, L. Spyracopoulos, B.D. Sykes, *Biochemistry* 38 (1999) 8289-8298.
[53] W.J. Dong, J. Xing, M. Villain, M. Hellinger, J.M. Robinson, M. Chandra, R.J. Solaro, P.K. Umeda, H.C. Cheung, *J Biol Chem* 274 (1999) 31382-31390.
[54] S. Ueki, M. Nakamura, T. Komori, T. Arata, *Biochemistry* 44 (2005) 411-416.
[55] G. De Nicola, C. Burkart, F. Qiu, B. Agianian, S. Labeit, S. Martin, B. Bullard, A. Pastore, *Structure* 15 (2007) 813-824.
[56] S.R. Martin, G. Avella, M. Adrover, G.F. de Nicola, B. Bullard, A. Pastore, *Biochemistry* 50 (2011) 1839-1847.
[57] G.M. Gasmi-Seabrook, J.W. Howarth, N. Finley, E. Abusamhadneh, V. Gaponenko, R.M. Brito, R.J. Solaro, P.R. Rosevear, *Biochemistry* 38 (1999) 8313-8322.
[58] W.J. Dong, J. Xing, M. Chandra, J. Solaro, H.C. Cheung, *Proteins* 41 (2000) 438-447.
[59] J.W. Howarth, J. Meller, R.J. Solaro, J. Trewhella, P.R. Rosevear, *J Mol Biol* 373 (2007) 706-722.
[60] G.M. Bou-Assaf, J.E. Chamoun, M.R. Emmett, P.G. Fajer, A.G. Marshall, *Int J Mass Spectrom* 302 (2011) 116-124.
[61] W.J. Dong, J. Xing, J.M. Robinson, H.C. Cheung, *J Mol Biol* 314 (2001) 51-61.
[62] W.J. Dong, J.M. Robinson, S. Stagg, J. Xing, H.C. Cheung, *J Biol Chem* 278 (2003) 8686-8692.
[63] W.J. Dong, J. An, J. Xing, H.C. Cheung, *Arch Biochem Biophys* 456 (2006) 135-142.
[64] R.M. Hoffman, T.M. Blumenschein, B.D. Sykes, *J Mol Biol* 361 (2006) 625-633.
[65] R.T. McKay, B.P. Tripet, R.S. Hodges, B.D. Sykes, *J Biol Chem* 272 (1997) 28494-28500.
[66] T.M. Blumenschein, D.B. Stone, R.J. Fletterick, R.A. Mendelson, B.D. Sykes, *Biophys J* 90 (2006) 2436-2444.
[67] J. Xing, M. Chinnaraj, Z. Zhang, H.C. Cheung, W.J. Dong, *Biochemistry* 47 (2008) 13383-13393.
[68] R.T. McKay, L.F. Saltibus, M.X. Li, B.D. Sykes, *Biochemistry* 39 (2000) 12731-12738.

[69] V. Gaponenko, E. Abusamhadneh, M.B. Abbott, N. Finley, G. Gasmi-Seabrook, R.J. Solaro, M. Rance, P.R. Rosevear, *J Biol Chem* 274 (1999) 16681-16684.

[70] N. Finley, M.B. Abbott, E. Abusamhadneh, V. Gaponenko, W. Dong, G. Gasmi-Seabrook, J.W. Howarth, M. Rance, R.J. Solaro, H.C. Cheung, P.R. Rosevear, *FEBS Lett* 453 (1999) 107-112.

[71] M.B. Abbott, W.J. Dong, A. Dvoretsky, B. DaGue, R.M. Caprioli, H.C. Cheung, P.R. Rosevear, *Biochemistry* 40 (2001) 5992-6001.

[72] O.K. Baryshnikova, M.X. Li, B.D. Sykes, *J Mol Biol* 375 (2008) 735-751.

[73] S. Sadayappan, N. Finley, J.W. Howarth, H. Osinska, R. Klevitsky, J.N. Lorenz, P.R. Rosevear, J. Robbins, *FASEB J* 22 (2008) 1246-1257.

[74] R.J. Solaro, P. Rosevear, T. Kobayashi, *Biochem Biophys Res Commun* 369 (2008) 82-87.

[75] T. Kobayashi, W.J. Dong, E.M. Burkart, H.C. Cheung, R.J. Solaro, *Biochemistry* 43 (2004) 5996-6004.

[76] S. Sakthivel, N.L. Finley, P.R. Rosevear, J.N. Lorenz, J. Gulick, S. Kim, P. VanBuren, L.A. Martin, J. Robbins, *J Biol Chem* 280 (2005) 703-714.

[77] B. You, G. Yan, Z. Zhang, L. Yan, J. Li, Q. Ge, J.P. Jin, J. Sun, *Biochem J* 418 (2009) 93-101.

[78] Y. Ouyang, R. Mamidi, J.J. Jayasundar, M. Chandra, W.J. Dong, *J Mol Biol* 400 (2010) 1036-1045.

[79] B.R. Nixon, A. Thawornkaiwong, J. Jin, E.A. Brundage, S.C. Little, J.P. Davis, R.J. Solaro, B.J. Biesiadecki, *J Biol Chem* 287 (2012) 19136-19147.

[80] S.M. Chong, J.P. Jin, *J Mol Evol* 68 (2009) 448-460.

[81] B. Wei, J.P. Jin, *Arch Biochem Biophys* 505 (2011) 144-154.

[82] J.P. Jin, Z. Zhang, J.A. Bautista, *Crit Rev Eukaryot Gene Expr* 18 (2008) 93-124.

[83] R. Mamidi, S.L. Mallampalli, D.F. Wieczorek, M. Chandra, *J Physiol* 591 (2013) 1217-1234.

[84] R. Mamidi, J.J. Michael, M. Muthuchamy, M. Chandra, *FASEB J* 27 (2013) 3848-3859.

[85] M.P. Sumandea, W.G. Pyle, T. Kobayashi, P.P. de Tombe, R.J. Solaro, *J Biol Chem* 278 (2003) 35135-35144.

[86] T. Kobayashi, L. Jin, P.P. de Tombe, *Pflugers Arch* 457 (2008) 37-46.

[87] A. Dvoretsky, E.M. Abusamhadneh, J.W. Howarth, P.R. Rosevear, *J Biol Chem* 277 (2002) 38565-38570.

[88] G.A. Olah, S.E. Rokop, C.L. Wang, S.L. Blechner, J. Trewhella, *Biochemistry* 33 (1994) 8233-8239.

[89] D.B. Stone, P.A. Timmins, D.K. Schneider, I. Krylova, C.H. Ramos, F.C. Reinach, R.A. Mendelson, *J Mol Biol* 281 (1998) 689-704.

[90] W.A. King, D.B. Stone, P.A. Timmins, T. Narayanan, A.A. von Brasch, R.A. Mendelson, P.M. Curmi, *J Mol Biol* 345 (2005) 797-815.

[91] Q. Kleerekoper, J.W. Howarth, X. Guo, R.J. Solaro, P.R. Rosevear, *Biochemistry* 34 (1995) 13343-13352.

[92] M.B. Abbott, A. Dvoretsky, V. Gaponenko, P.R. Rosevear, *FEBS Lett* 469 (2000) 168-172.

[93] W.J. Dong, J.M. Robinson, J. Xing, P.K. Umeda, H.C. Cheung, *Protein Sci* 9 (2000) 280-289.

[94] L. Schrödinger, The PyMOL Molecular Graphics System, Version 1.1r1, 2008.

[95] D.G. Ward, S.M. Brewer, C.E. Gallon, Y. Gao, B.A. Levine, I.P. Trayer, *Biochemistry* 43 (2004) 5772-5781.

[96] D.G. Ward, S.M. Brewer, M.J. Calvert, C.E. Gallon, Y. Gao, I.P. Trayer, *Biochemistry* 43 (2004) 4020-4027.

[97] W.T. Heller, E. Abusamhadneh, N. Finley, P.R. Rosevear, J. Trewhella, *Biochemistry* 41 (2002) 15654-15663.

[98] N.J. Greenfield, Y.J. Huang, G.V. Swapna, A. Bhattacharya, B. Rapp, A. Singh, G.T. Montelione, S.E. Hitchcock-DeGregori, *J Mol Biol* 364 (2006) 80-96.

[99] S.E. Hitchcock-DeGregori, Y. An, *J Biol Chem* 271 (1996) 3600-3603.

[100] F. Matsumoto, K. Makino, K. Maeda, H. Patzelt, Y. Maeda, S. Fujiwara, *J Mol Biol* 342 (2004) 1209-1221.

[101] A. Galinska-Rakoczy, P. Engel, C. Xu, H. Jung, R. Craig, L.S. Tobacman, W. Lehman, *J Mol Biol* 379 (2008) 929-935.

[102] J. Xing, J.J. Jayasundar, Y. Ouyang, W.J. Dong, *J Biol Chem* 284 (2009) 16432-16441.

[103] Z. Zhou, D. Rieck, K.L. Li, Y. Ouyang, W.J. Dong, *Arch Biochem Biophys* (2013).

[104] M. Nakamura, S. Ueki, H. Hara, T. Arata, *J Mol Biol* 348 (2005) 127-137.

[105] T. Burgoyne, F. Muhamad, P.K. Luther, *Cardiovasc Res* 77 (2008) 707-712.

[106] A. Narita, T. Yasunaga, T. Ishikawa, K. Mayanagi, T. Wakabayashi, *J Mol Biol* 308 (2001) 241-261.

[107] Y.B. Sun, M. Irving, *J Mol Cell Cardiol* 48 (2010) 859-865.

[108] X.E. Li, W. Lehman, S. Fischer, *J Struct Biol* 170 (2010) 313-318.

[109] R.F. Rayes, T. Kalai, K. Hideg, M.A. Geeves, P.G. Fajer, *PLoS One* 6 (2011) e21277.

[110] R.W. Kensler, *Biophys J* 82 (2002) 1497-1508.

[111] H.A. Al-Khayat, R.W. Kensler, J.M. Squire, S.B. Marston, E.P. Morris, *Proc Natl Acad Sci U S A* 110 (2013) 318-323.

[112] M. Suzuki, S. Ishiwata, *Biophys J* 101 (2011) 2740-2748.

[113] R. Craig, W. Lehman, *J Mol Biol* 311 (2001) 1027-1036.

[114] W.A. Mudalige, T.C. Tao, S.S. Lehrer, *J Mol Biol* 389 (2009) 575-583.

[115] X.E. Li, L.S. Tobacman, J.Y. Mun, R. Craig, S. Fischer, W. Lehman, *Biophys J* 100 (2011) 1005-1013.

[116] S.S. Lehrer, *J Muscle Res Cell Motil* 32 (2010) 203-208.

[117] G.P. Farman, D. Gore, E. Allen, K. Schoenfelt, T.C. Irving, P.P. de Tombe, *Am J Physiol Heart Circ Physiol* 300 (2011) H2155-2160.

[118] P.P. de Tombe, R.D. Mateja, K. Tachampa, Y. Ait Mou, G.P. Farman, T.C. Irving, *J Mol Cell Cardiol* 48 (2010) 851-858.

[119] S. Wu, J. Liu, M.C. Reedy, R.J. Perz-Edwards, R.T. Tregear, H. Winkler, C. Franzini-Armstrong, H. Sasaki, C. Lucaveche, Y.E. Goldman, M.K. Reedy, K.A. Taylor, *PLoS One* 7 (2012) e39422.

[120] R.L. Moss, M. Razumova, D.P. Fitzsimons, *Circ Res* 94 (2004) 1290-1300.

[121] J.D. Hannon, D.A. Martyn, A.M. Gordon, *Circ Res* 71 (1992) 984-991.

[122] D.A. Martyn, M. Regnier, D. Xu, A.M. Gordon, *Biophys J* 80 (2001) 360-370.

[123] M.X. Li, S.M. Gagne, S. Tsuda, C.M. Kay, L.B. Smillie, B.D. Sykes, *Biochemistry* 34 (1995) 8330-8340.

[124] X. Lin, G.A. Krudy, J. Howarth, R.M. Brito, P.R. Rosevear, J.A. Putkey, *Biochemistry* 33 (1994) 14434-14442.

[125] L. Spyracopoulos, M.X. Li, S.K. Sia, S.M. Gagne, M. Chandra, R.J. Solaro, B.D. Sykes, *Biochemistry* 36 (1997) 12138-12146.

[126] N.M. Cordina, C.K. Liew, D.A. Gell, P.G. Fajer, J.P. Mackay, L.J. Brown, *Biochemistry* (2013).

[127] K. Paakkonen, A. Annila, T. Sorsa, P. Pollesello, C. Tilgmann, I. Kilpelainen, P. Karisola, I. Ulmanen, T. Drakenberg, *J Biol Chem* 273 (1998) 15633-15638.

[128] K. Paakkonen, T. Sorsa, T. Drakenberg, P. Pollesello, C. Tilgmann, P. Permi, S. Heikkinen, I. Kilpelainen, A. Annila, *Eur J Biochem* 267 (2000) 6665-6672.

[129] J.M. Robinson, H.C. Cheung, W. Dong, *Biophys J* 95 (2008) 4772-4789.

[130] D.C. Rieck, K.L. Li, Y. Ouyang, R.J. Solaro, W.J. Dong, *Arch Biochem Biophys* 537 (2013) 198-209.

[131] A. Galinska, V. Hatch, R. Craig, A.M. Murphy, J.E. Van Eyk, C.L. Wang, W. Lehman, D.B. Foster, *Circ Res* 106 (2010) 705-711.

[132] W.J. Dong, C.K. Wang, A.M. Gordon, S.S. Rosenfeld, H.C. Cheung, *J Biol Chem* 272 (1997) 19229-19235.

[133] W.J. Dong, J.J. Jayasundar, J. An, J. Xing, H.C. Cheung, *Biochemistry* 46 (2007) 9752-9761.

[134] R. Stehle, J. Solzin, B. Iorga, C. Poggesi, *Pflugers Arch* 458 (2009) 337-357.

[135] K.L. Kreutziger, N. Piroddi, J.T. McMichael, C. Tesi, C. Poggesi, M. Regnier, *J Mol Cell Cardiol* 50 (2011) 165-174.

[136] D.R. Swartz, Z. Yang, A. Sen, S.B. Tikunova, J.P. Davis, *J Mol Biol* 361 (2006) 420-435.

[137] M.G. Bell, E.B. Lankford, G.E. Gonye, G.C. Ellis-Davies, D.A. Martyn, M. Regnier, R.J. Barsotti, *Biophys J* 90 (2006) 531-543.

[138] T.E. Gillis, D.A. Martyn, A.J. Rivera, M. Regnier, *J Physiol* 580 (2007) 561-576.

[139] M. Regnier, A.J. Rivera, C.K. Wang, M.A. Bates, P.B. Chase, A.M. Gordon, *J Physiol* 540 (2002) 485-497.

[140] V.L. Korman, V. Hatch, K.Y. Dixon, R. Craig, W. Lehman, L.S. Tobacman, *J Biol Chem* 275 (2000) 22470-22478.

[141] T.I. Bekyarova, M.C. Reedy, B.A. Baumann, R.T. Tregear, A. Ward, U. Krzic, K.M. Prince, R.J. Perz-Edwards, M. Reconditi, D. Gore, T.C. Irving, M.K. Reedy, *Proc Natl Acad Sci U S A* 105 (2008) 10372-10377.

[142] W.J. Dong, M. Chandra, J. Xing, R.J. Solaro, H.C. Cheung, *Biochemistry* 36 (1997) 6745-6753.

In: Troponin
Editor: J.-P. Jin

ISBN: 978-1-62948-320-7

Chapter VI

Kinetic Modeling of Troponin Dissociation in Striated Muscle

Asok K. Sen, PhD*

Richard G. Lugar Center for Renewable Energy
and Department of Mathematical Sciences
Indiana University

Abstract

Troponin (Tn) plays a vital role in the regulation of contraction in striated muscle. Troponin is made up of three subunits: TnC, TnI, and TnT. The subunit TnC binds to calcium, TnI inhibits the actin-activated myosin ATPase, and TnT binds to tropomyosin (Tm). Muscle contraction is initiated by binding of TnC to calcium, but because TnC does not interact directly with actin or tropomyosin, the signal is transmitted to the thin filament via the other two subunits. The purpose of this chapter is to describe our recent studies on the role of troponin in signal transduction in striated muscle. The signal transduction process is thought to involve three different states of actin-Tm resulting from changes in troponin's interaction with actin-Tm: (i) a blocked state (B) preventing myosin interaction, (ii) a closed state (C) allowing weak interaction, and (iii) an open state (M) allowing strong interaction. The apparent rate of troponin dissociation is often used to study thin filament regulation in striated muscle. The dissociation rate of Tn is extremely low so it is not intuitively clear that such a slow process would probe thin filament regulation. We investigate this issue using a kinetic model to explain the Tn dissociation rate measured by labeled Tn exchange in the myofibrils. In our model,

TnI's regulatory domain switches from actin-Tm to TnC followed by TnT dissociation from actin-Tm. On the basis of this model, we demonstrate that even though the apparent rate of Tn dissociation is low, it probes the blocked state to closed state equilibrium of the myofibrillar thin filament.

Keywords: Striated muscle, Troponin dissociation, Kinetic modeling

---

* 402 N. Blackford Street, Indianapolis, IN 46202, USA; Email: asen@iupui.edu.

Introduction

It has long been recognized that troponin plays an important role in the regulation of contraction in striated muscle (Ebashi et al., 1973). Within the half sarcomere, there is a region where the thick filaments and thin filaments overlap. In this overlap region, myosin interacts with actin to produce force leading to muscle contraction. Activation of contraction is a highly cooperative signal transduction process which involves binding of calcium by troponin and subsequent interactions between the thick and thin filaments (Gordon et al., 2000). The anatomical and biochemical features of the thin filaments are primarily responsible for this highly cooperative behavior. Troponin consists of three subunits: TnC, TnI, and TnT, with each subunit performing specific functions in the regulatory process. The subunit TnC binds to calcium, TnI inhibits the actin-activated myosin ATPase, and TnT binds to tropomyosin (Tm) (Gomes et al., 2002; Vingradova et al., 2005). Muscle contraction is initiated by binding of TnC to calcium, but because TnC does not interact directly with actin or tropomyosin, the signal is transmitted to the thin filament via the other two subunits TnI and TnT (Filatov et al., 1999). It is known that TnI operates like a switch, and moves away from action toward TnC, after calcium is bound. From the crystal structure of the regulatory domain of Tn, it has been found that there are complex interactions among the three subunits (Tao et al., 1999; Zhou et al., 2000). There are also interactions between TnT and actin, and between TnI and Tm. Myosin does not interact directly with TnC, but influences its binding with calcium.

Although troponin is found in both skeletal muscle and cardiac muscle, the specific subunits of troponin have different features depending on the type of muscle. For instance, for the TnC subunit of troponin in skeletal muscle, there are four calcium-ion binding sites, whereas in cardiac muscle there are three. In addition, no specific isoforms of TnC are known in human cardiac muscle. In human cardiac muscle tissue, TnC is present in an isoform typical of slow skeletal muscle, whereas a different form of TnC, known as the fast skeletal TnC isoform, is found in fast skeletal muscle. The TnI and TnT subunits may also be present in different forms. Two isoforms of TnI exist in human skeletal muscle, but only one TnI-isoform is found in human cardiac muscle. The TnT subunit is expressed in human cardiac tissue in several isoforms; in contrast, only two isoforms of TnT are present in human skeletal muscle tissue. There have been numerous studies describing the structure and function of the troponin subunits in both skeletal and cardiac muscle (Brown and Cohen, 2005; Dong et al., 2003a,b; Kowlessur and Tobacman, 2010; Lehman et al., 2009; Matsu et al., 2010; Murakami et al., 2005; Ngai et al., 2001; Sun et al., 2006, 2007; Takeda et al., 2003; Xing et al., 2008). Some of these studies have also investigated various aspects of the structure-function relationships of the troponin subunits including their differential regulatory patterns in skeletal and cardiac muscle (Clemmen et al., 2005; Maytum et al., 2002; Solaro et al., 2008; Yang et al., 2009).

The actin sites are exposed for myosin interaction by changes in the position of Tm on the surface of the actin filament. Earlier work suggested a two-state model where Tm sterically blocks the actin sites when calcium level is low, whereas high calcium level leads to a non-blocking position of Tm (Lehman et al., 1995; Vibert et al., 1997; Squire and Morris, 1998; Lehman and Craig, 2008). Recent studies, however, indicate that there are three different states of the thin filament in relation to the average position of Tm on the surface of the thin filament: a blocked state (B) where Tm is on the outer domain of the thin filament and blocks the actin site, a closed state (C) where Tm is mostly in the inner domain and allows weak interaction between actin and the motor domain, and an open state (M) in which Tm is displaced even further into the inner domain (McKillop and Geeves, 1993; Maytum et al., 1999). The M-state allows force-producing interactions. The calcium-saturated TnC site has a higher affinity such that the TnI domain switches from actin to TnC. This allows Tm to move from the outer domain to the inner domain. Tropomyosin (Tm) interacts with seven actin monomers along the thin filament, whereas TnI interacts with 1 or 2 actin monomers. The elastic properties of Tm facilitate the binding of Tm with actin. The elastic properties of Tm are also responsible for the cooperative nature of calcium activation (Smith et al., 2003a,b; Singh and DeGregori, 2003).

It has been shown in (Swartz et al., 1990) that the distribution of a fluorescently labeled, exogenous myosin motor domain (S1) within rigor myofibrils is modulated by calcium. This study probed the distribution of the M-state actin sites within the half sarcomere, and suggested that the intrinsic rigor cross-bridges shifted the actin-Tm sites into the M-state, and this was limited to about 14 actins adjacent to the rigor cross-bridge. More recent studies have demonstrated that calcium alone could not shift the actin sites into the M-state (Swartz et al., 1996; Zhang et al., 2000). However, a problem with this and many other studies is that the thin filament state was probed by changes in actin or Tm as a result of strong cross-bridge binding, which tends to alter the system. Furthermore, these studies did not probe the state of Tn in three different states. It is possible that Tn is present in only two states (B and C), whereas actin-Tm has three states (B, C, and M), in which the M-state is myosin-dependent and it does not alter the state of Tn on the thin filament.

The role of troponin in thin filament regulation can be studied by replacing the troponin complex in a muscle fiber with exogenous troponin (labeled troponin) (Brenner et al., 1999; She et al., 2000; Piroddi et al., 2003). This can be done using an exchange process in which the actin-tropomyosin complex is never devoid of full complement of troponin (Brenner et al., 1999). Using labeled troponin, it is found that the distribution of Tn exchange within the half sarcomere was not uniform. The exchange rate is limited by the intrinsic Tn dissociation rate, and when labeled Tn is present, the labeled protein gets incorporated into the myofibril. In our recent work (Swartz et al., 2006), we used labeled troponin (Tn) exchange in rigor myofibrils to probe Tn-subunit interactions on the thin filament. The location and rate of exchange of Tn or its subunits were measured by high-resolution fluorescence microscopy and image analysis. Three different rates of Tn exchange were observed that were dependent on calcium concentration and strong cross-bridge binding that strongly support the three-state model. The rate of Tn dissociation in the non-overlap region was 200-fold faster at pCa 4 (C-state region) than at pCa 9 (B-state region). When Tn contained engineered TnC mutants with weakened regulatory TnI interactions, the apparent exchange rate at pCa 4 in the non-overlap region increased proportionately with TnI-TnC regulatory affinity. This suggests that the mechanism of calcium enhancement of the rate of Tn dissociation is by favoring a TnI-TnC interaction over a TnI-actin-Tm interaction. At pCa 9, the rate of Tn dissociation in the overlap region (M-state region) was 100-fold faster than the non-overlap region (B-state region) suggesting that strong cross-bridges increase the rate of Tn dissociation. At pCa 4, the rate of Tn dissociation was twofold faster in the non-overlap region (C-state region) than the overlap region (M-state region) that likely involved a strong cross-bridge influence on TnT's interaction with actin-Tm. At sub-maximal calcium (pCa 6. 2-5. 8), there was a long-range influence of the strong cross-bridge on Tn to enhance its dissociation rate, tens of nanometers from the strong cross-bridge. These observations suggest that the three different states of actin-Tm are associated with three different states of Tn. They also support a model in which strong cross-bridges shift the regulatory equilibrium from a TnI-actin-Tm interaction to a TnC-TnI interaction that likely enhances calcium binding by TnC.

Swartz et al. (2006) estimated the Tn dissociation rate from the change in intensity ratio in the non-overlap region to overlap region, as a function of time (see Figure 5 of their paper) by fitting the data to the ratio of rates in each region. At early time points, the ratio is low at low calcium and high at high calcium levels. At infinite time it approaches one, independent of calcium level. The apparent rate of Tn dissociation in the absence of calcium is very low (k < $5\text{x}10^{-6}$ s^{-1}) corresponding with its high affinity for actin-Tm. The rate increases almost 200 fold ($1\text{x}10^{-3}$ s^{-1}) in the presence of calcium. In studies using Tn containing TnC mutants with altered TnI binding affinity, the apparent Tn-dissociation rate grades with TnC's affinity for TnI in the presence of calcium (Swartz et al., 2006). The dissociation rates are also strongly influenced by the TnC subunit isoform. These observations suggest that switching of the regulatory domain of TnI between actin-Tm and TnC is a major determinant of the apparent Tn dissociation rate.

A simple kinetic scheme to explain the observed influence of calcium and the TnC-TnI regulatory interaction on the Tn dissociation rate is shown in Figure 1, which contains two sequential steps (Swartz et al., 2006): step (1) represents dissociation of TnI from actin-Tm and its binding to the regulatory domain of TnC, and step, which is followed by TnT dissociation from actin-Tm in step (2). The first step is involved in the blocked-to-closed state transition of actin-Tm in the three-state model of thin filament regulation. Studies of the apparent rate of switching of TnI from actin-Tm to TnC (-Ca_2+ to +Ca_2+) shows that it is 500 s^{-1} while the reverse transition (+Ca_2+ to –Ca_2+) of switching from TnC back to actin-Tm is 60 s^{-1} (Shitaka et al., 2005). These observations suggest that the first step in the scheme can be six orders of magnitude faster than the overall process. The second step in the process, namely, TnT dissociation from actin-Tm, has an estimated rate of 2. 5x 10^{-3} s (Swartz et al., 2006) and can limit the rate of the overall process. It is not intuitively clear that such a fast upstream reaction can regulate the overall process as observed in the experiments.

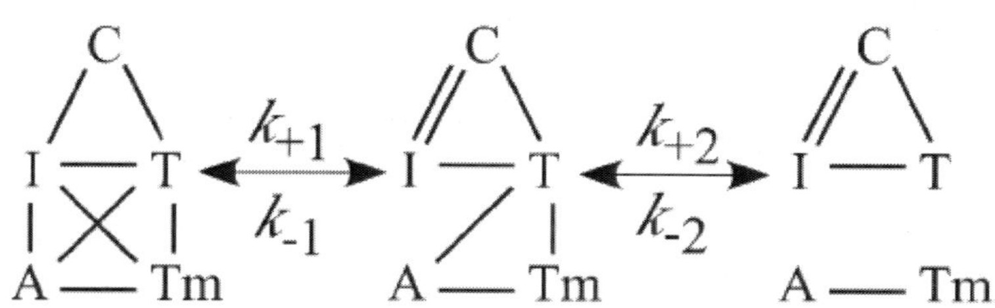

Figure 1. Pathway for Tn dissociation. Protein interactions are denoted by lines connecting the letters with A, Tm, T, I, and C, with the letters representing actin, tropomysosin, TnT, TnI and TnC, respectively.

Within the framework of the three-state system of thin filament regulation, we recently developed a kinetic model to determine if the apparent rate of Tn dissociation can be modulated by very fast upstream transitions (Sen et al., 2012). By integrating the governing rate equations, we computed the progress curves for labeled Tn exchange into the myofibrils, and derived the intensity ratio between the non-overlap and overlap regions. The computed progress curves for labeled Tn exchange and the derived intensity ratio well explain the intensity ratio progress curves observed experimentally. These numerical simulations and experimental observations reveal that even though the apparent rate of Tn dissociation is extremely low, it probes the calcium-dependent blocked-state to closed-state equilibrium of the myofibrillar thin filament.

The Kinetic Model

Our kinetic model is described by the following reaction scheme:

$$\begin{aligned} S + B &\underset{k_{-1}}{\overset{k_1}{\rightleftharpoons}} C \underset{k_{-2}}{\overset{k_2}{\rightleftharpoons}} D \\ S + E &\underset{k_{-3}}{\overset{k_3}{\rightleftharpoons}} F \underset{k_{-4}}{\overset{k_4}{\rightleftharpoons}} G \end{aligned} \qquad \text{(R1)}$$

Here *S*, *B*, *C*, *D*, *E*, *F* and *G* represent, respectively, actin-Tm (empty Tn site), unlabeled Tn, the thin filament in state 1 with unlabeled Tn, thin filament in state 2 with unlabeled Tn, labeled Tn, thin filament in state 1 with labeled Tn, thin filament in state 2 with labeled Tn (see appendix). State 1 is associated with the closed state of the thin filament, while state 2 is associated with the blocked state of the thin filament. It is appreciated that there are multiple protein and ligand interactions involved in this transition and thus more than two steps, but our initial modeling is to help explain how calcium influences troponin exchange kinetics in the sarcomere. The scheme is written as two parallel association reactions. At the beginning of the reaction, the most populated states are *C* and *D*, and as the reaction progresses, states *F* and *G* are populated. As such, in the kinetic model, step 2 is the calcium dependent switching of TnI from actin-Tm to TnC that is associated with the transition of the thin filament from the blocked to the open state. Step 1 is TnT dissociation from actin-Tm that allows for the exchange reaction to occur. Exhaustive exchange of labeled Tn for unlabeled Tn does not influence the high and low calcium ATPase activity suggesting that the label does not influence Tn interactions with actin-Tm (Swartz et al., 2006). Thus, the rate constants for the unlabeled and labeled Tn reactions are assumed the same. We estimate k_1 as being diffusion limited and thus 10^{-7} $M^{-1}s^{-1}$ whereas k_{-1} is estimated at 2. 5 x 10^{-3} s^{-1} from the observed values of TnT dissociation from myofibrils (Swartz et al., 2006). The values for k_2 and k_{-2} estimated from biochemical studies on TnI switching described in the introduction.

A schematic for the protein interactions in the kinetic model (R1) is shown in Figure 2, and the rate equations for the various components in (R1) are described in the appendix. The rate equations are solved numerically with appropriate initial conditions. We discuss below the results of the numerical simulation.

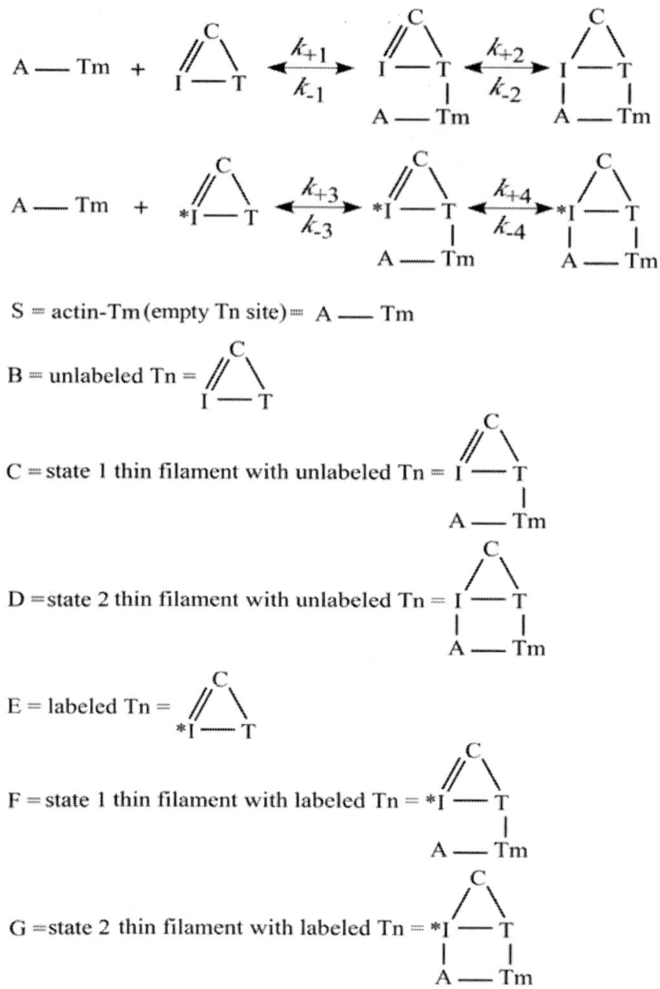

Figure 2. Schematic for the protein interactions in the kinetic model (R1).

Discussion

We experimentally measured the relative amount of labeled troponin bound to the thin filament. This is given by $F + G$. Accordingly, we will present the time course of $f + g$, which is the dimensionless form of $F + G$. We also measure the intensity ratio between the non-overlap and overlap regions. The rate in the overlap region is independent of calcium and thus $k_{-2} = k_{-4}$, and an appropriate value of these rate constants is $50\,\text{s}^{-1}$. The intensity in the non-overlap region is calcium sensitive and thus $k_{-2} = k_{-4}$ is low (on the order of $10\,\text{s}^{-1}$) at low calcium and high (500-1000 s^{-1}) at high calcium levels. Results for the time course of the intensity ratio for various values of $k_{-2} = k_{-4}$ in the non-overlap and overlap regions are presented below. By changing the values of the dimensionless parameter, δ, we have studied the effect of varying k_{-2} on the time course of labeled Tn bound to the thin filament. The initial amount (E_0) of labeled Tn can also be varied in the experiment. Since $\delta = k_{-2}/k_3E_0$, the effect of this variation on the time course of labeled Tn bound to the thin filament may also be examined by varying δ, with k_{-2} and k_3 fixed. We have obtained numerical solution of the IVP given by Eqn (9) for the following values of δ = 1, 2, 5, 10, 20, 50 and 100, corresponding to the values of k_{-2} = 10, 20, 50, 100, 500 and 1000 (s^{-1}). From the numerical solution, we have calculated the quantity $f + g$, which is the total (dimensionless) concentration of the labeled troponin bound to the thin filament. The progress curves of $f + g$ for various values of k_{-2} are depicted in Figure 3. This figure shows, as one would expect, that the progress curves rise as time increases and tend to level off at large times. As we increased the value of k_{-2} beyond $1000s^{-1}$, the differential equations became stiff and the computations required more than 40 minutes to converge. For values of $k_{-2} \geq 1000\ s^{-1}$, special algorithms for the solution of stiff differential equations must be used in order to enhance the rate of convergence and reduce computation time. The actual (dimensional) time in seconds can be obtained by multiplying the dimensionless time (τ) with k_3E_o. With our choice of $k_3 = 10^7\ M^{-1}s^{-1}$, and $E_o = 10^{-6}$ M, this amounts to multiplying the dimensionless time with 10 to obtain time in seconds. Accordingly, in figure 3, the progress curves tend to level off at times on the order of 25 to 30 hours. In our simulations, we varied the concentration of labeled troponin, E_0, and found that the results are qualitatively similar to those reported here.

We have computed the intensity ratio between the non-overlap and overlap regions, using a fixed level of Tn dissociation in the overlap region. The intensity ratio (R) for two different values of $\delta=\delta_1$ and $\delta=\delta_2$ is defined as:

$$R = \frac{(f+g)_{\delta=\delta_1}}{(f+g)_{\delta=\delta_2}}.$$

Figure 4 illustrates the variation of the intensity ratio with time for four values of δ_1 = 1, 2, 10, 20, 50 and 100, with $\delta_2 = 5$. Recall that the values of δ_1 correspond to the following values of $k_{-2} = k_{-4}$ = 10, 20, 100, 200, 500, and 1000 (s^{-1}), respectively, and the value $\delta_2 = 5$, corresponds to $k_{-2} = k_{-4}$ = 50 (s^{-1}). In this figure the various curves are labeled by the corresponding ratios of the values of $k_{-2} = k_{-4}$ in the non-overlap and overlap regions. The figure shows, as expected, that if $\delta_1 < \delta_2$, the intensity ratio decreases with time and approaches 1 from values less than 1, whereas if $\delta_1 > \delta_2$, the intensity ratio increases with time and approaches 1 from values greater than 1, at sufficiently large times. In the model, we varied k_{-2} and kept k_2 constant with the net result that we varied the calcium dependent equilibrium of switching of TnI between actin-Tm and TnC and thus the equilibrium between state 1 and state 2.

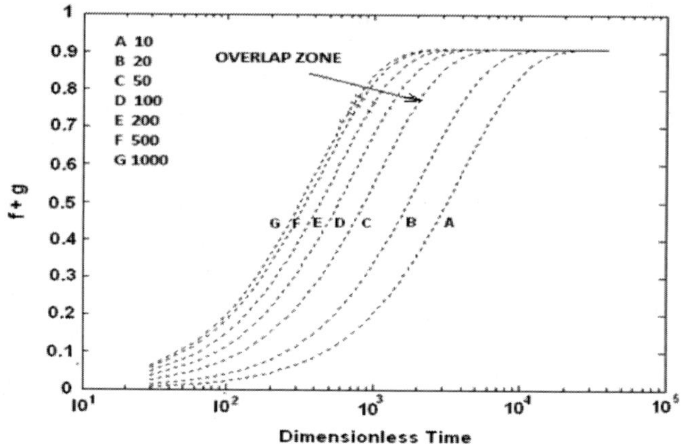

Figure 3. Progress curves for labeled Tn, (f + g), bound to myofibrils. The curves A through G correspond to the values of k_{-2} = 10, 20, 50, 100, 200, 500, and 1000, respectively. The actual time t is related to the dimensionless time τ using the relation $t = (1/k_3E_0)\tau$ (see appendix). With the values of $k_3 = 10^7\ \mathrm{M^{-1}s^{-1}}$ and $E_0 = 10^{-6}\ \mathrm{M}$ chosen in this study, the actual time t (in seconds) is given by: $t = 0.1\ \tau$.

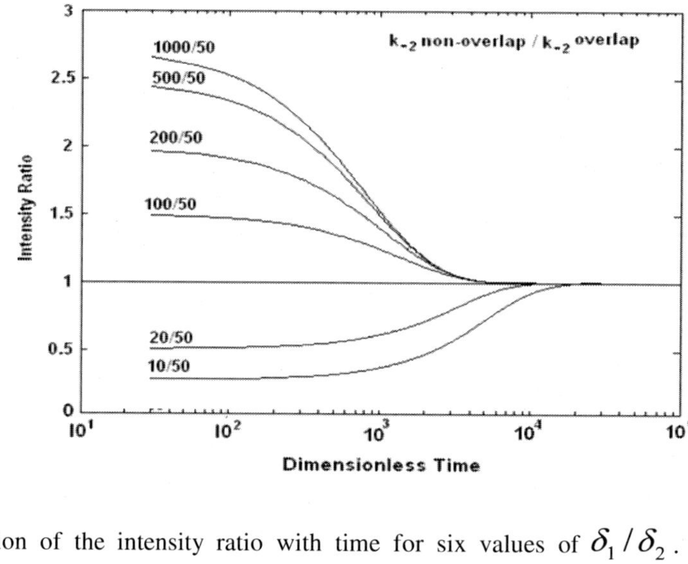

Figure 4. Variation of the intensity ratio with time for six values of δ_1/δ_2. The value of δ_2 corresponds to $k_{-2} = k_{-4} = 50\ \mathrm{s^{-1}}$, and the values of δ_1 correspond to $k_{-2} = k_{-4}$ = 10, 20, 50,100, 500, and 1000 $(\mathrm{s^{-1}})$, respectively. The actual time t is related to the dimensionless time τ using the relation $t = (1/k_3E_0)\tau$ (see appendix). With the values of $k_3 = 10^7\ \mathrm{M^{-1}s^{-1}}$ and $E_0 = 10^{-6}\ \mathrm{M}$ chosen in this study, the actual time t (in seconds) is given by: $t = 0.1\ \tau$.

The simulated curves match the general features observed experimentally for skeletal Tn in that at high calcium, the ratio goes from about two to one while at low calcium, the ratio starts from 0. 3 and approaches one with increasing time. More specifically, the curves that are most similar to data obtained from skeletal Tn are where k_{-2} = 200 and 10 (s^{-1}) for high and low calcium respectively for the non-overlap region. Thus, there is at least a 20-fold change in this rate in going from high to low calcium. This fold change is an underestimate considering that the intensity ratio value at low calcium is over-estimated because of imaging artifacts and the model does not include modest effects of calcium and cross-bridges on TnT dissociation as observed experimentally (Swartz et al., 2006). In terms of the effects of strong cross-bridges (comparison of the non-overlap to the overlap region), the fixed value of k_{-2} at 50 s^{-1} well approximates the initial values observed at high and low calcium. We did not vary this in the simulations to keep the modeling simple. However, changing this value within reasonable limits will change the initial intensity ratio value and the time required for the intensity ratio to obtain unity. For cardiac Tn, the modeled curves that best simulate the observed data are below one with low calcium being closest to the k_{-2} = 10 s^{-1} curve and the high calcium curve being closest to the k_{-2} = 20 s^{-1} or slightly higher. Here, the fold change in the rate is much smaller than observed with skeletal Tn. This is in accord with the observed difference in TnC and TnI interactions between the isoforms (Yang et al., 2009).

We would like to be able to better define the transition rates between the Tn-actin states, however, the complexity of the system in terms of the potential number of true steps involved makes this prohibitive. Our model uses just one step for the transition of the Tn-actin from state 2 to 1 (step 2) that is associated with the blocked to closed state transition of the thin filament. Mechanistically, this involves at least three steps: calcium binding to the regulatory domain of TnC and subsequent conformational changes, dissociation of the regulatory domain of TnI from actin-Tm, then binding of the regulatory domain of TnI to TnC. In our simple model, we assume that TnI dissociation from TnC then binding to actin-Tm (k_2) is independent while the reverse reaction is calcium dependent (k_{-2}). We use biochemical data to approximate k_2 and change the level of k_{-2} to approximate the low and high calcium data and use the biochemical data to approximate the maximum limit for k_{-2}. Even with these simple assumptions, the modeling shows that incorporation of the known fast kinetics of these reactions can explain the observed very slow overall reaction. Independent of the exact rates, the modeling allows for approximation of the step 2 equilibrium for Tn-actin at high and

low calcium and we feel that this is the major value of the modeling. We know that the governing rates are much faster than step one and show by the modeling that they can dictate the rates. More detailed modeling and incorporation of additional steps could allow for more explicit modeling of the transition rates, but some of the needed rate constants are currently not available in the literature.

The simulated curves match the general features observed experimentally for skeletal Tn in that at high calcium, the ratio goes from about two to one while at low calcium, the ratio starts from 0. 3 and approaches one with increasing time. More specifically, the curves that are most similar to data obtained from skeletal Tn are where k_{-2} = 200 and 10 (s^{-1}) for high and low calcium, respectively, for the non-overlap region. Thus, there is at least a 20-fold change in this rate in going from high to low calcium. This fold change is an underestimate considering that the intensity ratio value at low calcium is over-estimated because of imaging artifacts and the model does not include modest effects of calcium and crossbridges on TnT dissociation as observed experimentally (Swartz et al., 2006). In terms of the effects of strong crossbridges (comparison of the non-overlap to the overlap region), the fixed value of k_{-2} at 50 (s^{-1}) well approximates the initial values observed at high and low calcium. We did not vary this in the simulations to keep the modeling simple. However, changing this value within reasonable limits will change the initial intensity ratio value and the time required for the intensity ratio to obtain unity. For cardiac Tn, the modeled curves that best simulate the observed data are below one with low calcium being closest to the $k_{-2} = 10\ (s^{-1})$ curve and the high calcium curve being closest to the $k_{-2} = 20\ (s^{-1})$ or slightly higher. Here, the fold change in the rate is much smaller than observed with skeletal Tn. This is in accord with the experimentally observed difference in TnC and TnI interactions between the isoforms (Yang et al., 2009).

Conclusion

We have described the role of troponin in signal transduction in striated muscle. Our discussion is based on the three-state model of thin filament regulation. Within the framework of the three-state model, we used a simple kinetic scheme to investigate the relationship between the apparent rate of Tn dissociation and thin filament regulation. By solving the model rate equations numerically, the progress curves for labeled Tn exchange were computed, and the progress curves for the intensity ratio between the non-overlap and overlap regions were derived. The modeled progress curves for the intensity ratio are found to be qualitatively similar to those observed experimentally for skeletal Tn at physiological ionic strength (Swartz et al., 2006). In addition, studies comparing the cardiac isoform to the skeletal isoform show dramatic differences in the intensity ratio plots (Swartz et al, 2006) that are consistent with the model. Structural studies on cardiac TnC show that calcium saturation does not induce the same extent of conformational changes as in skeletal TnC and thus it has a less favored regulatory interaction with TnI (Li et al., 1999). Thus, the equilibrium for cardiac Tn in the presence of calcium is not fully favored for the closed state. The net result is that magnitude of the effect of calcium on Tn dissociation is less. Also, cTn dissociation at low calcium is not as slow as sTn and the comparative response is very similar to curves 20/50 and 10/50 respectively in Figure 4. Our model suggests that the apparent rate of Tn dissociation, measured by labeled Tn exchange, probes the blocked to closed state equilibrium of the thin filament. More detailed modeling of the signal transduction process using complex kinetic schemes will provide further insight into the regulatory functions of Tn in striated muscle contraction.

While this modeling is specific to muscle contraction, there are high affinity protein interactions involved in many other biological processes, and modulation of these processes can occur by differences in the dissociation rate. Gene silencing is mediated by some histones that have very slow dissociation rates (Misteli, 2001; Lele et al., 2006) and this is modulated by covalent modification (Elgin and Grewal, 2003). Growth factor binding proteins modulate growth factor function by binding with high affinity, and some of the isoform differences are mediated by differences in dissociation rates (18 Marinaro et al., 2003). Binding of calmodulin to calcium/calmodulin-dependent kinase II induces autophosphorylation that dramatically slows calmodulin dissociation from the enzyme, and this slow dissociation is thought to be involved in calcium signaling pathways in nerve (Singla et al., 2001). Lastly, protein turnover within the sarcomere of the myofibril of striated muscle may be limited by protein dissociation from the filaments or require limited proteolysis to enhance their dissociation from filaments before they can be processed by the proteasome (Goll et al., 2008). Thus, the general concepts and modeling methods developed here are of value in studying other high affinity protein multi-protein interactions and their modulation by post-translational modification or ligand binding.

Acknowledgment

I would like to thank Dr. Darl Swartz for stimulating discussions.

References

Brenner, B., Kraft, T., Yu, L. C., Chalovich, J. M., (1999). Thin filament activation probed by fluorescence of N-((2-(Iodoacetoxy)ethyl)-N-methyl)amino-7-nitrobenz-2-oxa-1, 3-diazole-labeled troponin I incorporated into skinned fibers of rabbit psoas muscle. *Biophys. J.* 77, 2677-2691.

Brown, J. H., Cohen, C., (2005). Regulation of muscle contraction by tropomyosin and troponin: how structure illuminates function. *Adv. Protein Chem.* 71, 121-159.

Clemmen, E. W., Entezari, M., Martyn, D. A., Regnier, M., (2005). Different effects cardiac versus skeletal muscle regulatory proteins on *in vitro* measures of actin filament speed and force. *J. Physiol.* 566, 737-746. .

Dong, W. J., An, J., Xing, J., Cheung, H. C., (2003a). Structural transition of the inhibitory region of troponin I within the regulated cardiac thin filament. *Arch. Biochem. Biophys.* 456, 135–142.

Dong, W. J., Robinson, J. M., Stagg, S., Xing, J., Cheung, H. C., (2003b). Kinetics of conformational transitions in cardiac troponin induced by Ca2+ dissociation determined by Forster resonance energy transfer. *J. Biol. Chem.* 278:42394–42402.

Ebashi, S., Ohtsuki, I., Mihashi, K., (1973). Regulatory proteins of muscle with special reference to troponin. *Cold Spring Harbor Symposia on Quantitative Biology* 37, 215-223.

Elgin, S. C., Grewal, S. I., (2003). Heterochromatin: silence is golden. *Curr. Biol.* 13, R895-898.

Filatov, V. L., Katrukha, A. G., Bulgarina, T. V., Gusev, N. B., (1999). Troponin, structure, properties, and mechanism of functioning. *Biochemistry* (Moscow) 64, 1155 -1174.

Goll, D. E., Neti, G., Mares, S. W., Thompson, V. F., (2008). Myofibrillar protein turnover: the proteasome and the calpains. *J. Anim. Sci.* 86, E19-35.

Gomes, A. V., Potter, J. D., Szczesna-Cordary, D., (2002). The role of troponin in muscle contraction, *Life* 54, 323-333.

Gordon, A. M., Homsher, E., Regnier, E., (2000). Regulation of contraction in striated muscle. *Physiol. Rev.* 80 853-924.

Kowlessur, D., Tobacman, L. S., (2010). Troponin regulatory function and dynamics revealed by H/D exchange and mass spectroscopy. *J. Biol. Chem.* 285, 2686-2694. .

Lehman, W., Craig, R., (2008). Tropomyosin and the steric mechanism of muscle regulation. *Adv. Exp. Med. Biol.* 644, 95-109.

Lehman, W., Galinska-Rakoczy, A., Hatch, V., Tobacman, L., Craig, R., (2009). Structural basis for activation of muscle contraction by troponin and tropomyosin. *J. Mol. Biol.* 388, 673-681.

Lehman, W., Vibert, P., Uman, P., Craig, R., (1995). Steric blocking by tropomyosin visualized in relaxed vertebrate muscle thin filaments. *J. Mol. Biol.* 251, 191-196.

Lele, T., Wagner, S. R., Nickerson, J. A., Ingber, D. E., (2006). Methods for measuring rates of protein binding to insoluble scaffolds in living cells: histone H1-chromatin interactions. *J. Cell. Biochem.* 99, 1334-1342.

Li, M. X., Spyracopoulos, L., Sykes, B. D., (1999). Binding of cardiac troponin-I147-163 induces a structural opening in human cardiac troponin-C. *Biochemistry* 38, 8289-8298.

Marinaro, J. A., Jamieson, G. P., Hogarth, P. M., Bach, L. A., (1999). Differential dissociation kinetics explain the binding preference of insulin-like growth factor binding protein-6 for insulin-like growth factor-II over insulin-like growth factor-I. *FEBS Lett.* 450, 240-244.

Matsuo, T., Iwamoto, H., Yagi, N., (2010). Monitoring the structural behavior of troponin and myoplasmic free Ca2+ concentration during twitch of frog skeletal muscle. *Biophys. J.* 99, 193-200.

Maytum, R., Lehrer, S. S., Geeves, M. A., (1999). Cooperativity and switching within the three-state model of muscle regulation. *Biochemistry* 38, 1102-1110.

Maytum, R., Westerdorf, B., Jaquet, K., Geeves, M. A., (2002). Differential regulation of the actomyosin interactin by skeletal and cardiac troponin isoforms. *J. Biol. Chem.* 278, 6696-6701.

McKillop, D. F., Geeves, M. A., (1993). Regulation of the interaction between actin and myosin subfragment 1: evidence for three states of the thin filament. *Biophys. J.* 65, 693-701.

Misteli, T., (2001). Protein dynamics: implications for nuclear architecture and gene expression. *Science* 291, 843-847.

Murakami, K., Yumoto, F., Ohki, S. Y., Yasunaga, T., Tanokura, M., (2005). Structural basis for Ca2+-regulated muscle contraction at interaction sites of troponin with actin and tropomyosin. *J. Mol. Biol.* 352, 178-201.

Ngai, S. -M., Hodges, R. S., (2001). Structural and functional studies on troponin I and troponin C interactions. *J. Cell. Biochem.* 83, 33-46.

Piroddi, N., Tesi, C., Pellegrino, M. A., Tobacman, L. S., Homsher, E., Poggesi, C., (2003). Contractile effects of the exchange of cardiac troponin for fast skeletal troponin in rabbit psoas single myofibrils. *J. Physiol.* 552, 917-731.

Sen, A. K., Swartz, D. R. . Gawalapu, R. K., (2012). A kinetic model of troponin dissociation in relation to thin filament regulation in striated muscle. *Math. Biosci.* 238, 32-37.

She, M., Trimble, D., Yu, L. C., Chalovich, J. M., (2000). Factors contributing to troponin exchange in myofibrils and in solution. *J. Muscle Res. Cell Motil.* 21, 737-745.

Shitaka, Y., Kimura, C., Miki, M., (2005). The rates of switching movement of troponin T between three states of skeletal muscle thin filaments determined by fluorescence resonance energy transfer. *J. Biol. Chem.* 280, 2613-2619.

Singh, A., DeGregori, S. E., (2003). Local destabilization of the tropomyosin coiled coil gives the molecular flexibility required for actin binding. *Biochemistry* 42, 14114-14121.

Singla, S. I., Hudmon, A., Goldberg, J. M., Smith, J. L., Schulman, H., (2001). Molecular characterization of calmodulin trapping by calcium/calmodulin-dependent protein kinase II. *J. Biol. Chem.* 276, 2953-2960.

Smith, D. A., Geeves, M. A., (2003b). Cooperative regulation of myosin-actin interactions by a continuous flexible chain II: actin-tropomyosin-troponin and regulation by calcium. *Biophys. J.* 84, 3168-3180.

Smith, D. A., Maytum, R., Geeves, M. A., (2003a). Cooperative regulation of myosin- actin interactions by a continuous flexible chain I: actin-tropomyosin systems. *Biophys. J.* 84, 3155-3167.

Solaro, R. J., Rosevear, P., Kobayashi, T., (2008). The unique functions of cardiac troponin I in the control of cardiac muscle contraction. *Biochem. Biophys. Res. Comm.* 369, 82-87.

Squire, J. M., Morris, E. P., (1998). A new look at thin filament regulation in vertebrate skeletal muscle. *FASEB J.* 12, 761-771.

Sun, Y. -B., Brandmeier, B., Irving, M., (2006). Structural changes in troponin in response to Ca2+ and myosin binding to thin filaments during activation of skeletal muscle. *Proc. Nat. Acad. Sci.* 103, 17771-17776.

Sun, Y. -B., Lou, F., Irving, M., (2007). Calcium- and myosin-dependent changes in troponin structure during activation of heart muscle. *J. Physiol.* 587, 155-163.

Swartz, D. R., Greaser, M. L., Marsh, B. B., (1990). Regulation of binding of subfragment 1 in isolated rigor myofibrils. *J. Cell Biol.* 111, 2989-3001.

Swartz, D. R., Moss, R. L., Greaser, M. K., (1996). Calcium alone does not fully activate the thin filament for S1 binding to rigor myofibrils. *Biophys. J.* 71, 1891-1004.

Swartz, D. R., Yang, Z., Sen, A. K., Tikunova, S. B., Davis, J. P., (2006). Myofibrillar troponin exists in three states and there is signal transduction along skeletal myofibrillar thin filaments. *J. Mol. Biol.* 361, 420-435.

Takeda, S., Yamashita, A., Maeda, K., maeda, Y., (2003). Structure of the core domain of human cardiac troponin in the Ca2+-saturated form. *Nature* 424, 35-41.

Tao, T., Gong, B. -J., Grabarek, Z., Gergely, J., (1999). Conformational changes induced in Troponin I by interaction with Troponin T and Actin/Tropomyosin. *Biochem. Biophys. Acta* 1450, 423-433.

Vibert, P., Craig, R., Lehman, W., (1997). Steric model for activation of muscle thin filaments. *J. Mol. Biol.* 266, 8-14.

Vingradova, M. V., Stone, D. B., Malanina, G. G., Karatzaferi, C., Cooke, R., Mendelson, R. A., Fletterick, R. J., (2005). Ca(2+)-regulated structural changes in troponin. *Proc. Nat. Acad. Sci.* 102, 5038-5043.

Xing, J., Chinnaraj, J., Zhang, Z., Cheung, H. C., Dong, W. J., (2008). Structural studies of interactions between cardiac troponin I and actin in regulated thin filament using Förster resonance energy transfer. *Biochemistry* 47, 13382-13393.

Yang, Z., Yamazaki, M., Shen, Q. W., Swartz, D. R., (2009). Differences between cardiac and skeletal troponin interaction with the thin filament probed by troponin exchange in skeletal myofibrils. *Biophys. J.* 97, 183-194.

Zhang, D., Yancey, K. W., Swartz, D. R., (2000). Influence of ADP on cross-bridge dependent activation of myofibrillar thin filaments. *Biophys. J.* 78. 3103-3111.

Zhou, X., Morris, E. P., Lehrer, S. S., (2000). Binding of troponin I and troponin I- troponin C complex to actin-tropomyosin: Dissociation by myosin subfragment. *Biochemistry* 30, 1128-1132.

Appendix

The time courses of the various components in the kinetic scheme shown in Figure 2 can be described by the following rate equations.

$$\frac{dS}{dt} = -k_1 SB + k_{-1} C - k_3 SE + k_{-3} F, \tag{1a}$$

$$\frac{dB}{dt} = -k_1 SB + k_{-1} C, \tag{1b}$$

$$\frac{dC}{dt} = k_1 SB - (k_{-1} + k_2)C + k_{-2} D, \tag{1c}$$

$$\frac{dD}{dt} = k_2 C - k_{-2} D, \tag{1d}$$

$$\frac{dE}{dt} = -k_3 SE + k_{-3} F, \tag{1e}$$

$$\frac{dF}{dt} = k_3 SE - (k_{-3} + k_4)F + k_{-4} G. \tag{1f}$$

$$\frac{dG}{dt} = k_4 F - k_{-4} G. \tag{1g}$$

Here each of the letters represents the concentration of the corresponding component in the kinetic scheme. These equations are subject to the initial conditions:

$$S(0) = 0,\ B(0) = 0,\ C(0) = 0,\ D(0) = D_0,$$
$$E(0) = E_0,\ F(0) = 0,\ G(0) = 0, \tag{2}$$

D_0 and E_0 being the initial concentrations of *D* and *E*, respectively.

Note that in our experiment we measure the amount of labeled troponin bound to the thin filament in blocked and closed states of the non-overlap region relative to the open state in the overlap region of the sarcomere. The last two initial conditions, $F(0) = 0$ and $G(0) = 0$, follow from the fact that initially there is no labeled troponin in the thin filament in either blocked or closed state. By combining the various equations in (1), integrating the results and applying the initial conditions, the following conservation relations are obtained (Sen et al., 2012).

$$B + C + D = D_0. \tag{3}$$

$$E + F + G = E_0. \tag{4}$$

$$S + C + D + F + G = D_0. \tag{5}$$

From Eqns (4) and (5), we have

$$S = B - F - G. \tag{6}$$

Using the relations (3), (4) and (6), the variables, C, E and S in Eqns (1b), (1d) and (1e), respectively, can be replaced in terms of the remaining variables, resulting in the following equations.

$$\frac{dB}{dt} = -k_1 B(B-F-G) + k_{-1}(D_0 - B - D), \tag{7a}$$

$$\frac{dD}{dt} = k_2(D_0 - B \quad D) - k_{-2}D, \tag{7b}$$

$$\frac{dF}{dt} = k_3(B - F - G)(E_0 - F - G) - (k_{-3} + k_4)F + k_{-4}G. \tag{7c}$$

Equations (7a), (7b) and (7c), in conjunction with Eqn (1g), form a system of four differential equations for the variables B, D, F and G. These four equations with the appropriate initial conditions given in Eqn (2) constitute an initial value problem (IVP). We nondimensionalize this IVP by introducing the dimensionless variables b, d, f and g as follows.

$$B = E_0\, b, \quad D = E_0\, \bar{d}, \quad F = E_0\, f, \quad G = E_0\, g. \tag{8}$$

We also introduce a dimensionless time τ with $t = (1/k_3 E_0)\tau$. With these variables, the dimensionless IVP is given by

$$\frac{db}{d\tau} = -\alpha b(b - f - g) + \beta(\sigma - b - \bar{d}), \tag{9a}$$

$$\frac{d\bar{d}}{d\tau} = \gamma(\sigma - b - \bar{d}) - \delta \bar{d}, \tag{9b}$$

$$\frac{df}{d\tau} = (b - f - g)(1 - f - g) - (\varepsilon + \mu)f + \eta g, \tag{9c}$$

$$\frac{dg}{d\tau} = \mu f - \eta g, \tag{9d}$$

with

$$b(0) = 0, \ \bar{d}(0) = \sigma, \ f(0) = 0, \ g(0) = 0. \tag{9e}$$

Equations (9) contain the following dimensionless parameters:

$$\alpha = \frac{k_1}{k_3}, \ \beta = \frac{k_{-1}}{k_3 E_0}, \ \gamma = \frac{k_2}{k_3 E_0}, \ \delta = \frac{k_{-2}}{k_3 E_0}, \tag{10}$$

$$\varepsilon = \frac{k_{-3}}{k_3 E_0}, \ \mu = \frac{k_4}{k_3 E_0}, \ \eta = \frac{k_{-4}}{k_3 E_0}, \ \sigma = \frac{D_0}{E_0}.$$

In our computations we will use the following numerical values for the rate constants.

$$k_1 = k_3 = 10^7\,\mathrm{M}^{-1}\mathrm{s}^{-1}, \ k_2 = k_4 = 100\,\mathrm{s}^{-1}, \ k_{-1} = k_{-3} = 2.5x10^{3}\,\mathrm{s}^{-1} \tag{11}$$

In addition, the rate constants k_{-2} and k_{-4} will be taken equal, and will be varied from 1 to 1000. The initial concentrations of D and E are chosen as

$$D_0 = 10^{-7}\,\mathrm{M}, \ E_0 = 10^{-6}\,\mathrm{M}. \tag{12}$$

These choices lead to the following numerical values for the dimensionless parameters given in Eqn (9).

$$\alpha = 1, \ \beta = 2.5x10^{-4}, \ \gamma = 10, \ \varepsilon = 2.5x10^{-4}, \ \mu = 10, \ \sigma = 0.1, \tag{13}$$

with $\delta = \eta$ varied from 1 to 100.

Note that in the dimensional form, there are seven rate equations, given by Eqn (1a-g), one for each of the variables in the kinetic model. These equations contain eight parameters, which are the rate constants, k_1 through k_4, and k_{-1} through k_{-4}. Additionally, the initial conditions given by Eqn (2) contain two parameters, *D*o and *E*o. In other words, the dimensional initial value problem (IVP) contains seven differential equations and ten parameters. In contrast, the dimensionless IVP, given by Eqns (9) and (10), consists of four differential equations and eight parameters. Thus the nondimensionalization procedure has, in effect, provided a significant reduction in the dimensionality of the problem, improving calculation efficiency. Once the dimensionless IVP is solved for the four variables, *b*, $\bar{d}$, *f*, and g, their dimensional forms, B, D, F, and G can be obtained. The remaining three variables, C, E, and S can then be found from the conservation relations, given by Eqns (3), (4) and (6), respectively.

Using the above parameter values, the dimensionless IVP given by Eqns (9a-e) is numerically solved with MATLAB. The MATLAB routine ode45 was used for this purpose. This routine implements a Runge-Kutta method with a variable step size for efficient computation.

In: Troponin
Editor: J.-P. Jin

ISBN: 978-1-62948-320-7

Chapter VII

Developmental and Adaptive Regulations of Troponin

Han-Zhong Feng,MD and J.-P. Jin, MD, PhD
Department of Physiology, Wayne State University School of Medicine,
Detroit, Michigan, US

Abstract

Genes encoding muscle type isoforms of troponin subunits are expressed in a fiber type-specific manner during the development of heart and skeletal muscles with functional significances. The main structural differences of TnI and TnT isoforms are in their N-terminal region that is a regulatory structure conferring different functions in altering muscle contractility. Alternative splicing further generates TnT protein variants with N-terminal variations during development and adaptations. Modifications of N-terminal variable region of TnI and TnT provide additional fine tune on muscle contraction and relaxation. This chapter introduces troponin gene transcriptional, posttranscriptional and posttranslational regulations focusing on developmental and functional adaptations and the understanding of their structure-function relationships.

Introduction

Homologous genes have evolved encoding isoforms of troponin subunits. Expression of the troponin isoform genes is muscle fiber type-specific and regulated during embryonic and postnatal development. Troponin T (TnT) further has alternatively spiced isoforms/variants, which are also regulated during development and muscle adaptation to physiological and pathological conditions. This chapter discusses the regulations and relevant functional significances of these isoforms and splice forms. The regulation and function of cardiac troponin I (TnI) and cardiac TnT function through restrictive proteolysis is also discussed.

Troponin C Isoforms and Their Expression in Cardiac and Skeletal Muscles

There are two TnC isoform genes in the vertebrate genomes: fast skeletal muscle TnC (*TNNC2*) and slow skeletal muscle/cardiac TnC (*TNNC1*) [1, 2] The main difference between the two TnC isoforms is in their N-domain, in which the fast TnC contains two metal ion-binding sites and the slow cardiac TnC has only one [3, 4]. The N-domain metal ion-binding sites (sites I and II) are the regulatory sites in muscle contraction. During the activation of contraction, transient increase of cytoplasmic Ca^{2+} concentration results in binding to the N-domain of TnC to initiate the activation of actomyosin ATPase [5]. The metal ion-binding site I in the slow/cardiac TnC is inactive, which resulted in functional differences from the fast TnC, in which metal ion-binding sites I and II are both active [6-8].

The fast TnC gene is specific to fast skeletal muscle but transiently expresses in embryonic cardiac muscle [9]. The slow/cardiac TnC gene is expressed in both cardiac and slow skeletal muscles and transiently expresses in embryonic tongue fast muscle fibers [10]. The physiological significance of the developmental expression of TnC isoforms remains to be investigated.

Ca^{2+} affinity of the fast TnC site II is higher than that of cardiac/slow TnC [11, 12], Replacing the fast TnC with cardiac/slow TnC in skinned rabbit psoas muscle fibers resulted in a slower rate of Ca^{2+} activation and a lower maximum Ca^{2+}-activated force [12]. The exposure size of COOH-terminal hydrophobic patch of TnC following Ca^{2+} activation is different in cardiac/slow TnC and fast TnC [13-15], corresponding to a less interaction strength between cardiac/slow TnC and cardiac TnI switch peptide than that between fast TnC and the fast TnI switch peptide [13]. These data indicate that the differences in the structure/function of cardiac/slow TnC and fast TnC contribute to the differences in the physiological characteristics of different muscle types.

Muscle Type-Specific Expression of Troponin I Genes and Developmental Regulation

Three homologous TnI genes are present in higher vertebrates encoding three muscle fiber type-specific TnI isoforms: cardiac TnI (*TNNI3*), fast skeletal muscle TnI (*TNNI2*), and slow skeletal muscle TnI (*TNNI1*) [16-19]. The fast TnI gene is expressed exclusively in fast skeletal muscle fiber [20] and the cardiac TnI gene is expressed only in the heart [21]. During the slow to fast fiber type transition in muscle adaptation to unloading, synchronized slow to fast TnI and TnT isoform switching occurs [22, 23].

In addition to slow skeletal muscle fibers, the slow TnI gene is expressed in the embryonic and neonatal heart [24, 25]. The embryonic cardiac muscle expresses exclusively slow skeletal muscle TnI [26]. The expression of slow TnI is down-regulated during development after cardiac TnI begins to express, for example, around embryonic day 18 in the rat [24]. Timing of this TnI isoform switch during heart development in avian and mammalian species appears corresponding to the functional adaptation of the postnatal heart with the cardiac TnI completely replacing slow TnI approximately 10 days after birth [27].

The expression of different TnI isoforms in different types of muscle cells may play a role in the Ca^{2+} sensitivity, cooperativity, and pH tolerance of force development [28]. In transgenic mice over-expressing slow TnI in the cardiac muscle, the heart displayed an increased Ca^{2+} sensitivity in force development [29]. Slow TnI confers higher resistance to acidosis than that of cardiac TnI, as shown in the observation that neonatal heart muscles have more resistance to the deactivating effect compared to adult heart muscles [30]. Ca^{2+} sensitivity of tension development increased and myofilament cooperativity decreased in cardiac myocytes transfectively expressing slow TnI, further supporting functional significance of TnI isoform regulation during cardiac development [31].

Restrictive N-Terminal Truncation of Cardiac Troponin I in Cardiac Adaptation

Proteolytic degradation and fragmentation are rapid mechanisms to alter the function of proteins. Cardiac TnI is known to be sensitive to intracellular proteases, such as μ-calpain [32], and proteolytic modifications of cardiac TnI have been found during myocardial adaptation to stress conditions.

A proteolytic truncation of 19 amino acids at the COOH terminus of cardiac TnI was found during myocardial ischemia and reperfusion injury [33]. Experimental over-expression of this cardiac TnI fragment, $cTnI_{1-192}$, in transgenic mice produced myocardial stunning with systolic and diastolic dysfunctions [33]. In reconstituted troponin, $cTnI_{1-192}$ caused a significant increase in the maximum Ca^{2+}-activated actomyosin ATPase activity and increased Ca^{2+} sensitivity of both the actomyosin-ATPase activity and the Ca^{2+}-dependent sliding velocity compared with that of intact cardiac TnI [34]. When exchanged into cardiac myofibrils or permeabilized cardiomyocytes, $cTnI_{1-192}$ increased the Ca^{2+} sensitivity and decreased the kinetics of activation [35]. Consistent with the observation that the COOH terminus of TnI is a Ca^{2+}-regulated allosteric site in the function of troponin [36, 37], the C-terminal truncation of cardiac TnI is a destructive proteolysis.

A proteolytic N-terminal truncation of cardiac TnI occurs in normal cardiac muscle with increased levels during chronic hemodynamic adaptations [38]. This restrictive proteolytic N-terminal modification of cardiac TnI selectively removes the cardiac specific N-terminal extension (amino acid residues 1-30), including the two serine residues that are PKA substrates, but preserves the conserved TnI core structure. Studies of a similarly truncated recombinant cardiac TnI fragment showed no functional defect resulting from the deletion of the N-terminal extension [39]. Transgenic expression of cTnI-ND effectively rescued the lethality of mice in which the endogenous cardiac TnI gene was deleted [40]. Therefore, the selective removal of the N-terminal extension of cardiac TnI through restrictive proteolysis is not destructive but a posttranslational mechanism that regulates cardiac function in physiological and pathological adaptations.

Transgenic mouse cardiac muscle containing the N-terminal truncated cardiac TnI (cTnI-ND) showed lower affinity for Ca^{2+} in actomyosin ATPase assay, similar to the effect of isoproterenol treatment. The transgenic mouse hearts showed a faster rate of relaxation and lower left ventricular end diastolic pressure compared with controls both *in vivo* and *ex vivo* in isolated working hearts. The higher baseline relaxation rate of cTnI-ND hearts was also similar to that of β-adrenergic stimulated wild type mouse hearts [41]. These findings indicate that this restrictive N-terminal truncation is an adaptive regulation that enhances diastolic function of the heart.

cTnI-ND is up-regulated in mouse hearts with β-adrenergic deficiency due to the deletion of the α subunit of Gs (Gsα-KO) [42]. Transgenic over-expression of cTnI-ND in Gsα-KO failing mouse hearts increased the relaxation velocity and left ventricular end diastolic volume, which produced higher left ventricle maximum systolic pressure and stroke volume [42]. In the meantime, the adaptive expression of β-myosin heavy chain in adult Gsα-KO failing hearts was diminished with the over-expression of cTnI-ND [42]. The increased myocardial relaxation mimics the effect of PKA phosphorylation [41]. Since phosphorylation of cTnI at Ser^{23}/Ser^{24} by PKA is decreased in heart failure [43, 44], removal of cTnI N-terminal region provides an alternative mechanism of cardiac adaptation. Consistent with improved diastolic function, in the heart of double transgenic mice, cTnI-ND corrected the Ca^{2+} sensitization-caused diastolic dysfunction caused by a restricted cardiomyopathy mutation in cardiac TnI (cTnI-R193H) [45]. These results support the observation that up-regulation of cTnI-ND is a compensatory rather than a destructive myocardial adaption to impaired β-adrenergic signaling.

Studies have indicated that the functions of cardiac TnI may contribute to the Frank-Starling mechanism of the heart. Slow skeletal muscle TnI lacks the cardiac specific N-terminal extension [46], and replacing cTnI with slow TnI in adult transgenic mouse cardiac muscle decreased the length-dependent activation and Frank-Starling response [47]. In contrast, the removal of the N-terminal extension of cardiac TnI produces enhanced Frank-Starling response [48]. Therefore, the adaptive regulation of cardiac function through restrictive proteolysis of the N-terminal extension of cardiac TnI is based on other specific features of cardiac TnI.

Muscle Type Specific Troponin T Isoforms and Alternative Spliced Variants

Three homologous genes have evolved encoding the cardiac (*TNNT2*), slow (*TNNT1*) and fast (*TNNT3*) skeletal muscle TnTs [49-51]. By transcriptional regulation, the three TnT isoform genes are expressed in cardiac, slow and fast skeletal muscles of adult animals in a fiber type-specific manner. More complex than the regulation of the TnC and TnI genes, the expression of all three TnT isoform genes further involves alternative RNA splicing, which is regulated during muscle and heart development and/or adaptation and generates multiple protein isoforms from the transcript of each of the muscle type-specific TnT genes [46].

Encoded by constitutively expressed exons, the middle and C-terminal regions of the TnT polypeptide chain are highly conserved among the three muscle type-specific TnT isoforms and across species. Alternative RNA splicing generates an N-terminal variable region in all three TnT isoforms [49, 50, 52-54] and a C-terminal variable region specifically in fast skeletal muscle TnT [50, 55, 56]. A short segment (2-3 amino acids) connecting the middle and C-terminal regions of cardiac TnT is also encoded by an alternatively spliced exon [57, 58]. The size of vertebrate TnTs varies from 223 to 305 amino acids, largely due to the variable length of the N-terminal segment encoded by various combinations of multiple alternative exons [53].

Alternatively Spliced Exons in the Fast Skeletal Muscle TnT Gene

The mammalian fast TnT gene contains 19 exons. Exons 4, 5, 6, 7, 8 and a fetal exon are alternatively spliced encoding the highly variable N-terminal region [50, 52, 55]. Exons 16 and 17 are alternatively spliced in a mutually exclusive manner encoding the C-terminal variable region [55, 59]. Theoretically, the 8 alternative exons of fast TnT gene, among which exons 16 and 17 are mutually exclusive [50], could generate as many as 256 splicing variants. However, these alternative exons are not included randomly and not all possible splicing combinations are at a significant level detectable by cDNA cloning. For example, only 13 mouse fast TnT mRNA variants and 12 chicken fast TnT mRNA variants have been identified as the products of alternative splicing at the levels of likely physiological significance [55, 56, 59].

In addition to exons 4 – 8, avian fast skeletal muscle TnT gene has evolved with unique alternatively spliced exons. A w exon and a y exon are found between exons 4 - 5 and 7 - 8, respectively [60]. In the fast TnT gene of birds in the orders of *Galliformes* and *Craciformes*, four to seven P (pectoral) exons located between exons 5 and 6 encode a Tx segment [56, 61, 62] that is a cluster of 4-7 tandem repeats of pentapeptides (AHH[A/E]E) forming 4-7 His pairs in an α-helix, which bind transition metal ions Cu(II), Ni(II), Co(II) and Zn(II) with high affinity [63]. No homologous counterpart was found in mammalian fast TnT gene and avian slow skeletal muscle and cardiac TnT genes.

Alternatively Spliced Exons in the Cardiac TnT Gene

The mammalian cardiac TnT gene contains 14 constitutively expressed exons and 3 alternatively spliced exons [57, 58, 64]. Exon 5 encoding 9 or 10 amino acids in the N-terminal variable region is included in embryonic but not adult cardiac TnT [54]. Independent of developmental stages, exon 4 in the N-terminal region and exon 13 in between the middle and C-terminal regions are alternatively spliced in mammalian cardiac TnT [57, 58]. The avian cardiac TnT gene contains 16 constitutively spliced exons and only 1 alternative exon (the embryonic exon 5), whereas exon 4 was constitutively expressed and no counterpart of mammalian exon 13 was found [49]. Altogether, eight mammalian and two avian cardiac TnT alternative splicing variants have been found in normal cardiac muscle by cDNA cloning and sequencing.

Alternatively Spliced Exons in the Slow Skeletal Muscle TnT Gene

The slow TnT gene contains 14 exons, fewer than that of the fast TnT and cardiac TnT genes. Exon 5 encoding an N-terminal segment is alternatively spliced to generate two protein isoforms [51, 53, 65]. Splicing at alternative acceptor sites in the intron 5 of mouse slow TnT gene generates another one amino acid variation in the exon 6-encoded segment [51]. With an exon-intron organization same as that of the mammalian slow TnT genes, chicken slow TnT gene is significantly smaller (3-kb versus 9-kb) by containing shorter intron sequences [51,66]. Exon 5 of chicken slow TnT gene is alternatively spliced and two alternative acceptor sites in intron 4 are utilized [66].

The above alternative splicing patterns represent protein variants expressed at significant amount and thus with functional significance [46]. Splicing errors of TnT transcript have also been observed. The two mutually exclusive C-terminal exons in fast TnT was found both missing in fast TnT mRNA at low frequency, which results in a reading frame change after the exon 16 encoded segment, producing a non-functional COOH terminus in the fast TnT protein [67]. A human slow TnT cDNA with an additional 48 nucleotides inserted between the exon 11 and exon 12 encoded segments was cloned [65]. Genomic sequence analysis indicated that an alternative receptor site in intron 11 of the slow TnT gene is responsible for the inclusion of a portion of the 3'-region of intron 11 (exon 12') into the mRNA [51]. However, no corresponding slow TnT protein was detectable in muscles of multiple species examined [53]. Therefore, this alternative splicing product suggests a splicing error with minimum functional impact.

Expression of TnT Isoform Genes in Developmental Heart and Skeletal Muscle

The expression of TnT isoform genes in embryonic striated muscle is not as restricted to fiber types as that in the adult. Cardiac TnT is expressed at significant levels in embryonic skeletal muscles [27, 68-70]. *In situ* hybridization showed that the expression of cardiac TnT in the developing heart begins at day 7.5 postcoitum and in skeletal muscles at day 11.75 postcoitum [71]. The cardiac TnT gene is down-regulated in skeletal muscle during perinatal development and ceases expression in adult skeletal muscle [68, 72, 73]. The developmentally regulated expression of cardiac TnT in embryonic skeletal muscle occurs in both avian and mammalian species [27, 70, 7 4, 75].

Transient expression of slow TnT, but not fast TnT, was found in the embryonic heart [71, 76]. Slow TnT mRNA is detected in all skeletal muscles at 10-14 days of mouse gestation and was then rapidly down-regulated, whereas it is also detected in fetal mouse heart at 10 days of gestation and peaks at 15 days [76]. Similarly, transient regional expression of slow TnT is found in human fetal and rat fetal hearts [77]. At day 13.5 postcoitum, expressions of all three TnT genes were detected by *in situ* hybridization in the developing mouse tongue and this co-expression pattern continued to day 16.5 postcoitum with the fast TnT being predominant [71]. Cardiac TnT transcript was detected in embryonic mouse urinary bladder [71]. The significance of this low level expression of cardiac TnT in embryonic bladder smooth muscle remains to be investigated. microRNAs may play a role in regulating striated muscle development and pathophysiological remodeling [78-81]. In adult mouse heart, the deletion of miR-208a increased the expressions of fast TnI and fast TnT, which could be corrected by over-expression of miR-499 [81]. In skeletal muscle, deletion of both miR-208b and miR-499 reduced the number of slow fibers [81].

The expressions of TnI and TnT isoform genes are coupled in a muscle fiber type-specific manner [23, 82]. Cardiac TnC was co-expressed with cardiac TnT at early developmental stages in chicken skeletal muscle [74]. During the development of avian skeletal muscle, cardiac TnT and cardiac TnC are down-regulated and skeletal muscle troponin subunits are up-regulated, a process dependent on diffusible neurohumoral factors but not innervation [69].

An interesting exception is that the cardiac muscle of toads (**Bufo**) that live in dry land environment expresses exclusively slow skeletal muscle TnT together with cardiac forms of TnI and myosin. In contrast, the heart of frogs (*Rana*), a closely related amphibian species living in water environment, expresses normal cardiac TnT [83]. Functional comparison of *ex vivo* working hearts of toad and frog revealed that toad hearts have higher contractile and relaxation velocities and are able to pump against a significantly higher afterload than that of frog hearts [83]. The selective utilization of slow TnT in an evolutionary adaptation of the toad cardiac muscle demonstrates a role of the TnT isoform diversity in determining muscle contractility. The function of slow versus cardiac TnT for the heart of dry land toads to sustain circulation in a wide range of body fluid changes indicates a molecular mechanism to improve systolic function and suggests a potential target for the development of new treatment of heart failure.

Developmental Regulation of TnT Expression Through Alternative RNA Splicing

Many of the alternatively spliced exons of the three TnT isoform genes exhibit a developmentally regulated expression. Combinations of inclusion or exclusion of these exons produce a large number of splice-forms of TnT proteins during heart and skeletal muscle development with functional significance [84].

The N-terminal coding exon 5 of cardiac TnT is alternatively spliced during the perinatal development of avian and mammalian hearts [49, 57]. The inclusion or exclusion of 9 or 10 amino acids encoded by exon 5 in the N-terminal variable region results in a developmental switching from a larger embryonic cardiac TnT to a smaller adult cardiac TnT [49, 54, 57, 58]. The exon 5-encoded segment is highly acidic (negatively charged at physiological pH) and, therefore, this alternative splicing regulation also generates a more acidic to less acidic switch of cardiac TnT [85].

The time course of this developmental switch has been described for chicken, mouse and rat hearts. Cardiac TnT cDNAs with the same embryonic and adult splicing patterns are also found in human hearts [86] and the same protein isoform switch was seen in developing human skeletal muscles where cardiac TnT is transiently expressed [68]. It was intriguing that the embryonic to adult cardiac TnT switching occurs earlier in mouse than that in rat (transition time at 16-17 days gestation versus 5-6 days postnatal), while the perinatal development and functional switching of mouse and rat hearts are largely similar [27]. In contrast, the transition time of slow TnI to cardiac TnI switching is at 5-6 days postnatal in the hearts of both rat and mouse, reflecting a correlation to the postnatal changes in cardiac function. The notion is that a genetically determined program, other than functional demands, determines the alternative splicing of cardiac TnT during development. This observation is supported by the fact that when cardiac TnT gene is transiently expressed in neonatal skeletal muscle, the alternative splicing pattern is synchronized with that in the heart [27]. The developmentally synchronized alternative splicing is also found between cardiomyocytes residing in mouse pulmonary and azygos veins and the heart, although they are anatomically isolated [87].

Significantly more complex alternative splicing patterns are present for fast skeletal muscle TnT during development, involving multiple coding exons for the N-terminal and C-terminal variable regions. Similar to that of cardiac TnT, a fetal exon located between the alternative exon 8 and constitutive exon 9 is found in mammalian fast TnT genes [52]. Involving the fetal exon and other N-terminal alternative exons (4, 6, 7 and 8), which all encode mainly acidic residues, the expression of fast TnT also shows a developmental switching from high molecular weight acidic isoforms to low molecular weight basic isoforms [55].

The alternative splicing of the two mutually exclusive exons (exons 16 and 17) of the fast TnT gene encoding a 14 amino acids segment in the C-terminal region is also under developmental regulation. cDNA cloning of mouse and chicken fast TnT mRNA found that exon 16 was expressed only in adult muscles, whereas all embryonic and neonatal fast skeletal muscles TnT mRNA contained exon 17 [55, 67]. The amino acid sequence of the fast TnT exon 17-encoded segment is more similar to that of the counterparts in cardiac TnT and slow skeletal TnT than that of exon 16 [67].

While the developmental switches of cardiac TnT and mammalian fast TnT both represent splice-out of embryonic/fetal exons in adult isoforms, a unique phenomenon of the developmental regulation of fast TnT gene is the post-hatching up-regulated splice-in of multiple P exons encoding a unique Tx segment in the N-terminal region of pectoral but not leg muscles of avian species in the orders of *Galliformes* and *Craciformes* [59, 71]. In the meantime, the developmental splice-out of the other 5'-alternative exons of avian fast TnT is retained similar to that of mammalian fast TnT.

Functional significance of the specific splice-in of the Tx segment in adult pectoral muscles of those avian species is not fully understood. While the large number of Glu residues encoded by the P exons may serve as a calcium reservoir in the thin filaments [88], and the negative charges added to the N-terminal variable region of Tx-positive fast TnT correlated to an increased tolerance to acidosis [59], the biological significance of its binding to transition metal ions remains to be investigated.

Along with the developmentally regulated expression of TnT isoform genes, alternative splicing adds a major mechanism for functional modifications of the contractility of cardiac and skeletal muscles [46]. The most significant difference between the embryonic and adult isoforms is in the N-terminal variable region (Table 1) that plays a role in modulating the overall conformation of TnT and the interactions with TnI, TnC and tropomyosin. When reconstituted into skinned cardiac muscle strips, embryonic cardiac TnT with more negative N-terminal charge increased Ca^{2+} sensitivity of myosin ATPase and force development in comparison to that of the adult cardiac TnT with less negatively N-terminal charge [89]. The co-existence of slow skeletal muscle TnI in embryonic heart may also contribute to this effect [93]. Consistently, studies using reconstituted myofilaments showed that the embryonic cardiac TnT produced higher Ca^{2+} sensitivity as compared with that of adult cardiac TnT [54, 90]. Similarly, experiments using skinned chicken skeletal muscle fibers showed that adult pectoral muscle containing fast TnT with more negatively charged residues in the N-terminal variable region due to the inclusion of the Tx segment exerted higher myofilament calcium sensitivity than control muscles containing fast TnT with less N-terminal negative charges [91-93].

Table 1. The N-terminal hypervariable region of mammalian and avian TnT

Exons alternatively spliced under physiological conditions:
A fetal exon in mammalian embryonic fast TnT
Exons 4, 5, 6, 7 and 8 in avian and mammalian fast TnT
P exons (up to 7) in avian fast TnT (encoding the Tx segment)
Exons w and y in avian fast TnT
Exon 5 in avian and mammalian cardiac TnT
Exon 4 in mammalian cardiac TnT
Exon 5 in mammalian slow TnT
Alternative acceptor sites in intron 5 of mouse and intron 4 of chicken slow TnT genes
Developmental regulations:
Alternative splicing of exon 5 in cardiac TnT results in isoform switch in heart
Alternative splicing of exons 4, 6, 7, 8 and fetal in fast TnT results in isoform switches in skeletal muscles
Postnatal inclusion of P exons in avian pectoral muscle fast TnT
Posttranslational modifications:
Constitutive phosphorylation of Ser_2 at the N terminus
Selective removal of the N-terminal variable region by restrictive proteolysis for adaptation to stress conditions
Functional significance
Variable N-terminal negative charges determine the overall charge of TnT
N-terminal structures modulate the conformation and function of the middle and C-terminal regions of TnT
TnT N-terminal variation alters thin filament Ca^{2+}-sensitivity and force production
N-terminal acidic residues of TnT provide a binding capacity for Ca^{2+}
Deletion of the N-terminal variable region of cardiac TnT increases heart efficiency
Pathological relevance:
Exclusion of exon 4 in adult cardiac TnT in human heart failure
Increased exclusion of exon 4 in adult cardiac TnT in diabetic or hypertrophic rat hearts

Inclusion of the fetal exon-encoded segment in the N-terminal variable region of embryonic fast TnT had an inhibitory effect on the level of myosin ATPase activity in reconstituted myofilaments [94]. Embryonic and neonatal cardiac muscle containing embryonic cardiac TnT exhibited higher tolerance to acidosis [95]. In contrast, overexpression of fast skeletal muscle TnT that has a less negatively charged N-terminal segment than that of cardiac TnT decreased the tolerance to acidosisin transgenic mouse cardiac muscle [96].

The molecular and cellular mechanisms that regulate TnT alternative splicing during development remain to be established. Both *cis* and *trans* regulatory factors have been implicated to affect the alternative splicing of cardiac TnT [97]. Alternative splicing of fast TnT was found in myogenesis. Muscle-specific *trans* regulatory factors were required for appropriate splicing and incorporation of constitutive and alternative exons of fast TnT during myotube differentiation in culture [98].

The Functional Importance of Slow Skeletal Muscle TnT

A nonsense mutation at codon Glu^{180} in the slow skeletal muscle TnT gene (*TNNT1*) causes an inherited nemaline myopathy in the Amish people with a recessive postnatal lethal phenotype [99]. Due to the premature termination of the slow TnT polypipetide chain, the binding sites for TnI and TnC as well as the T2 region tropomyosin binding site are lost. This destruction causes inability to form troponin complex and drastically decreased binding affinity for tropomyosin [100]. The truncated fragment of slow TnT was not detectable in patient's muscles, indicating the loss of the T2 binding site for tropomyosin abolished myofilament incorporation. It has been demonstrated that non-incorporated or over synthesized myofilament proteins are degraded in muscle cells [101]. This finding is consistent with the recessive phenotype of the Amish nemaline myopathy, in which the neonatal lethality of the loss of only one isoform of TnT demonstrates the non-redundant nature and functional importance of slow TnT in skeletal muscle.

The TnI and TnT genes are closely linked in vertebrate genomes [102] and the deletion of the upstream cardiac TnI in a gene knockout mouse line [103] results in the loss of a portion of the enhancer region of the slow TnT gene [104]. Using transgenic expression of a cardiac TnI transgene to rescue the postnatal lethality, the double transgenic mice demonstrated an effective knockdown of slow TnT expression [40]. The deficiency of slow TnT caused a slow-to-fast muscle fiber type switch in the diaphragm muscle and significantly decreased fatigue tolerance and hindered recovery [103]. The data demonstrate the critical role of slow TnT in the slow fiber-based function of skeletal muscles.

Alternative Splicing of TnT in Cardiomyopathy and Adaptation

Aberrant alternative splicing of cardiac TnT is found in animal cardiomyopathies. Abnormal skipping of exon 8 in cardiac TnT occurs in turkey hearts with inherited dilated cardiomyopathy [105]. Counterpart of this exon (exon 7) in mammalian cardiac TnT was abnormally spliced out in dog hearts with dilated cardiomyopathy [106]. This N-terminal region coding exon is constitutively included in normal cardiac TnT [46]. Its aberrant splice-out in dilated cardiomyopathy turkey and dog hearts indicates a causal relationship to the pathogenesis. Over-expression of exon 7-deleted cardiac TnT to replace normal cardiac TnT in transgenic mouse hearts severely decreased systolic function [107]. In addition to the splice-out of exon 7, dilated cardiomyopathy dog hearts showed an abnormal inclusion of the embryonic exon 5 in cardiac TnT in the adult cardiac muscle [106].

Alternative splicing of exon 4 that encodes 4-5 amino acids in the N-terminal variable region of cardiac TnT has also been correlated to disease conditions. Significant expression of low molecular weight cardiac TnT excluding exon 4 was found in failing human hearts [108, 109], diabetic rat hearts [110], and hypertrophic rat hearts [43] (the abnormally increased exclusion of exon 4 in cardiac TnT was mistakenly quoted in some old cardiology textbooks as re-expression of fetal cardiac TnT in failing human hearts).

These aberrant splicing of the N-terminal variable region of cardiac TnT do not abolish the core function of TnT. Skeletal muscles normally contain multiple alternatively spliced variants of fast and slow TnT. Therefore, the mechanism for the aberrantly spliced cardiac TnT to contribute to the pathogenesis of dilated cadiomyopathy in turkeys and dogs, and potentially in failing human hearts, required an explanation. An important feature of vertebrate hearts is the synchronized uniform ventricular contraction activated as an electrical syncytium. It is logical to expect that uniform myofilament activation including uniform effect of TnT on Ca^{2+}-sensitivity would be critical to the ventricular function of rhythm pumping. This is very different from the function of skeletal muscles, in which multiple TnT isoforms corresponding to a widened range of myofilament Ca^{2+}-sensitivity would be beneficial for producing broader twitches to fuse into tetanic contractions.

Therefore, the abnormality resulting from the presence of aberrantly spliced cardiac TnT is not a loss of function but is the chronic presence of more than one class of TnT in the thin filaments of adult cardiac muscle [111]. In this hypothesis, desynchronized activation of ventricular muscle at the myofilament level due to the co-existence of TnT variants that produce split Ca^{2+} sensitivity would decrease the efficiency of ventricular pumping. This mechanism was first demonstrated using a model of transgenic mouse heart that co-express a wild type fast skeletal muscle TnT together with the endogenous cardiac TnT. The co-existence of two non-mutant variants of TnT in adult cardiac muscle altered the overall cooperativity of Ca^{2+} activated force production [112], decreased cardiac function and produced chronic myocardial degeneration [113]. Further studies in transgenic mouse hearts that co-expressing one or two of the cardiac TnT splicing variants found in turkey and canine dilated cardiomyopathy with endogenous wild type adult cardiac TnT showed decreased cardiac efficiency proportional to the degree of TnT heterogeneity [111].

Abnormal inclusion of the embryonic exon 5 in adult cardiac TnT was also found in cat and Guinea pig hearts [106]. In addition, the Guinea pig hearts expresses cardiac TnT with an exclusion of the segment encoded by exon 6 in the N-terminal region [106]. Cat and Guinea pigs are both reported to have high incidence of inherited cardiomyopathy and heart failure [114, 115]. Therefore, aberrant splicing of N-terminal exons of cardiac TnT might be a common pathogenic mechanism. In contrast to the above cases of chronic TnT heterogeneity, the co-existence of embryonic and adult cardiac TnT in neonatal developing hearts is only a transient brief event with no obvious negative impact.

The functional effect of the alternative splicing-generated decreases of size and negative charge in the N-terminal variable region of cardiac TnT suggests a mechanism of functional adaptation to pathological conditions. An example is that the low molecular weight slow TnT with exon 5 excluded was up-regulated in type 1 (demyelination) but not type 2 (axon loss) Charcot-Marie-Tooth disease, suggesting a role of TnT alternative splicing in skeletal muscle adaptation to neuromuscular disorders [116].

Restrictive N-Terminal Truncation of Cardiac TnT in Myocardial Adaptation

Several experimental models of TnT and TnI over-expression or under-expression demonstrated that myofilament incorporation determines the stoichiometry of troponin subunits in cardiac myocytes [40]. It was documented in dog hearts that TnT and TnI both have rapid turnover rates in cardiac myocytes with half-life of approximately 3.5 days, which was shorter than the 5.3 days half-life of TnC [117]. The effective removal of surplus TnT and TnI over-expressed in transgenic mouse hearts under the strong alpha myosin heavy chain promoter suggests that a potent proteolytic clearance of non-myofilament-incorporated TnT and TnI is critical to maintain the integrity and protein stoichiometry of the thin filament regulatory system [40]. Consistently, no cytoplasmic pool of non-myofilament-incorporated cardiac TnT was detected [117].

Proteolytic fragmentation of TnT has been reported under pathological conditions. Hypoxia in canine diaphragm muscle produced a truncated 28-kDa TnT fragment [118]. A cleavage of cardiac TnT by caspase 3 was found in apoptotic rat cardiomyocytes to generate a 25-kDa fragment with a loss of the N-terminal variable region plus a partial deletion of the middle conserved region. This destructive proteolysis of cardiac TnT decreased the maximum myosin ATPase activity and myofibril force generation [119].

A restrictive proteolysis was found to produce an N-terminal truncation of cardiac TnT. Different from the destructive cleavage by caspase 3 [119], the restrictive proteolysis selectively removes only the N-terminal variable region (amino acid 1-71 in mouse cardiac TnT) and preserves the conserved regions of cardiac TnT. The restrictive N-terminal truncation of cardiac TnT has been identified in mouse, rat and pig hearts during acute ischemia-reperfusion [120] and pressure overload [121]. Myofilament-associated calpain I [122] has been indicated with a role in this restrictive modification of cardiac TnT [120].

The selective deletion of the N-terminal variable region does not destroy the function of TnT but altered the binding affinities for TnI, TnC and tropomyosin [123]. Studies by several laboratories showed that such selective removal of the N-terminal variable region of TnT slightly but significantly decreased the maximum myosin ATPase activity and myofibril force generation without affecting thin filament calcium sensitivity and cooperativity [124-126]. Transgenic mouse hearts over-expressing cardiac TnT lacking the N-terminal region exhibited a moderate but clearly detectable decrease in the velocity of ventricular contraction without decreasing the maximum left ventricular pressure [121]. This reduced contractile velocity consequently extended the rapid ejection phase of ventricular pumping cycle to effectively increase the stroke volume. This mechanism provides a plausible functional benefit to compensate for imbalances between systolic power and workload such as that occur in myocardial ischemia and ventricular pressure overload [121].

The cellular mechanism that produces the restrictive N-terminal truncation of cardiac TnT remains to be investigated. A recent study demonstrated that fatigue-stress conditions induced restrictive N-terminal truncation of fast skeletal muscle TnT [127]. Experimental treatment of transgenic mouse cardiac muscle expressing fast skeletal muscle TnT induced restrictive N-terminal truncation of both fast TnT and endogenous cardiac TnT despite their different amino acid sequences at the cleavage sites[120]. Therefore, the restrictive calpain I cleavage of cardiac TnT under stress conditions is likely regulated by the accessibility of calpain and/or the molecular conformation of TnT other than an increase in the overall elevation of calpain activity.

Conclusion

Troponin plays a central role in the Ca^{2+}-mediated regulation of muscle contraction and relaxation. The expression of muscle fiber type-specific troponin isoform genes is regulated during heart and skeletal muscle development and functional adaptations. A transcriptional regulation of TnI during heart development is the switch from slow skeletal muscle isoform in the embryo to the cardiac isoform in the adult. The major difference between the two TnI isoforms is the adult heart-specific N-terminal extension of cardiac TnI, which adds mechanisms for modifying cardiac function under PKA phosphorylation as well as restrictive proteolysis during physiological and pathological adaptations. Another major mechanism of troponin regulation during development and adaptation is the N-terminal variable region of TnT, which differs among the muscle type-specific TnT isoforms, undergoes complex alternative splicing, and is regulated through restrictive proteolysis. By altering the overall conformation of TnT with functional impacts, the developmental and adaptive regulation of TnT N-terminal structure confers fine-tuning of the function of troponin and the contractility of muscle. To further understand the function and molecular mechanisms of troponin regulation during development and adaptation will help the development of novel approaches for the treatment of various muscle diseases and heart failure.

References

[1] Parmacek, M. S., Karpinski, B. A., Gottesdiener, K. M., Thompson, C. B., and Leiden, J. M. (1989) *Nucleic acids research* 17, 1915-1931

[2] Parmacek, M. S., Bengur, A. R., Vora, A. J., and Leiden, J. M. (1990) *The Journal of biological chemistry* 265, 15970-15976

[3] Johnson, J. D., Collins, J. H., Robertson, S. P., and Potter, J. D. (1980) *The Journal of biological chemistry* 255, 9635-9640

[4] Holroyde, M. J., Robertson, S. P., Johnson, J. D., Solaro, R. J., and Potter, J. D. (1980) *The Journal of biological chemistry* 255, 11688-11693

[5] McKay, R. T., Tripet, B. P., Hodges, R. S., and Sykes, B. D. (1997) *The Journal of biological chemistry* 272, 28494-28500

[6] Dedman, J. R., Potter, J. D., Jackson, R. L., Johnson, J. D., and Means, A. R. (1977) *The Journal of biological chemistry* 252, 8415-8422

[7] van Eerd, J. P., and Takahashi, K. (1975) *Biochemical and biophysical research communications* 64, 122-127

[8] Leavis, P. C., and Kraft, E. L. (1978) *Archives of biochemistry and biophysics* 186, 411-415

[9] Stoutamyer, A., and Dhoot, G. K. (2005) *Journal of muscle research and cell motility* 26, 237-245

[10] Prigozy, T. I., Dalrymple, K., Shuler, C., and Kedes, L. (1997) *Developmental dynamics : an official publication of the American Association of Anatomists* 209, 36-44

[11] Li, M. X., Gagne, S. M., Spyracopoulos, L., Kloks, C. P., Audette, G., Chandra, M., Solaro, R. J., Smillie, L. B., and Sykes, B. D. (1997) *Biochemistry* 36, 12519-12525

[12] Moreno-Gonzalez, A., Gillis, T. E., Rivera, A. J., Chase, P. B., Martyn, D. A., and Regnier, M. (2007) *The Journal of physiology* 579, 313-326

[13] Li, M. X., Spyracopoulos, L., and Sykes, B. D. (1999) *Biochemistry* 38, 8289-8298

[14] Gagne, S. M., Tsuda, S., Li, M. X., Smillie, L. B., and Sykes, B. D. (1995) *Nature structural biology* 2, 784-789

[15] Sia, S. K., Li, M. X., Spyracopoulos, L., Gagne, S. M., Liu, W., Putkey, J. A., and Sykes, B. D. (1997) *The Journal of biological chemistry* 272, 18216-18221

[16] Hallauer, P. L., Bradshaw, H. L., and Hastings, K. E. (1993) *Development* 119, 691-701

[17] Zhu, L., Lyons, G. E., Juhasz, O., Joya, J. E., Hardeman, E. C., and Wade, R. (1995) *Developmental biology* 169, 487-503

[18] Guenet, J. L., Simon-Chazottes, D., Gravel, M., Hastings, K. E., and Schiaffino, S. (1996) *Mammalian genome : official journal of the International Mammalian Genome Society* 7, 13-15

[19] Hastings, K. E. (1997) *Cell structure and function* 22, 205-211

[20] Koppe, R. I., Hallauer, P. L., Karpati, G., and Hastings, K. E. (1989) *The Journal of biological chemistry* 264, 14327-14333

[21] Ausoni, S., De Nardi, C., Moretti, P., Gorza, L., and Schiaffino, S. (1991) *Development* 112, 1041-1051

[22] Stevens, L., Bastide, B., Kischel, P., Pette, D., and Mounier, Y. (2002) *American journal of physiology. Cell physiology* 282, C1025-1030

[23] Yu, Z. B., Gao, F., Feng, H. Z., and Jin, J. P. (2007) *American journal of physiology. Cell physiology* 292, C1192-1203

[24] Saggin, L., Gorza, L., Ausoni, S., and Schiaffino, S. (1989) *The Journal of biological chemistry* 264, 16299-16302

[25] Sabry, M. A., and Dhoot, G. K. (1989) *Journal of muscle research and cell motility* 10, 85-91

[26] Sasse, S., Brand, N. J., Kyprianou, P., Dhoot, G. K., Wade, R., Arai, M., Periasamy, M., Yacoub, M. H., and Barton, P. J. (1993) *Circulation research* 72, 932-938

[27] Jin, J. P. (1996) *Biochemical and biophysical research communications* 225, 883-889

[28] Westfall, M. V., Albayya, F. P., and Metzger, J. M. (1999) *The Journal of biological chemistry* 274, 22508-22516

[29] Fentzke, R. C., Buck, S. H., Patel, J. R., Lin, H., Wolska, B. M., Stojanovic, M. O., Martin, A. F., Solaro, R. J., Moss, R. L., and Leiden, J. M. (1999) *The Journal of physiology* 517 (Pt 1), 143-157

[30] Westfall, M. V., Albayya, F. P., Turner, II, and Metzger, J. M. (2000) *Circulation research* 86, 470-477

[31] Westfall, M. V., and Metzger, J. M. (2001) *News in physiological sciences : an international journal of physiology produced jointly by the International Union of Physiological Sciences and the American Physiological Society* 16, 278-281

[32] Di Lisa, F., De Tullio, R., Salamino, F., Barbato, R., Melloni, E., Siliprandi, N., Schiaffino, S., and Pontremoli, S. (1995) *The Biochemical journal* 308 (Pt 1), 57-61

[33] McDonough, J. L., Arrell, D. K., and Van Eyk, J. E. (1999) *Circulation research* 84, 9-20

[34] Foster, D. B., Noguchi, T., VanBuren, P., Murphy, A. M., and Van Eyk, J. E. (2003) *Circulation research* 93, 917-924

[35] Narolska, N. A., Piroddi, N., Belus, A., Boontje, N. M., Scellini, B., Deppermann, S., Zaremba, R., Musters, R. J., dos Remedios, C., Jaquet, K., Foster, D. B., Murphy, A. M., van Eyk, J. E., Tesi, C., Poggesi, C., van der Velden, J., and Stienen, G. J. (2006) *Circulation research* 99, 1012-1020

[36] Pirani, A., Vinogradova, M. V., Curmi, P. M., King, W. A., Fletterick, R. J., Craig, R., Tobacman, L. S., Xu, C., Hatch, V., and Lehman, W. (2006) *Journal of molecular biology* 357, 707-717

[37] Jin, J. P., Yang, F. W., Yu, Z. B., Ruse, C. I., Bond, M., and Chen, A. (2001) *Biochemistry* 40, 2623-2631

[38] Yu, Z. B., Zhang, L. F., and Jin, J. P. (2001) *The Journal of biological chemistry* 276, 15753-15760

[39] Guo, X., Wattanapermpool, J., Palmiter, K. A., Murphy, A. M., and Solaro, R. J. (1994) *The Journal of biological chemistry* 269, 15210-15216

[40] Feng, H. Z., Hossain, M. M., Huang, X. P., and Jin, J. P. (2009) *Archives of biochemistry and biophysics* 487, 36-41

[41] Barbato, J. C., Huang, Q. Q., Hossain, M. M., Bond, M., and Jin, J. P. (2005) *The Journal of biological chemistry* 280, 6602-6609

[42] Feng, H. Z., Chen, M., Weinstein, L. S., and Jin, J. P. (2008) *The Journal of biological chemistry* 283, 33384-33393

[43] McConnell, B. K., Moravec, C. S., and Bond, M. (1998) *The American journal of physiology* 274, H385-396

[44] Zakhary, D. R., Moravec, C. S., Stewart, R. W., and Bond, M. (1999) *Circulation* 99, 505-510

[45] Li, Y., Charles, P. Y., Nan, C., Pinto, J. R., Wang, Y., Liang, J., Wu, G., Tian, J., Feng, H. Z., Potter, J. D., Jin, J. P., and Huang, X. (2010) *Journal of molecular and cellular cardiology* 49, 402-411

[46] Jin, J. P., Zhang, Z., and Bautista, J. A. (2008) *Critical reviews in eukaryotic gene expression* 18, 93-124

[47] Nowak, G., Pena, J. R., Urboniene, D., Geenen, D. L., Solaro, R. J., and Wolska, B. M. (2007) *Journal of muscle research and cell motility* 28, 415-419

[48] Feng, H. Z., Huang, X. P., and Jin, J. P. (2010) *Biophys J* 98, 357a

[49] Cooper, T. A., and Ordahl, C. P. (1985) *The Journal of biological chemistry* 260, 11140-11148

[50] Breitbart, R. E., and Nadal-Ginard, B. (1986) *Journal of molecular biology* 188, 313-324

[51] Huang, Q. Q., Chen, A., and Jin, J. P. (1999) *Gene* 229, 1-10

[52] Briggs, M. M., and Schachat, F. (1993) *Developmental biology* 158, 503-509

[53] Jin, J. P., Chen, A., and Huang, Q. Q. (1998) *Gene* 214, 121-129

[54] Jin, J. P., and Lin, J. J. (1989) *The Journal of biological chemistry* 264, 14471-14477

[55] Wang, J., and Jin, J. P. (1997) *Gene* 193, 105-114

[56] Smillie, L. B., Golosinska, K., and Reinach, F. C. (1988) *The Journal of biological chemistry* 263, 18816-18820

[57] Jin, J. P., Huang, Q. Q., Yeh, H. I., and Lin, J. J. (1992) *Journal of molecular biology* 227, 1269-1276

[58] Jin, J. P., Wang, J., and Zhang, J. (1996) *Gene* 168, 217-221

[59] Ogut, O., and Jin, J. P. (1998) *The Journal of biological chemistry* 273, 27858-27866

[60] Schachat, F., Schmidt, J. M., Maready, M., and Briggs, M. M. (1995) *Developmental biology* 171, 233-239
[61] Miyazaki, J., Jozaki, M., Nakatani, N., Watanabe, T., Saba, R., Nakada, K., Hirabayashi, T., and Yonemura, I. (1999) *Journal of muscle research and cell motility* 20, 655-660
[62] Jin, J. P., and Samanez, R. A. (2001) *Journal of molecular evolution* 52, 103-116
[63] Jin, J. P., and Smillie, L. B. (1994) *FEBS letters* 341, 135-140
[64] Farza, H., Townsend, P. J., Carrier, L., Barton, P. J., Mesnard, L., Bahrend, E., Forissier, J. F., Fiszman, M., Yacoub, M. H., and Schwartz, K. (1998) *Journal of molecular and cellular cardiology* 30, 1247-1253
[65] Gahlmann, R., Troutt, A. B., Wade, R. P., Gunning, P., and Kedes, L. (1987) *The Journal of biological chemistry* 262, 16122-16126
[66] Hirao, C., Yonemura, I., and Miyazaki, J. (2004) *Gene* 338, 243-256
[67] Jin, J. P., Wang, J., and Ogut, O. (1998) *Biochemical and biophysical research communications* 242, 540-544
[68] Jin, J. P., Brotto, M. A., Hossain, M. M., Huang, Q. Q., Brotto, L. S., Nosek, T. M., Morton, D. H., and Crawford, T. O. (2003) *The Journal of biological chemistry* 278, 26159-26165
[69] Toyota, N., and Shimada, Y. (1983) *Cell* 33, 297-304
[70] Cooper, T. A., and Ordahl, C. P. (1984) *Science (New York, N.Y.)* 226, 979-982
[71] Wang, Q., Reiter, R. S., Huang, Q. Q., Jin, J. P., and Lin, J. J. (2001) *The Anatomical record* 263, 72-84
[72] Saggin, L., Gorza, L., Ausoni, S., and Schiaffino, S. (1990) *Development* 110, 547-554
[73] Sabry, M. A., and Dhoot, G. K. (1991) *Journal of muscle research and cell motility* 12, 447-454
[74] Toyota, N., and Shimada, Y. (1981) *The Journal of cell biology* 91, 497-504
[75] Swiderski, R. E., and Solursh, M. (1990) *Developmental biology* 140, 73-82
[76] Krishan, K., Morgan, M. J., Zhao, W., and Dhoot, G. K. (2000) *Journal of muscle research and cell motility* 21, 527-536
[77] Barton, P. J., Felkin, L. E., Koban, M. U., Cullen, M. E., Brand, N. J., and Dhoot, G. K. (2004) *Molecular and cellular biochemistry* 263, 91-97
[78] Tatsuguchi, M., Seok, H. Y., Callis, T. E., Thomson, J. M., Chen, J. F., Newman, M., Rojas, M., Hammond, S. M., and Wang, D. Z. (2007) *Journal of molecular and cellular cardiology* 42, 1137-1141
[79] van Rooij, E., and Olson, E. N. (2007) *J Clin Invest* 117, 2369-2376
[80] van Rooij, E., Sutherland, L. B., Qi, X., Richardson, J. A., Hill, J., and Olson, E. N. (2007) *Science (New York, N.Y.)* 316, 575-579
[81] van Rooij, E., Quiat, D., Johnson, B. A., Sutherland, L. B., Qi, X., Richardson, J. A., Kelm, R. J., Jr., and Olson, E. N. (2009) *Dev Cell* 17, 662-673
[82] Brotto, M. A., Biesiadecki, B. J., Brotto, L. S., Nosek, T. M., and Jin, J. P. (2006) *American journal of physiology. Cell physiology* 290, C567-576
[83] Feng, H. Z., Chen, X., Hossain, M. M., and Jin, J. P. (2012) *The Journal of biological chemistry* 287, 29753-29764
[84] Wei, B., and Jin, J. P. (2011) *Archives of biochemistry and biophysics* 505, 144-154
[85] Jin, J. P., and Lin, J. J. (1988) *The Journal of biological chemistry* 263, 7309-7315

[86] Townsend, P. J., Barton, P. J., Yacoub, M. H., and Farza, H. (1995) *Journal of molecular and cellular cardiology* 27, 2223-2236
[87] Kracklauer, M. P., Feng, H. Z., Jiang, W., Lin, J. L., Lin, J. J., and Jin, J. P. (2013) *The FEBS journal* 280, 880-891
[88] Zhang, Z., Jin, J. P., and Root, D. D. (2004) *Biochemistry* 43, 2645-2655
[89] Gomes, A. V., Guzman, G., Zhao, J., and Potter, J. D. (2002) *The Journal of biological chemistry* 277, 35341-35349
[90] Gomes, A. V., Venkatraman, G., Davis, J. P., Tikunova, S. B., Engel, P., Solaro, R. J., and Potter, J. D. (2004) *The Journal of biological chemistry* 279, 49579-49587
[91] Reiser, P. J., Greaser, M. L., and Moss, R. L. (1992) *The Journal of physiology* 449, 573-588
[92] Reiser, P. J., Greaser, M. L., and Moss, R. L. (1996) *The Journal of physiology* 493 (Pt 2), 553-562
[93] Ogut, O., Granzier, H., and Jin, J. P. (1999) *The American journal of physiology* 276, C1162-1170
[94] Chaudhuri, T., Mukherjea, M., Sachdev, S., Randall, J. D., and Sarkar, S. (2005) *Journal of molecular biology* 352, 58-71
[95] Solaro, R. J., Lee, J. A., Kentish, J. C., and Allen, D. G. (1988) *Circulation research* 63, 779-787
[96] Nosek, T. M., Brotto, M. A., and Jin, J. P. (2004) *Archives of biochemistry and biophysics* 430, 178-184
[97] Ladd, A. N., and Cooper, T. A. (2002) *Genome biology* 3, reviews0008
[98] Breitbart, R. E., and Nadal-Ginard, B. (1987) *Cell* 49, 793-803
[99] Johnston, J. J., Kelley, R. I., Crawford, T. O., Morton, D. H., Agarwala, R., Koch, T., Schaffer, A. A., Francomano, C. A., and Biesecker, L. G. (2000) *American journal of human genetics* 67, 814-821
[100] Wang, X., Huang, Q. Q., Breckenridge, M. T., Chen, A., Crawford, T. O., Morton, D. H., and Jin, J. P. (2005) *The Journal of biological chemistry* 280, 13241-13249
[101] Robbins, J. (2000) *Annual review of physiology* 62, 261-287
[102] Chong, S. M., and Jin, J. P. (2009) *Journal of molecular evolution* 68, 448-460
[103] Feng, H. Z., Wei, B., and Jin, J. P. (2009) *The Journal of biological chemistry* 284, 31798-31806
[104] Huang, Q. Q., and Jin, J. P. (1999) *Journal of molecular evolution* 49, 780-788
[105] Biesiadecki, B. J., and Jin, J. P. (2002) *The Journal of biological chemistry* 277, 18459-18468
[106] Biesiadecki, B. J., Elder, B. D., Yu, Z. B., and Jin, J. P. (2002) *The Journal of biological chemistry* 277, 50275-50285
[107] Wei, B., Gao, J., Huang, X. P., and Jin, J. P. (2010) *The Journal of biological chemistry* 285, 27806-27816
[108] Anderson, P. A., Greig, A., Mark, T. M., Malouf, N. N., Oakeley, A. E., Ungerleider, R. M., Allen, P. D., and Kay, B. K. (1995) *Circulation research* 76, 681-686
[109] Mesnard-Rouiller, L., Mercadier, J. J., Butler-Browne, G., Heimburger, M., Logeart, D., Allen, P. D., and Samson, F. (1997) *Journal of molecular and cellular cardiology* 29, 3043-3055
[110] Akella, A. B., Ding, X. L., Cheng, R., and Gulati, J. (1995) *Circulation research* 76, 600-606

[111] Feng, H. Z., and Jin, J. P. (2010) *American journal of physiology. Heart and circulatory physiology* 299, H97-H105

[112] Huang, Q. Q., Brozovich, F. V., and Jin, J. P. (1999) *The Journal of physiology* 520 Pt 1, 231-242

[113] Huang, Q. Q., Feng, H. Z., Liu, J., Du, J., Stull, L. B., Moravec, C. S., Huang, X., and Jin, J. P. (2008) *American journal of physiology. Cell physiology* 294, C213-222

[114] Tilley, L. P., Liu, S. K., Gilbertson, S. R., Wagner, B. M., and Lord, P. F. (1977) *The American journal of pathology* 86, 493-522

[115] Hasenfuss, G. (1998) *Cardiovascular research* 39, 60-76

[116] Larsson, L., Wang, X., Yu, F., Hook, P., Borg, K., Chong, S. M., and Jin, J. P. (2008) *American journal of physiology. Cell physiology* 295, C722-731

[117] Martin, A. F. (1981) *The Journal of biological chemistry* 256, 964-968

[118] Simpson, J. A., van Eyk, J. E., and Iscoe, S. (2000) *J Appl Physiol* 88, 753-760

[119] Communal, C., Sumandea, M., de Tombe, P., Narula, J., Solaro, R. J., and Hajjar, R. J. (2002) *Proceedings of the National Academy of Sciences of the United States of America* 99, 6252-6256

[120] Zhang, Z., Biesiadecki, B. J., and Jin, J. P. (2006) *Biochemistry* 45, 11681-11694

[121] Feng, H. Z., Biesiadecki, B. J., Yu, Z. B., Hossain, M. M., and Jin, J. P. (2008) *The Journal of physiology* 586, 3537-3550

[122] Goll, D. E., Thompson, V. F., Li, H., Wei, W., and Cong, J. (2003) *Physiological reviews* 83, 731-801

[123] Biesiadecki, B. J., Chong, S. M., Nosek, T. M., and Jin, J. P. (2007) *Biochemistry* 46, 1368-1379

[124] Fujita, S., Maeda, K., and Maeda, Y. (1992) *Journal of biochemistry* 112, 306-308

[125] Chandra, M., Montgomery, D. E., Kim, J. J., and Solaro, R. J. (1999) *Journal of molecular and cellular cardiology* 31, 867-880

[126] Pan, B. S., Gordon, A. M., and Potter, J. D. (1991) *The Journal of biological chemistry* 266, 12432-12438

[127] Feng, H. Z., Wang, Q., Reiter, R. S., Lin, J. L., Lin, J. J., and Jin, J. P. (2013) *American journal of physiology. Cell physiology* 304, C1002-1012

In: Troponin
Editor: J.-P. Jin

ISBN: 978-1-62948-320-7

Chapter VIII

Post-Translational Modification of Troponin

Margaret Westfall, PhD
Department of Cardiac Surgery
University of Michigan
Ann Arbor, Michigan, US

Abstract

An overview of post-translational troponin modifications is presented in this chapter. A number of Ser/Thr kinases are known to phosphorylate the troponin complex in the myofilament, including multiple members of the AGC, CaMK-related, STE, TKL, and TK kinase families. The troponin residues targeted by each kinase, and the contribution of each phosphorylation site to the modulation of contractile function are discussed in this chapter. Relatively less is known about the specificity of the phosphatase regulatory components for the troponin complex compared to the many kinase pathways. However, the importance of phosphorylation turnover is increasingly recognized as an important tool for fine-tuning the modulation of cellular processes such as contractile performance. A second section covers advances in understanding phosphatases and their contribution to the cyclic turnover of phosphorylation. Additional post-translational modifications observed in troponin and their known or putative roles in modulating contractile function also are examined in this chapter.

Introduction

Signaling pathways play an important role in phosphorylating cardiac troponin (cTn) and modulating contractile function. The targeting of cTn subunits for phosphorylation and modulation of contractile performance was recognized over 30 years ago [43, 176, 191, 223]. Interest in phosphorylation of cTn continues in present research for a variety of reasons. We now know multiple residues may be targets for phosphorylation within a given cTn subunit, the impact of cTn phosphorylation on function may not be straight forward, and changes in cTn phosphorylation are often associated with cardiac remodeling and/or contractile dysfunction. Significant advancements also have emerged for identifying novel signaling pathways targeting cTn, multiple residues within cTn targeted by these pathways, as well as an increasing variety of post-translational modifications. A more dynamic view of signaling processes also continues to unfold [230, 250]. Signaling cascades operating as mini-processors to modulate cardiac structure and performance in response to multiple environmental inputs are increasingly replacing "on-off" signaling models. Incorporation of these recent insights into future work is anticipated to continue advancing our understanding of cTn as a post-translational target for "tuning" cardiac performance.

Perhaps the most studied and best described signaling modulation of a myofilament protein is the protein kinase A (PKA) -mediated phosphorylation of cardiac troponin I (cTnI), the inhibitory subunit of the troponin heterotrimer. A significant group of classic studies fundamentally advanced our understanding of the cTnI residues phosphorylated by PKA and their impact on relaxation during β-adrenergic activation. While several detailed reviews have focused on this collective group of work [109, 110, 222, 225, 229], many novel post-translational modifications in cardiac troponin also have recently been identified. The goals of the present review are to provide basic overviews of both established and emerging post-translational modifications of troponin, and discuss important questions that remain to be addressed about the role and/or contribution of each modification in health and disease. The initial part of this chapter covers kinases and post-translational phosphorylation, followed by brief summaries describing the role of phosphatases, and additional post-translational modifications of troponin. References to amino acid residues in this chapter are based on rat isoforms and Uniprot KB numbering, as shown in Table 1.Motifs in this chapter are designated by single letter amino acid abbreviations and hyphens between residues, with * indicating the phosphorylated amino acid, Φ for hydrophobic, β for basic, and X for non-specified amino acid residues.

Table 1. UniprotKB identification and total amino acid residues for each rat troponin subunit

Rat Troponin Subunit	Cardiac [total amino acids]	Slow Skeletal [total amino acids]	Fast Skeletal [total amino acids]
Troponin I (TNNI)	P23693 [211]	P13413 [187]	P27768 [182]
Troponin T (TNNT)	P50753 [289; Adult-Isoform 2]	Q7TNB2 [261;Isoform 1]	P09739 [259; βIsoform]
Troponin C (TNNC)	Q4PP99 [161]	See cardiac TNNC	Q304F3 [160]

Phosphorylation: Signaling Pathways and Targeted Residues

AGC Kinases

β-Adrenergic activation of PKA and downstream influences on cardiac troponin. Troponin phosphorylation was first described in the early 1970's [176, 191] and work on the signaling pathways, target residues, and impact on function continues today. The physiological relevance of this phosphorylation was established in studies showing cTnI phosphorylation closely correlated in time with the *in vivo* response to β-adrenergic receptor (β-AR) activation of protein kinase A (PKA), and the ensuing increase in contractile performance [43, 223]. Later work demonstrated PKA phosphorylated Ser23 and Ser24 residues on cTnI [242], and this cTnI phosphorylation increased Ca^{2+} dissociation from troponin C (TnC) to cause a reduction in the Ca^{2+} sensitivity of myofilament force [199]. These changes in myofilament function were later shown to play an important role in the accelerated relaxation observed in response to PKA activation [101]. In addition, both Ser23 and Ser24 phosphorylation were found to be required for the cTnI contribution to PKA – mediated accelerated relaxation [287]. PKA phosphorylated the Ser24 residue more quickly than Ser23 [96,151], and Ser24 was predicted and later verified to be constitutively phosphorylated in the myocardium [96, 151, 287].

Transgenic animal models provided direct evidence that the shift in myofilament Ca^{2+} sensitivity produced by cTnI Ser23/24 phosphorylation contributed significantly to accelerated relaxation during β_1-AR-mediated activation of PKA in the heart [245, 276]. Down regulation of β-AR signaling during human and animal models of heart failure, the associated decrease in cTnISer23/24 phosphorylation [70, 141, 164, 272], and slowing of myocardial relaxation [10, 15, 282] were in agreement with this mechanism. Phosphorylation of these residues also can increase the Ca^{2+} sensitivity of velocity [77], although work in genetic animal models showed PKA primarily influenced cardiac contractile kinetics via thick filament phosphorylation targets [70, 231].

Additional experiments demonstrated PKA also phosphorylates cTnI residues Ser40 and Ser167 [262]. More extensive Ser167 phosphorylation is detected compared to Ser40, although PKA-mediated phosphorylation of both residues developed more slowly than Ser23/24 phosphorylation. In addition, top-down and electron capture analysis identified partial phosphorylation of residues equivalent to rat cTnI Ser78 and Thr79 in commercially available human heart tissue [280]. Consensus phosphorylation motifs for PKA and PKC are present around Ser78 (R-X-S/T*, R-X-X-S*/T* and S*/T*X-K/R), and PKC phosphorylated this site under *in vitro* conditions [227]. PKA also could phosphorylate cTnI Thr79, Thr124, Thr144 and Ser151, although the motifs around Thr79 and Thr124 conform more closely to a PKG consensus phosphorylation motif (R/K-$(X)_{1\text{-}3}$-S*/T*). A proteomic analysis recently confirmed the residues equivalent to cTnI Ser78, Thr79, and Ser167 are phosphorylated in failing human hearts [286].

Phosphorylation of cardiac troponin T (cTnT) by cAMP-activated PKA is not detectable when analyzed in the cTn complex and/or in more complex preparations [232]. Consensus phosphorylation sites for PKA are predicted for cTnT (Ser69, Ser181, Thr195, Ser199, Thr204, Ser276, Thr285), and when purified, cTnT phosphorylation is detected at the site equivalent to Ser199 [243]. However, in other studies, the detection of phosphorylation in cTnT is more often attributed to other kinases [232, 285]. More recent work suggests splice variants of cTnT may be targeted for phosphorylation by PKA in response to length changes, based on the inhibition of cTnT phosphorylation by H-89 [156]. However, the H-89 concentration used for this most recent work also inhibits other AGC kinases along with members of the CaMK and STE kinase families [36]. Thus, other kinases may contribute to this response. Currently, there is no direct evidence cAMP-dependent PKA activation directly targets cTnT in the myocardium.

Activation of PKC and targeting cTn for phosphorylation. Work has focused on the relationship between protein kinase C (PKC) and myofilament targets for over 25 years. PKC signaling and modulation of contractile function are significantly more complex in comparison to β-AR and PKA-mediated myofilament phosphorylation. Our fundamental understanding of the PKC pathway is primarily based on work demonstrating G-protein receptors (GPR) coupled to $G\alpha_q$ activate phospholipase C (PLC) to convert phosphatidylinositol 4,5 bisphosphate (PIP_2) into inositol 1,4,5 trisphosphate (IP_3) and diacylglycerol (DAG). Additional upstream events also activate and/or inactivate PKC, which may include more complex signaling crosstalk [227]. The release of DAG from membranes activates most PKC, isoforms, which then phosphorylate and/or modulate downstream kinases/ phosphatases in addition to end target proteins [162, 230]. In the heart, activated PKC modulates contractile function by targeting key proteins in excitation-contraction coupling, and phosphorylates transcription factors to modulate cellular remodeling of cardiac structure and function [123, 227, 230]. The identification of multiple isoforms at each step in this activation process dramatically increases the flexibility of this signaling cascade, but also contributes to the large and somewhat confusing body of literature. More detailed analyses of individual steps in this pathway are available in several reviews, including the different forms of PLC, DAG, and PKC [146, 195, 196, 220, 227, 230].

The relationship between PKC activation and cTnI phosphorylation is complex [110]. A series of biochemical studies initially showed activated PKC phosphorylates multiple cTnI residues in the presence and absence of Ca^{2+} and phospholipid cofactors [92, 166, 169, 271]. These studies also identified the PKC isoforms capable of phosphorylating cTnI [87, 168], and their functional impact on myocardial actomyosin ATPase activity [87, 165-167]. Taken together, this work showed PKC activation targets 3-4 clusters of residues with phosphorylation initially detected at Ser43/45, Thr144 and Ser78 followed by Ser23/24 [169, 255]. Additional PKC consensus phosphorylation motifs also are present within cTnI. While established PKC sites are phosphorylated in recent proteomic studies using human myocardium [280, 286], phosphorylation also is detected in these additional PKC consensus sites (Thr79, Ser167, Ser181; Note: Ser181 is only present in human cTnI) [175, 280]. More recent work indicates several other kinases also may target these residues (see below). The identification of 3 PKC-targeted clusters (Ser23/24, Ser43/45 and Thr144) spawned substantial research to determine the impact of PKC-mediated cTnI phosphorylation on cardiac function, the PKC isoform(s) responsible for cTnI phosphorylation, and the conditions responsible for PKC activation.

The functional consequences of cTnI phosphorylation at PKC targeted sites has been examined extensively using phospho-mimetic (Glu or Asp) and phospho-null (Ala) substitutions in cTnI Ser23/24, Ser43/45 and/or Thr144. Biochemical analysis of cTnI with a phosphonull Ser43/45Ala substitution showed that phosphorylation of these residues reduced maximum actomyosin ATPase activity. In addition, phosphorylation of Ser43/45, Thr144 or Ser23/24 each diminished myofilament Ca^{2+} sensitivity and decreased the affinity of cTnI for cTnC [11, 118, 166, 168]. Both myofilament Ca^{2+} sensitivity and peak tension also were reduced after extraction and replacement with cTn containing phosphomimetic Glu substitutions at cTnISer43/45 in permeabilized papillary bundles [23], and comparable changes were observed in *in vitro* motility experiments [23]. In contrast, phosphomimetic cTnI Thr144Glu reduced Ca^{2+} sensitivity in the *in vitro* motility studies but had little impact on Ca^{2+} sensitivity of tension [23]. These experiments also showed phosphorylation of more than one cluster within cTnI produced specific, but not necessarily additive, functional responses [23, 102, 166, 168].

Based on the rightward shift in myofilament Ca^{2+} sensitivity of tension produced by cTnI with Ser43/45 and Ser23/24 (see PKA) substitutions, PKC-mediated phosphorylation of these residues should accelerate cardiac relaxation, while Thr144 likely has less influence on relaxation rate under loaded conditions. Alternatively, a Ser43/45-induced reduction in maximum tension could decrease peak pressure in the heart and, if severe enough, there may be no net change in relaxation rate. Non-phosphorylatable substitutions in this cluster also are not expected to change basal contractile function, but could slow relaxation in response to PKC activation if cTnI plays a significant role in the PKC response. These predictions were tested in myocytes and genetic animal models expressing phospho-null or phospho-mimetic cTnI substitutions. Mice expressing phospho-mimetic cTnISer23/24Asp developed the anticipated acceleration of relaxation detected as dP/dt_{min} [205, 276]. Basal contractile function also is unchanged in myocytes and myocardium expressing non-phosphorylatable cTnISer23/24Ala or cTnIThr144Pro [245, 264, 276]. PKC activation in myocytes expressing phospho-null cTnI Ser23/24Ala or Thr144Pro also produced the expected slowing of relaxation in response to PKC activation [178, 179, 264]. Overall, the cTnIThr144Pro and cTnISer23/24Ala results indicate cellular and/or organ-level relaxation responds in a manner predicted by earlier biochemical and biophysical studies and are consistent with Ser23/24 and Thr144 phosphorylation modulating relaxation.

A more complex literature emerged from animal models with Ser43/45 substitutions, and controversy remains about the functional consequences when PKC phosphorylates this cluster. In one animal model, the substitution of all 5 PKC-targeted sites with phospho-mimetic Asp (cTnIAllP mouse) completely replaced endogenous cTnI [205]. Systolic pressure development slowed without changing peak pressure or relaxation rate in this model [205]. In contrast, very low level (<10%) expression is achieved with cTnI containing Ser43/45Glu plus Thr144Glu (PKC-P mouse; 102). However, even low level cTnIPKC-P expression decreased maximum tension without changing Ca^{2+} sensitivity of tension in permeabilized papillary muscles, and this translated into reduced systolic pressure and slowed rates of contraction and relaxation [102]. The greater level of cTnIAllP replacement and modest cardiac phenotype compared to cTnIPKC-P could result from differences in cTnI conformation, intra- and inter- protein interactions within the myofilament, and/or the amount and type of compensatory adaptations in response to expression of each cTnI. Despite these differences, the direction of change in myofilament and cardiac contractile function observed with the phospho-mimetic cTnI substitutions are generally consistent with the idea that Ser43/45 phosphorylation acts as a brake to modulate peak pressure and/or contraction and relaxation rates in the intact heart. This brake may be transiently beneficial for modulating heart function, but sustained elevations in Ser43/45 phosphorylation levels also appear to contribute to cardiac dysfunction during chronic stress and heart failure [39, 118, 211].

Studies utilizing phospho-null substitutions should add further insight into the relationship between Ser43/45 phosphorylation and contractile function, but instead indicate this region has a complex role in thin filament function. A knock-in cTnI mouse expressing Ala substitutions at cTnI Ser23/24, Ser43/45 and Thr144 ($cTnIAla_5$) was compared to wild type and cTnISer23/24Ala-expressing (cTnIAla$_2$) mice [179]. Endothelin- or phorbol ester-mediated PKC activation was expected to reduce peak tension/ATPase and Ca^{2+} sensitivity in cardiac preparation expressing cTnIAla$_2$ or wild type cTnI, and this response should be blunted in samples expressing cTnIAla$_5$.While agonists did not cause a change in peak tension with cTnIAla$_5$, PKC activation enhanced rather than reduced the Ca^{2+} sensitivity of actomyosin ATPase and tension in wild type and cTnIAla$_2$ preparations [178, 179]. Further work also showed cTnIAla$_5$ blunted the functional response to myosin light chain 2 (MLC$_2$) phosphorylation by myosin light chain kinase (MLCK) [179]. These results led to the conclusion that activated PKC targets cTnI Thr144 and/or Ser43/45 phosphorylation, which synergize with MLC$_2$ phosphorylation to enhance myofilament Ca^{2+} sensitivity. However, compensatory phosphorylation of cTnI on another residue (e.g., Ser78) and/or on other myofilament proteins is evident in cTnIAla$_5$myocardium compared to wild type or cTnIAla$_2$ samples [178,179], and could contribute to different responses. A recent gene transfer study determined whether substantial replacement with cTnISer43/45Ala in myocytes produced similar results. In this study, sarcomere expression of cTnISer43/45Ala decreased myofilament Ca^{2+} sensitivity of tension, reduced shortening amplitude, and slightly accelerated re-lengthing rate, which are similar to the outcomes anticipated from earlier work with Glu or Asp substitutions at cTnISer43/45 [107]. This unexpected functional change observed with extensive cTnISer43/45Ala replacement of endogenous cTnI shows this substitution is not functionally conservative in intact myofilaments. A similar rightward shift in Ca^{2+} sensitivity also is evident on inspection of basal ATPase and tension results from cTnIAla$_5$ mice [178, 179]. In addition, the same dose of endothelin used for cTnIAla$_5$ studies [179] slightly accelerated re-lengthening in myocytes expressing cTnI while re-lengthening slowed in cTnISer43/45Ala-expressing myocytes. While cTnI Ser43/45Ala is not identical to cTnIAla$_5$, the divergent response to PKC may be in part explained by the presence of a phosphatase inhibitor in the earlier work [179] versus its absence in the gene transfer study [107]. This phosphatase inhibitor enhances cTnISer78 phosphorylation independently from PKC [255], which may explain the divergent outcomes in studies with Ser43/45Ala substitutions. Future studies are needed to test this idea and determine whether Ser78 increases myofilament Ca^{2+} sensitivity in the presence or absence of Ser43/45 phosphorylation.

The role of Ser43/45 phosphorylation is more perplexing when considering results from a mouse expressing cTnISer43/45Ala, which developed minimal changes in the cardiac phenotype [133]. At first glance this lack of change is not expected based on either the cTnISer43/45Ala gene transfer experiments or the knock-in cTnIAla$_5$ mouse. This transgenic mouse is predicted to replace ~50% of total cTnI in the myofilament with cTnISer43/45Ala [160]. If cTnISer43/45 residues are typically phosphorylated *in vivo,* enhanced systolic and diastolic function could be expected in hearts from this mouse, and yet the phenotype is comparable to wild type. The minimal cardiac phenotype observed in this mouse was attributed to a compensatory reduction in the cellular Ca^{2+} transient and/or compensatory changes in cTnI and cTnT phosphorylation levels [134, 160, 187]. In keeping with this phenotype, a modest decrease in maximum tension and blunted negative inotropic response to PKC activation are observed in papillary muscles from these mice [187]. The minimal cardiovascular phenotype in this mouse also is consistent with results obtained in myocytes after partial replacement (<50%) with cTnISer43/45Ala via gene transfer [107]. Based on current experimental results, high level sarcomere expression of cTnI with Ser43/45Ala unexpectedly reduces myofilament function. It is unclear whether the minimal cardiac phenotype observed with lower level Ser43/45Ala expression is due to a concentration effect and/or due to compensatory adaptations. The Ser43/45Ala substitution provides contradictory evidence about the role this cluster plays in the PKC response, and yet also indicates this region plays a previously unrecognized but important functional role in the thin filament. Further studies are clearly necessary to better understand the role cTnI Ser43/45 plays under basal conditions in addition to its role during PKC activation.

Another important focus in this field is to define the PKC isoform(s) responsible for phosphorylating cTnI residues in intact myocytes and the myocardium. The PKC family is classified into conventional, novel, and atypical subclasses, with adult myocardium expressing the conventional α isoform, novel δ and ε isoforms, and to a lesser extent, the atypical ζ isoform. In some mammalian species such as mouse, expression of the conventional, alternatively spliced β_I and β_{II} isoforms also are detected [124], although these isoforms are expressed at very low levels or are not detectable until heart failure develops in humans and other species [80]. The most abundant isoform in adult myocardium is PKCα [16, 184], and in solution biochemical studies, this isoform phosphorylates all 3 clusters of PKC-targeted cTnI residues [168]. However, in animal models and intact myocytes PKCα produces diminished cardiac function, which is attributed to the activation of downstream phosphatase activity [19] and is independent from cTnI phosphorylation.

Under *in vitro* conditions, the conventional PKCβ_{II} isoform also phosphorylates the same cTnI residues as PKCα [104, 127, 203, 260]. In addition, PKCβ_{II} up-regulation enhances cardiac cTnI phosphorylation in different animal models [75, 244]. This consistent phosphorylation could indicate cTnI is an important target for PKCβ_{II}, and yet reduced cardiac function is observed in the animal model expressing wild type PKCβ_{II} [244], while enhanced cardiac function develops in a second constitutively active, inducible PKCβ_{II} expression model [75]. This functional divergence suggests signaling via PKCβ_{II} is complex, and led to recent experiments aimed at identifying the impact of PKCβ_{II} signaling on contractile function in myocytes. Wild type PKCβ_{II} in this experiment acted more as a signaling node to activate downstream kinase and phosphatase activity, which in turn modulated contractile performance and targeted cTnI for phosphorylation [80]. Moreover, receptor activated PKC shifted the balance between phosphatase and kinase activation [79]. Thus, a different balance of kinase versus phosphatase activation may be present in the two genetic animal models, which produce opposing cardiac phenotypes.

In contrast to the conventional PKC isoforms, there is solid evidence that novel class PKCε directly phosphorylates cTnI to modulate contractile function, and less direct evidence PKCδ also targets cTnI. PKCε and δ each phosphorylate cTnI Ser23/24, Ser43/45 and Thr144 and sometimes Ser78 under *in vitro* conditions [87, 168]. Sustained PKCε up-regulation produces systolic and diastolic cardiac dysfunction which is ameliorated by the phospho-null cTnISer43/45Ala [211]. Most importantly, imaging studies show PKCε specifically translocates to the sarcomeric z-band [76, 200, 201], although this localization depends on the receptor and/or environmental conditions responsible for PKCε activation [42, 89]. While PKCε binds to F-actin, its docking to the myofilament does not use this site [76] and instead depends, at least in part, on the actin capping protein, CapZ [186]. In contrast to PKCε, PKCδ activation is often linked to hypertrophic responses involving nuclear rather than myofilament translocation [228]. PKCδ may typically target other functions, and yet there are studies showing stress-induced Src-mediated phosphorylation of PKCδ at residue Tyr311 significantly increases cTnI phosphorylation and shifts phosphorylation from Ser23/24 to Thr144 in cTnI [239]. Up-regulation of PKCδ in response to hypertension also is associated with increased cTnISer43/45 phosphorylation [39]. Although there is direct imaging evidence to show PKCδ translocation to the myofilament, these results indicate PKCδ may target cTnI residues as part of a coordinated response to stress and/or under pathophysiological conditions.

Mammalian myocardium also expresses detectable levels of the atypical isoform, PKCζ [16, 184]. This isoform shows less ability to phosphorylate cTnI relative to other PKC isoforms in the heart [87]. It also became associated with a protein phosphatase 2A (PP2A) complex in myocytes, which reduced Thr phosphorylation in cTnI [274]. While this work provides a function of PKCζ, future studies are needed to understand the circumstances and importance of this PKC in healthy and diseased myocardium. This summary demonstrates the substantial progress made in dissecting out the role of individual PKC isoforms in modulating contractile function, and yet continued efforts are needed to understand the relationship between PKC activation, cTnI phosphorylation and cardiac function.

The relationship between PKC isoforms, cTnT, and cardiac function is equally complex. The discovery that PKC targets cTnT phosphorylation, reduces cTnT affinity for actin-tropomyosin (Tm) and decreases both maximum actomyosin ATPase and Ca^{2+} sensitivity in response to cTnT phosphorylation emerged from the same biochemical studies with cTnI [92, 243]. Residues in cTnT targeted for phosphorylation by PKC include Thr195, Ser199, Thr204, and Thr285, with kinetic data suggesting Thr285 is the preferred target in intact myocytes (bovine cTnT Thr191, Ser195, Thr200, Thr281; uniprotKB P13789; 87,169). *In vitro* studies demonstrated PKC isoforms α, δ, and ε phosphorylated each of these residues. PKC α and ε primarily targeted Thr204 and Thr285 [87], while PKCδ showed a lower overall phosphorylation response compared to other PKC isoforms. Both the peak actomyosin ATPase activity and Ca^{2+} sensitivity dramatically decreased in response to PKCα treatment. In hearts expressing ~50% of total TnT as fast skeletal TnT (fsTnT), phorbol ester treatment attenuated the reduction in papillary muscle peak tension compared to wild type. This improvement in the functional response to PKC activation was attributed to the absence of sites equivalent to cTnT Thr195 and Thr285 in fsTnT [159]. However, in work with phospho-mimetic substitutions, Ser204 phosphorylation appeared to be largely responsible for the reduction in peak tension and most of the TnT-dependent rightward shift in Ca^{2+} sensitivity [237, 238]. Phosphorylation of this residue increases in hearts expressing a mutation linked to inherited cardiomyopathies [212], and yet decreases after myocardial infarction [40]. Compensatory changes in cTnT phosphorylation also develop in response to cTnI phospho-substitutions [187]. Collectively, the PKC-related results suggest a better understanding of the conditions responsible for PKC-dependent cTnT and cTnI phosphorylation and the contribution to functional modulation of cardiac performance are needed. This type of insight will continue to require integration of multiple approaches ranging from biochemistry to systems biology.

Fewer published studies have examined PKC-mediated Tn phosphorylation and modulation of skeletal muscle function, in part due to early biochemical work indicating PKC is unlikely to modulate skeletal muscle function via Tn phosphorylation. This early study showed phosphatidylserine and Ca^{2+}-activated PKC phosphorylated Tn purified from mixed muscle, presumably composed of fast and slow isoforms of TnI and TnT [140]. The Thr11 and Ser13 residues in slow skeletal troponin I (ssTnI) or Thr12 in fast skeletal TnI (fsTnI) are equivalent to the Ser43 and/or Ser45 sites in cTnI and are candidate residues for phosphorylation. Alternatively, there is a PKC phosphorylation motif at Ser46 in ssTnI, which is analogous to cTnISer78, although this site is absent in fsTnI. Based on the much higher amount of Ser versus Thr phosphorylation in skeletal Tn, the most likely targets for PKC phosphorylation in the skeletal muscle samples are ssTnI Ser13 or Ser46. In contrast, equal amounts of Ser and Thr residues are phosphorylated by PKC in fsTnT, and yet the sites identified as PKC targets in cTnT are absent from slow skeletal TnT (ssTnT) and fsTnT. Instead, PKC phosphorylation motifs are found around Ser42, Thr117, Thr176, Ser186 and Ser242 of ssTnT, with homologous residues at Thr45, Ser168 and Ser243 in fsTnT. Interestingly, the presence of TnC or Tm substantially attenuated Tn phosphorylation, which suggests activated PKC may not target Tn residues in intact skeletal muscle. This possibility is supported by work showing activated PKC fails to phosphorylate ssTnI after gene transfer and expression of this isoform in cardiac myocytes [263].

Protein kinase G (PKG)-mediated phosphorylation of cTn. The Ser23/24 cluster also is a target for PKG activated in response to cGMP [14], although PKG phosphorylation of cTnI is slower than cAMP-activated PKA [120]. The Ser23/24 cluster conforms to the consensus phosphorylation motif for PKG of R/K-$(X)_{1\text{-}2}$-S/T [175]. Natriuretic peptides [21, 180, 273] and eNOS (NOS3) formation [93, 109] each increase cGMP-dependent PKG activity, although high levels of cGMP are needed for PKG to modulate contractile function [206]. More recently, studies found an important arm of β_3-adrenergic receptor (β_3-AR) depended on PKG activation and downstream cTnI Ser23/24 phosphorylation to modulate contractile function [111]. Activation of β_3-AR integrates cAMP- and NOS-mediated modulation of contractile function, and loss of this integration contributes to heart failure [152, 158]. Pathophysiological conditions producing inflammation, such as myocardial ischemia, also activate PKG [51].

Inflammatory processes also stimulate cGMP and soluble guanylate cyclases (GCs) to increase PKG activity and cTnI phosphorylation [51]. Currently, there is no direct evidence to link PKG to cTnT phosphorylation in the heart. However, decreased PKG expression along with diminished cTnI and cTnT phosphorylation are observed in a dog model of diastolic heart failure with preserved ejection fraction [63]. Experiments designed to show PKG is responsible for these changes are not yet available, and it is also possible other kinases may play a role in this animal model. For example, PKCα up-regulation is associated with heart failure, and is expected to increase phosphatase activity, which could then produce the observed decrease in cTn phosphorylation [19]. As with PKC, future studies utilizing integrated approaches are needed to better understand the relationship between PKG and cTn phosphorylation under physiological and pathophysiological conditions.

RhoA-dependent protein kinase (ROCK-II) phosphorylation of cTn. Multiple myofilament proteins are targeted for phosphorylation by the activation of ROCK-II, which is found in the sarcomere [50, 188]. RhoA, the upstream effector of ROCK-II also is detected in sarcomeres and can be activated by obscurin [50, 188], as well as stressors such as rapid pacing and chronic pressure overload [207, 234]. In permeabilized fiber bundles, constitutively active ROCK-II targeted cTnI and cTnT for phosphorylation [252]. The

ROCK-II phosphorylation targets identified in these experiments included Ser23/24 and Thr144 of cTnI along with Ser276 and Thr278 in cTnT. However, ROCK-II also is associated with downstream activation of protein phosphatases, as well as phosphatase-associated kinases such as mitogen activated protein kinases (MAPK), and p90RSK [108, 114]. In the future, it will be important to establish whether specific conditions cause ROCK-II to target cTn instead of, or in addition to, activating downstream phosphatases in the intact myocardium.

Additional AGC kinases: Ribosomal S6 kinases (RSK or p90S6K), protein kinase B (Akt), and ribosomal protein S6 kinase (S6K or p70S6K). The first evidence one of these kinases phosphorylates cardiac and skeletal TnI came from studies on p90S6K published in 1988 [44]. Activation of p90S6K contributes to cell survival and growth, and the 4 isoforms of this kinase are now classified as RSKs, which are found in the cytosol and the nucleus [202]. Each RSK is composed of 2 kinases, with a CaMK in the carboxyl-half linked to an AGC kinase in the amino-terminal half of the protein. RSK kinases are primary downstream targets for MAPKs, which is an upstream activator of the RSK CaMK in the canonical signaling pathway. In turn, the CaMK activates the RSK AGC kinase to phosphorylate downstream targets [202]. RSKs play a significant role in cardiac myocyte hypertrophy and cell survival in response to stress [116, 202]. Itoh and colleagues reported RSK1 directly phosphorylates cTnI, and oxidative stress increased RSK auto-phosphorylation in addition to cTnISer23/24 phosphorylation [82].

Phosphorylation of RSK and cTnISer23/24 also are both elevated in a PKCβ transgenic mouse model and in response to oxidative stress [82]. Oxidative stress-induced cTnI phosphorylation is blocked by gene transfer and expression of a kinase dead RSK [82]. Gene transfer of RSK1 or 2 into myocytes also confirmed RSK phosphorylates cTnISer23/24 [33]. The earlier finding that RSK phosphorylates skeletal TnI [44] indicates RSKs may target more than just Ser23/24 in cTnI, as skeletal TnI isoforms lack the extension containing Ser23/24. An additional consensus phosphorylation motif is present at cTnI Ser78 and ssTnI Ser46, but is absent in the fsTnI isoform. There is little understanding of the integrated relationship between RSK, cTn phosphorylation and contractile function beyond this current information. Less information is available about the potential of p70S6Ks (S6K) and Akt to target cTn subunits. Akt recognizes the same R/K-X-R-X-X-S*/T* phosphorylation motif as RSK [82, 202, 250], and both cTnI and Akt phosphorylation increase in response to the nitric oxide precursor, L-arginine [254]. However, there is no direct evidence Akt and p70S6K target the sarcomere and/or cTnI [8], and L-arginine activation of PKG may be responsible for the increased cTnI phosphorylation.

Collectively, this brief summary touched on some of the significant advances made in understanding of the relationship between AGC kinases and cTn phosphorylation. In addition to providing insight into the modulatory role cTn plays in the myofilament, previous work on AGC kinases also established the fundamental approaches for investigating kinase-mediated phosphorylation of cTn and modulation of contractile function. There is evidence additional cellular signaling pathways target cTnI, including kinases classified in the CaMK (PKD, AMPK, PHK, CaMKII), STE (PAK, STK4/MST, ASK1), TKL (TNNI3, CK1, Raf-1), and possibly TK (SRC) kinome families. Our current understanding of the relationship between these additional signaling cascades and cTn phosphorylation are described below. This summary also shows there are still significant gaps in our understanding of signaling modulation of cTn phosphorylation and contractile function. Future work is expected to fill in these gaps for each of these kinases, and could perhaps lead to therapeutic strategies for treating cardiac dysfunction.

CaMK-Related Kinases

PKD phosphorylation of cTnI. Protein kinase D (PKD) integrates multiple upstream regulators with an equal variety of downstream targets and thus acts as an important signaling node within myocytes [5]. The contribution of PKD to the modulation of cardiac performance under physiological and pathophysiological conditions continues to evolve. In heart, PKD can be directly and indirectly activated by several agonists which bind to Gq-coupled receptors, and upstream PKC appears to be a dominant pathway for indirect PKD activation [5, 59, 66, 67]. High levels of PKD expression are detected in neonatal cardiac myocytes, where PKD1 is the predominant isoform along with lower expression of PKD 2 and 3 [5]. Initially, the expression of PKD1 in adult hearts appeared to be restricted to non-myocytes [66], but there is now direct evidence this isoform is expressed in adult cardiac myocytes [18, 34, 80]. In addition, studies analyzing a cardiac-specific PKD1 null mouse showed this kinase plays a significant role in stress-induced cardiac remodeling [49].

Just after re-classification of protein kinase Cμ as PKD1 [135], a yeast 2-hybrid screen revealed PKD interacts with cTnI [65]. Studies in ventricular myofibrillar preparations showed cTnI phosphorylation in response to the catalytic domain of PKD. Peptide mass fingerprinting identified Ser23, Ser24, and Ser198 as the targeted residues in recombinant cTnI, with PKD preferring Ser24 over Ser23 [65]. The Ser23/24 sites are part of a PKD consensus phosphorylation motif (ϕ-X-R-X-X-S*/T*), and PKD targeted this cluster in myofibrils and Tn subunits reconstituted with cTnISer23/24Ala. In functional studies, Ser23/24 phosphorylation by PKD reduced myofilament Ca^{2+}-sensitivity of tension in permeabilized preparations and attenuated ET-1-induced increases in myocyte shortening [34, 65]. Activated PKD also increased the rate of tension redevelopment (k_{tr}), although this response is attributed to the phosphorylation of other myofilament proteins [9]. In at least one study, PKD failed to phosphorylate cTnI [59], although this work used neonatal rat myocytes, which primarily express slow skeletal TnI (ssTnI) lacking the phosphorylatable amino-terminal extension found in cTnI [192, 265]. Less is known about the Ser198 site, which does not conform to a consensus PKD phosphorylation motif, but does contain the phosphorylation motif for CKII [100]. The recent identification of Ser198 phosphorylation in human myocardium [286], and its location in the COOH-terminal tail of all TnI isoforms suggests further work is needed to understand both upstream signaling and the role this residue may play in modulating contractile function.

AMPK and cTnI phosphorylation. The heterotrimeric 5'-AMP-activated protein kinase (AMPK) is present in the cytosol and acts as a master energy gauge and metabolic regulator [281]. Multiple inputs modulate the expression and activity of this kinase. The activation of this kinase in response to pathophysiological stress such as ischemia and pressure overload is thought to be cardio protective. As expected for a metabolic regulator, AMPK activation maintains energy balance by increasing fatty acid and glucose uptake and metabolism, produces a more sustained influence via actions on nuclear transcription factors and histones, reduces protein synthesis and increases autophagy. The more recent recognition that AMPK activation modulates cytoskeletal organization by targeting actin polymerization [150] raised the possibility that AMPK also may target myofilament proteins. Indeed, cTnI alone, cTnI within the cTn complex, and cTnI in permeabilized myocytes served as a substrate for the catalytic domain of AMPK [224]. Phosphorylation of cTnI is detected in multiple species and in fsTnI [163, 224]. Thorough analysis by ^{32}P-incorporation, Western detection and top down analysis demonstrated rat cTnI Ser23 and Ser151 are targets for AMPK phosphorylation, with the preferred site at Ser151. These 2 sites also contain the AMPK phosphorylation motif (ϕ-X/β–$(X)_{1\text{-}3}$-S/T-X-X-X-ϕ; 224). In fsTnI, AMPK targeted Ser118, which is homologous to cTnISer151.Subsequent studies demonstrated activation of the AMPK heterotrimer and/or endogenous AMPK phosphorylates cTnISer151 in intact myocytes, and Ser151 phosphorylation increases during myocardial ischemia [63, 171]. In contrast to known influences of phosphorylation at other residues in cTnI, these more recent studies demonstrate cTnI Ser151 phosphorylation enhances myofilament sensitivity to Ca^{2+} without changing cooperativity or peak force.

Interestingly, studies with cTnI containing phospho-mimetic Asp substitutions indicate there may be cross-talk between phosphorylated Ser151 and Ser23/24 in cTnI [163]. Permeabilized fibers reconstituted with Tn containing cTnISer151Asp attenuated the known ability of cTnISer23/24Asp phospho-mimetic to decrease myofilament Ca^{2+} sensitivity, without changing the heightened myofilament cooperativity produced by Ser23/24Asp [163]. The maintenance of heightened cooperativity could explain why enhanced myofilament Ca^{2+} sensitivity in permeabilized preparations did not translate into consistent increases in peak contractile function and/or slowing of relaxation in intact cells [171]. These results suggest AMPK phosphorylation of cTnI could stimulate force production without causing elevated Ca^{2+} transients and/or significant diastolic dysfunction in intact hearts. The ability of Ser23/24Asp to reduce Ca^{2+} sensitivity also is attenuated by increases in sarcomere length [163], and this cross-talk was diminished by Ser151Asp, which decreased the ability of increased length to override the impact of Ser23/24Asp on myofilament Ca^{2+} sensitivity. Future work is needed to understand the molecular basis for this cross-talk, the conditions leading to cross-talk, and to define the functional impact of AMPK phosphorylation on cTnI and other myofilament proteins.

CaMKII and cTnT phosphorylation. Bovine cTnT is a substrate for Ca^{2+}/Calmodulin-dependent kinase II (CaMKII) [85]. CaMKII targeted the residue homologous to Thr195 in rat cTnT, and this residue lies within the consensus phosophorylation motif for this kinase (K/R-X-X-S*/T*; 250). The region surrounding Thr195 is present in the cTnT of multiple species, and yet CaMKII-mediated phosphorylation of this residue is not significant when cTnT is part of the Tn-Tm complex [83]. Instead, PKC likely targets this residue. In addition, the CaMKII phosphorylation motif in cTnT is disrupted by an extra Glu in mammalian hearts such as rabbit and humans. While it remains possible CaMKII could target cTnT in some species, the evidence to date suggests cTnT is not a typical substrate for CaMKII in the myocardium.

Phosphorylase kinase and Troponin T phosphorylation. Phosphorylase kinase (PHK) is one of the first kinases found to phosphorylate TnT, and it phosphorylates both skeletal and cardiac TnT isoforms [176, 153, 189, 233]. The Ser2 and Ser181 residues in rat cTnT are targets for PHK [189]. The PHK phosphorylation motif of K-X-K-X-X-S*/T* is present at Ser181 in cTnT, Ser149 in fsTnT, and Ser154 in ssTnT. A consensus motif is absent from

Ser2 in cTnT and ssTnT, and is more likely phosphorylated by CKII (see below; 197). *In vitro* studies show PHK also phosphorylates cTnI Ser78 followed by Ser23, Thr144, and Thr167 [154]. Consensus phosphorylation motifs are present at Ser78 and Thr144 in cTnI, and the Ser78 equivalent residues also are present in ssTnI and fsTnI (Ser46). However, the functional importance of PHK-induced cTnI phosphorylation remains to be verified *in vivo*. PHK-mediated phosphorylation of myofilament proteins such as cTn subunits is postulated to synchronize glycogen breakdown with contractile activity [283]. If this idea is correct, increased stimulation frequency should increase TnT phosphorylation, and yet the TnT phosphorylation state is unchanged with tetanic stimulation [153]. This idea remains to be fully tested in experimental studies.

*SPEG (Striated muscle-specific Ser/Thr protein kinase).*SPEG (or APEG) is an additional kinase in the CaMKII family and is an ortholog of the MLCK in obscurin [241]. As with obscurin, this kinase is critical for myofibrillogenesis, and disruption of the gene encoding SPEG leads to cellular hypertrophy, myofibrillar degeneration, and neonatal mortality in mice [126]. Based on this work, SPEG is postulated to play an important role during myocardial development. SPEG also is detected in adult sarcomeres [277], and the expression during cardiac dysfunction is not known. The strong localization of this kinase in the sarcomere may indicate myofilament proteins are targets in adult hearts. While MLC2 is a likely substrate, cTn may be a target, although there is no experimental evidence supporting this possibility.

STE Kinases

Rac-1/p21-activated kinase (PAK) and Tn phosphorylation. Over a decade ago, constitutively active PAK3 was found to phosphorylate cTnISer151 in permeabilized cardiac fiber bundles [24]. PAKs were shown to be Ser/Thr kinases activated by small GT Pases, such as cdc42 and rac1 [24]. In mammals, 6 PAK isoforms were classified into 2 sub-groups, with the first subgroup composed of isoforms 1-3 [84]. Work in myocardium has focused almost exclusively on the actions of this first sub-group of PAK. Different upstream pathways activate PAK isoforms 1-3, and they target an array of downstream proteins [see 84, 136, 98, 99]. Rac1 expression and PAK1 activation contributed to cardiac hypertrophy in myocardium [240], PAK2 plays a role in apoptosis, and PAK3 activation rapidly reorganizes the cytoskeleton, but is expressed primarily in brain [98].

The PAK3-mediated phosphorylation of cTnISer151 increases myofilament Ca^{2+} sensitivity via changes in cTnI-cTnC interactions [24, 172]. The initial studies identifying cTnISer151 phosphorylation also indicate at least 2 residues maybe phosphorylated in cTnT [24]. However, only cTnT Ser181 lies within residues similar to the PAK consensus phosphorylation sequence (R/K-R/K-X-S*/T*-X-Y/ϕ-ϕ; 193). In contrast to the findings with PAK3, cTnI phosphorylation is significantly decreased in the absence versus the presence of PAK1-Raf-1 inhibitors [177]. The circumstances and extent PAK3 influences cTn phosphorylation in intact myocytes and/or myocardium remain unknown.

While the role played by PAK3 in the heart remains unclear, more recent work has focused more on the impact of PAK1 on cTn phosphorylation. Gene transfer of active PAK1 reduced cTnI phosphorylation at Ser23/24 via PP2A activation, and increased myofilament Ca^{2+} sensitivity of tension [95, 214]. In further studies, cTnI dephosphorylation in response to

PAK1 activation of PP2A could be stimulated by bradykinin treatment of myocytes [94].Improved functional outcomes after myocardial ischemia/repefusion (I/R) injury also are associated with PAK1-mediated dephosphorylation of cTnT [157]. The residues targeted for dephosphorylation on cTnT remain to be defined, although the increase in Ca^{2+} sensitivity observed after dephosphorylation indicates residues such as Thr195, which diminish Ca^{2+} sensitivity when phosphorylated would appear to be likely candidates. Interestingly, *in vitro* studies show PAK-1 enhances the ability of the small GTPase, Raf-1, to directly phosphorylate cTnTThr204 [177]. Overall, work in intact myocardium and cells suggest PAK1 activation of PP2A targets residues on both cTnI and cTnT in addition to other myofilament and Ca^{2+} cycling proteins [213]. There remains a need to define the spatiotemporal activation pattern for PAK in myocytes, which are most likely activated under pathophysiological conditions, such as pressure overload and myocardial ischemia. PAK3 is primarily expressed in brain [99], and yet if cTn is verified as a downstream target for PAKs 1 and 3, cTn phosphorylation state may serve as an indicator of PAK activation in myocardium.

Serine/Threonine kinase 4 (STK4/MST-1) targets TnI. Activation of STK4/MST-1 in response to ischemia/re-perfusion (I/R) injury induced production of reactive oxygen species (ROS) initiates a potent apoptotic response in the heart [278]. A yeast two-hybrid screen identified targets for STK4/MST-1 phosphorylation, and cTnI alone and in a cTn complex proved to be a key phosphorylation target for this kinase [279]. STK4/MST-1 failed to phosphorylate other cTn subunits during a 30 min incubation [279]. MALDI-TOF analysis revealed STK4/MST-1 phosphorylated Thr residues 32, 52, 130 and 144 in recombinant human cTnI [279]. Each of these residues lies within conserved regions of cTnI and a putative STK4/MST-1 phosphorylation motif. Phospho-null Ala substitutions at each site indicated STK4/MST-1 preferentially phosphorylates cTnI Thr32 in isolated cTnI and in the cTn complex. This residue is within a sequence which is similar to STK4/MST-1 auto-phosphorylation sites [54]. A phospho-specific antibody showed direct evidence cTnIThr32 is phosphorylated in neonatal myocytes treated with hydrogen peroxide to induce oxidative stress. This peroxide-induced increase in cTnI phosphorylation is blocked by a STK4/MST-1 specific siRNA, and this potent phosphorylation of cTnI is detected at a point in time when these myocytes are only beginning to express adult cTnI [192]. Further work is needed to define the time course and magnitude of phosphorylation at this cTnI site, along with determining the functional consequences of this phosphorylation, and whether peroxide-induced injury also changes phosphorylation on other sites within cTn in animal models.

Apoptosis signal-regulating kinase (ASK1) phosphorylation of cTnT. Inflammatory signaling via tumor necrosis factor and oxidative stress are both associated with ASK1 activation [20]. A yeast 2-hybrid screen revealed ASK1 interacted with cardiac troponin T [68]. Further work demonstrated ASK1 co-localized with cTnT in adult myocytes. The ASK1-induced phosphorylation targeted cTnT Thr195 and Ser199, which conforms to the ASK1 phosphorylation consensus motif S*/T*-X-X-X-S*/T* [68]. ASK1 phosphorylation of cTnT coincided with a significant decrease in the peak amplitude of myocyte shortening in electrically paced myocytes. ASK1 also acts as a mitogen activated kinase kinase or MEKK (MAP3K5 or MEKK5), which ultimately targets MAPKs and contributes to apoptosis [125]. However, observations such as the ASK1 translocation to a detergent insoluble fraction of myocytes in response to oxidative stress [20] indicate ASK1 also may target additional proteins such as the myofilaments during stress. These findings suggest further work is needed to understand the importance of cTnT phosphorylation during ASK1 activation should be investigated in more detail in the future.

TKL Kinases

Cardiac Troponin-I-interacting-Kinase (TNNI3K) and cTnI. A putative kinase targeting cTnI was identified using a yeast two-hybrid screen [289]. Bioinformatic analysis of this novel kinase, TNNI3K (or cardiac ankyrin repeat kinase, CARK) indicates this kinase is a mitogen activated kinase kinase kinase or MAPKKK in the TKL family of kinases, is most closely related to integrin-like kinase (ILK) [129, 259], and may have dual specificity as a Ser/Thr and Tyr kinase [47]. The predicted consensus phosphorylation motif for Ser/Thr residues is ϕ-$(X)_{2\text{-}4}$-S*/T* (ϕ is often L) and E-X-E/X-X-X-X-Y*-$(X)_{1\text{-}4}$-P/ϕ.A strong correlation between cardiac remodeling and/or dysfunction in response to the expression of this kinase is reported in multiple studies [210, 246, 270]. TNNI3K is reportedly localized to the sarcomeric z-band [246], and gene transfer into adult myocytes produces increased cTnISer43 and Thr144 phosphorylation [261]. While Ser45 phosphorylation was not directly examined, this site, but not Ser43 and Thr144 minimally conforms to the consensus phosphorylation motif of other TNNI3K targets [74]. Phosphorylation of cTnI at these residues also enhanced contractile function [261] rather than producing the decrease in function observed with PKC-mediated phosphorylation of and/or phospho-mimetic substitutions at these residues [23, 255]. The functional importance of cTnI in the remodeling response to TNNI3K may be complex, as other members of the TKL family are linked to bifurcative downstream modulation [219]. At least one group predicted TNNI3K may be linked to detection of environmental conditions by ILK [129]. However, the neurohormones, receptors, and/or stimuli responsible for TNNI3K activation are only vaguely understood in the heart, and involve multiple end targets [246]. More work is needed to define the relative importance of cTnI phosphorylation in this response as additional targets for phosphorylation by this kinase are identified in experimental studies.

Raf-1 and TnT. Recent studies show an additional TnT kinase is present in myocytes, which is activated by oxidative stress and can be inhibited by GW5074 [177]. This inhibitor primarily targets Pim and Raf kinases, although it can also influence other kinases known to target cTnT phosphorylation such as phosphorylase kinase and CK2 at the same concentration [7]. To focus in on the potential kinase responsible for cTnT phosphorylation, *in vitro* radiolabeling showed cTn phosphorylation in response to c-Raf-1 in the presence and absence of PAK1.In this work, c-Raf-1 increased cTnT phosphorylation, which is further enhanced by the addition of PAK1. The Raf-1-dependent phosphorylation of cTnT is absent when a cTnT containing Thr204Glu is reconstituted into the cTn complex [177]. The Thr204 residue is located within a consensus phosphorylation site for Raf-1 (R/P-X-R-S/R/P-X-S/T*-X-E/P).As with the studies of PAK1 (see above), additional work is needed to understand the cellular environment responsible for c-Raf-1-dependent cTnT phosphorylation, along with verification that Raf-1 targets cTnT in intact cells, and not a different kinase.

Other Kinases

Casein kinase 2 (CK2) and TnT. There is sporadic information suggesting that TnT may be a target for CK2.This ubiquitous kinase is constitutively active, targets a substantial range of proteins, and functions as a regulatory node for a range of cellular functions including apoptosis and cell development [145, 181]. In animal models, diminished CK2 and/or diminished phosphorylation of downstream targets produced myofibril derangements and hypertrophy [81, 117, 121]. However, skeletal TnT followed by cTnT were among the first targets identified for a kinase which was later identified as CK2 [62, 153, 198, 256]. These earlier studies also showed CK2 phosphorylates the Ser2 residue of cTnT, which contains the CK2 S*/T*-D/E-X-E/D phosphorylation motif [250, 145], and cTnT Ser2 is phosphorylated in mammalian myocardium [285]. There also is *in vitro* evidence CK2 phosphorylates the fsTnT residue equivalent to cTnT Thr181, although it has not been detected in proteomic studies designed to analyze cTnT phosphorylation [62, 256, 285]. The conditions leading to and the extent of cTnT phosphorylation, and the role in modulating myofilament function under physiological and pathophysiological conditions remains to be determined. It should be noted that both CK1 and CK2 phosphorylate skeletal TnT, although these kinases are clustered into different kinome families and they appear to target different groups of residues [218].

TK Kinases

While recent mass spectroscopy evidence suggests Tyr26 is phosphorylated in cTnI [286], the tyrosine kinase responsible for this phosphorylation is not known. The residues surrounding this tyrosine (RRRSSNYRAYA) also are not a match for interacting domains in tyrosine kinases, although there is an atypical proline rich domain (P-X-P-X-P-X-R) analogous to the SH2 or SH3 sites in Src kinase [250]. This proline rich region is located upstream from the Tyr26 phosphorylation site. While TK kinases such as Src recognize E/D-X-X-Y/F motifs for phosphorylation, it is possible phosphorylated cTnISer23 could act as a substitute for the acidic residue in this motif. The Tyr26 phosphorylation detected in failing human hearts may signal an increase in myofilament degradation, although this idea is purely speculative. Interestingly, stretch-induced activation is associated with increased expression of the ring finger, ubiquitin ligase BRCA1 [22]. BRCA1 typically repairs DNA breaks, and phosphorylation at Tyr26 could allow BRCA1 to bind to the S-X-X-Y motif on cTnI, and then regulate degradation of cTnI in a manner similar to the ring finger protein MuRF [97].

In Silico Predictions

While there is direct evidence the signaling pathways described above are capable of phosphorylating cTnI, there are also algorithms [13, 149, 170] which predict these same residues as well as additional residues in cTnI may be targeted by other signaling pathways. Phosphorylation of some of these additional residues was recently demonstrated in human cardiac tissue [286]. Five predicted residues in cTnI are phosphorylated (Ser4, Ser5, Tyr25,

Thr50, and Thr181; Note: Thr181 is in human sequence, not present in rat cTnI) in addition to 7 of the 12 known residues discussed above (Ser23, Ser24, Ser43, Ser45, Ser78, Thr144, Ser198). The Thr32, Ser40, Thr52, Ser78, Thr130, and Ser151 sites previously shown to be targets for phosphorylation were not detected in this study. Algorithm predictions also included Tyr111 ($p>0.5$), which also is not yet detected experimentally. Instead, phosphorylation of Tyr 25 is detected, which is not a predicted kinase target in the *in silico* analysis. A similar analysis also identified an additional 4 predicted phosphorylation sites in cTnT, which also have not been experimentally validated. Investigators also propose there may be other unidentified post-translational modifications (PTMs), such as myristoylation present in cTnT [225, 285].A site conforming to the consensus sequence for myristoylation is present at cTnTAla39 in both the rat and human sequence [138, 251]. Although these *in silico* predictions may not be completely correct, further work is needed to understand the influence of potentially novel as well as known PTM sites, the possible functional response when multiple sites are phosphorylated, and the kinases responsible for their phosphorylation.

Phosphatases and Phosphorylation Turnover

Phosphatase activity is an equally important component of phosphorylation-mediated signaling pathways. Protein phosphatase I (PP1) and PP2A are ubiquitously expressed Ser/Thr phosphatases, and they act as the primary phosphatases responsible for phosphate cleavage from Tn subunits in heart and skeletal muscle [31, 132]. Several in depth reviews cover the structure and function of these phosphatases [31, 32, 139, 216, 217]. Four highly homologous mammalian genes produce PP1 catalytic subunits (PP1c), which combine with more than 100 regulatory proteins to localize and target specific proteins for phosphate cleavage in the cell (reviewed in 32,216). Regulatory subunits generally contain a PP1c consensus recognition motif. Separate groups of chaperones, activity modulators, and/or inhibitors such as inhibitor-1 (I-1) bind and further regulate PP1 activity [32, 216]. PP2A phosphatases are composed of a dimeric core containing a scaffolding subunit A and catalytic C subunit. Two genes encode isoforms of each A and C subunit. The specificity, activity, and localization of PP2A are determined by a dozen known regulatory B subunits classified into multiple subfamilies [139, 217]. The B subunit binds to the core to form a heterotrimeric holoenzyme, which results in a specific spatial/temporal localization of PP2A. Reversible methylation of the C subunit enhances heterotrimer formation for many but not all B subunits [45, 112, 248, 275], and can be used as an index of PP2A activation during cardiac dysfunction [38, 137, 208].

Our initial understanding of PP1 and PP2A came from studies in skeletal muscle [see 31]. Later work demonstrated these phosphatases also are present in human atria and ventricles [132]. More recently, expression of each C and A subunit and all of the known B subunits, except PPP242C in the PP2A complex are detected in human myocardium [37]. Moreover, immunohistochemistry demonstrated the A and C subunits along with at least 3 B subunits are co-localized with the myofilaments. This study also provided some evidence that distinct cardiac PP2A expression and cellular distribution fingerprints develop with each type of cardiomyopathy during heart failure. In addition, a proteomic analysis confirmed PP2A and PP1 are present in adult rat cardiac myofilaments [277]. This analysis detected the PP1β catalytic subunit along with the MYPT1 and 2 regulatory subunits of PP1. Components of PP2A present in the same samples included the regulatory B'α along with the Aα scaffolding subunit of PP2A [277].

The relative contribution of PP1 versus PP2A to cTn dephosphorylation and the role phosphatases play in contractile function continue to be the focus of ongoing research. In biochemical studies, PP1 contributed more than PP2A to the de-phosphorylation of cTnI S23/24. Over a longer time period, PP2A rather than PP1 contributed more to de-phosphorylating these residues [86]. Permeabilized myocyte tension generation also is more sensitive to Ca^{2+} after the addition of PP1c, and this sensitization is absent when cTnI is replaced by phospho-mimetic cTnISer23/24Asp [161]. Comparable *in vivo* studies evaluating the contribution of these phosphatases to dephosphorylation of other residues in cTnI and/or cTnT are not available. Defining the relative contributions of each phosphatase in intact myocytes and under *in vivo* conditions also remains a challenge, due in part to the potential overlap between PP1 and PP2A on downstream targets. However, in myocytes expressing an I-1 polymorphism, PP1 is prevented from responding to the PKA signal to reduce PP1 activity.In this study, cTnI phosphorylation remained unchanged at the same point in time when diminished phospholamban phosphorylation is present [28]. These findings suggest PP1 targets cTn subunits *in vitro* but not necessarily *in vivo*, although there may be alternative explanations. For example, PKA-mediated phosphorylation of cTnI could adjust to the heightened PP1 activity and/or the I-1 polymorphism modifies interactions between I-1 and PP1, which in turn reduces the ability of PP1 to compete for myofilament targets.

A number of studies have focused on defining PP2A end-targets and the impact on contraction. Enhanced cTnI phosphorylation developed with increasing phosphatase inhibitor levels using an inhibitor thought to favor PP2A over PP1, although cTnI phosphorylation state did not correlate well with changes in contractile function [17]. Investigators also studied genetically altered mice expressing a 1.6-fold increase in PP2A holoenzyme expression. Phosphorylation is significantly reduced in cTnI and other proteins in this model and coincided with elevated end diastolic pressure and slowed rates of contraction and relaxation [53]. At the same time, cardiac remodeling in the form of hypertrophy, fibrosis, and necrosis accompanied this functional phenotype. Thus, reduced cTnI phosphorylation may not completely account for the results in this study. However, outcomes in other studies are consistent with this work and suggest cTnI may be a primary myofilament target for PP2A under physiological conditions. For example, PP2A enhanced myofilament Ca^{2+} sensitivity in non-failing but not failing heart tissue [266], due in part to the lower level of endogenous cTnI phosphorylation present in failing hearts. Moreover, gene transfer and expression of a dominant negative PP2A dimer in myocytes led to a reduced affinity for regulatory B subunits and to enhanced cytoskeleton protein phosphorylation in cardiac myocytes [208]. Although this later study did not examine cTnI phosphorylation levels, a separate study demonstrated enhanced cTnI phosphorylation levels coincided with reductions in PP2A activity [38]. Taken together, these results are consistent with PP2A-mediated cTn dephosphorylation in myocytes, but more direct experiments are clearly needed to adequately test this idea.

Phosphatase-induced cleavage from a target protein is important because it returns the functional response to baseline. Phosphatase activity also may change the balance between kinase and phosphatase activity to produce allosteric regulation of a structural or functional response [32]. In support of this idea, a number of kinases modulate cTn phosphorylation indirectly via their influence on phosphatases and/or their simultaneous influence on phosphorylation and phosphatase pathways. For example, β_2-adrenergic-mediated PKA activation in cardiac myocytes phosphorylates end targets such as cTnI and reduces PP1 activity [61]. In contrast, myocardial PKCα increases PP1 activity and reduces overall end target phosphorylation [19]. Increased phosphatase activity and diminished end-target phosphorylation also are associated with the Akt and MAPK pathways [122, 128, 253]. In other signaling cascades, the initial kinase activation acts as a node to coordinate activation of downstream kinases and phosphatases [26, 27, 106]. This node can produce highly specific responses, influence the gain of the response, and/or modulate more complex outcomes [61]. In other studies, the focus on target protein phosphorylation and phosphatase activity produces a different conclusion than studies focused exclusively on target phosphorylation by kinases such as PAK [213]. In addition, the evaluation of protein phosphorylation status for a single time point may depend on the environment as well as the protocol used for tissue collection, especially for human tissue [257, 258]. Based on our current understanding of phosphatases, they are an important contributor to the dynamic signaling networks present in the cardiac myocyte and skeletal muscle fiber. While many previous studies on cTn phosphorylation in health and disease focused on the specific level of cTn phosphorylation, there is little understanding of dynamic turnover and its impact on cTn modulation of myofilament function. Thus, studies are needed to examine the relative contribution of kinases versus phosphatases to cTn phosphorylation and turnover rate in response to physiological as well as pathophysiological conditions.

Additional Post-Translational Modifications (PTMs)

Experimental work on additional PTMs identified in cellular proteins has dramatically increased in recent years. To date, only a few of many possible modifications have been identified in the Tn complex. Some PTMs, such as methylation, are predicted from algorithms such as ACEVIEW and are used as a modification to improve signal detection [2]. Other contractile proteins and Tn genes also are known to be methylated [35, 147, 194], but the presence of Tn methylation has not been verified in experimental work. Other PTMs have been shown *in vitro* and yet remain to be investigated *in vivo,* such as TnC biotinylation [209]. In another group of PTMs, the protein modification is present in intact tissue or myocytes, and efforts are underway to understand the environment responsible for activating the pathway, the components in the pathway, and functional influence of the modification on troponin function. Troponin PTMs in this category are often linked to environmental stress and include oxidation, nitration and nitrosylation, glycosylation and acetylation, as well as other modifications such as SUMOylation. Each of these PTMs are described in more detail below.

PTMs Associated with Cellular/Oxidative Stress and/or Cell Dysfunction

Some Tn subunit PTMs develop in response to cellular stress and the generation of reactive oxygen species (ROS) caused by a variety of chronic cardiac diseases, such as ischemia and diabetes [183, 229, 235]. Both enzymatic and non-enzymatic processes produce ROS, and detoxification pathways normally keep ROS metabolites at low levels in cardiac myocytes [247]. TnI was first identified as a target for modification in response to oxidative stress over 15 years ago, with the observation that H_2O_2 treatment of cardiac myocytes caused alterations in cTnI phosphorylation [236]. Additional PTMs developed with cardiac ischemia and I/R injury, including the formation of proteolytic cTn subunit cleavage products [144, 249] as well as covalently linked complexes containing cTnI complexed with cTnC or cTnT [55, 144]. Protein cleavage, degradation and processing significantly impact contractile function, although this PTM is not universally detected in animal models associated with elevated ROS levels [143]. In addition, several recent reviews provide in depth coverage of ROS production and the relationship to downstream PTMs [143, 229].

Crosslinking and Transglutamination. The crosslinked products detected after I/R injury are predicted to result from transglutaminase activation [144]. However, some *in vitro* data are not consistent with this conclusion [55, 226]. Alternatively, subunits such as TnC and TnI also are known to spontaneously form intra- and inter-molecular disulfide bridges in response to mild oxidation [131, 185]. While investigators predict crosslinked products may protect against further degradation and/or anchor Tn subunits [144, 182], modifications also can significantly influence contractile function [57, 64, 173]. Based on these findings, further studies are needed to determine the relative impact of Tn crosslinking on cardiac and/or skeletal muscle dysfunction in response to oxidative stress, and identify/verify the mechanism(s) responsible for crosslinking.

Oxidation/S-Glutathionylation. While there is increasing evidence indicating myofilament proteins are oxidized in response to ROS and/or associated cellular stresses, there is little direct evidence showing the presence and time course of *in vivo* post-translational oxidation of troponin. Myofilament protein oxidation develops in human and animal models of cardiac dysfunction [25, 41], although Tn oxidation is either not analyzed or not present in several of these studies. Commercially available human heart tissue shows evidence of cTnI oxidation, although it was unclear whether this modification is present in intact hearts [280]. There is more direct data showing TnI oxidation in skeletal muscle studies. For example, oxidation is detected in residue Trp161 in fsTnI (homologous to cTnITrp192) after stimulating ROS production by UV irradiation [46]. The fsTnI also is a target for oxidation of Cys133 via glutathionylation, which produces increased myofilament sensitivity to Ca^{2+} and is linked to exercise [155]. Comparable residues in slow skeletal and cardiac TnI are substituted with Ser, and thus, would not be similarly modified. A rodent myocardial ischemia model also failed to produce detectable changes in troponin glutathionylation [6]. Collectively, these findings suggest further work is needed to identify the pathways responsible for Tn oxidation/glutathionylation in skeletal muscle and understand environmental conditions associated with this PTM.

Nitrosylation and Nitration. Stress and/or ROS activation also produce PTMs via S-linked nitrosylation (SNO) of cysteine residues and nitration of tryptophan residues on cellular proteins [1, 48, 119]. A SNO modification may result from cGMP-dependent or independent processes [119]. Database analysis [113] indicates 2-3 sites within all TnI isoforms are predicted targets for SNO modification, and a single Cys residue is targeted for SNO in the both cTnC and skeletal TnC (rat cTnC Cys35). However, TnT isoforms are not predicted to be a SNO target. A recent study demonstrated the presence of SNO modification in one of the cTnI sites (equivalent to rat cTnICys97) and a cTnC site (rat cTnCCys35 equivalent; 105), although only modest to low SNO occupancy of cTn subunits is reported in this work. While SNO is expected to protect against further protein oxidation during cell stress, this PTM competes against signaling pathways producing oxidation, glutathionylation, and cross-linking [119], and there may be additional roles in thin filament proteins such as Tn. In future work, it will be important to identify the environmental conditions favoring this PTM in cardiac and skeletal myofilaments along with determining the impact of SNO modifications on cardiac and skeletal muscle function and the role played by Tn and other myofilament proteins.

While the formation of nitric oxide (NO) and downstream N-linked nitrosylation of proteins in response to cellular stress are most often positive adaptations, the presence of protein nitration reduces contractile function and increases protein degradation [48]. Most cardiovascular studies on nitration focus on peroxynitrite production in response to nitric oxide (NO) synthesis and oxygen derived free radical or ROS production [1, 48], although more than one signaling pathway can produce post-translational nitration [1]. The conversion of tyrosine to 3-nitrotyrosine is a common marker of nitration, but several amino acid residues can be targeted for modification [1]. Currently, there is disagreement about the presence and role Tn nitration plays in heart and skeletal muscle. The direct addition of peroxynitrite significantly impairs myofilament contractile function in permeabilized trabeculae and in an *in vitro* motility assay, but yielded only small changes in cardiac Tn nitration [148, 221]. However, nitration develops more slowly than oxidation and is often linked to chronic disease states such as aging [73, 90, 91]. Indeed, one study showed increased cTnI and cTnC nitration levels in the myocardium with age [90]. This outcome requires further validation, as increased cardiac Tn subunit nitration is not detected by other approaches and/or with advanced age [73]. In addition, there are no detected changes in Tn subunit nitration in aging skeletal muscle [91]. In contrast, there is evidence oxidative stress caused by repeated use of 3,4-methylenedioxymeth-amphetamine increases myocardial protein and cTnT nitration at a point in time when cardiac diastolic dysfunction is present [215]. Thus, studies are needed to verify and/or determine whether significant troponin nitration develops in response to cellular stress and/or ROS as well as the function nitration plays in myofilament function.

O-linked N-acetyl-D-glucosaminylation (O-GlcNAc). The reversible addition of monosaccharides to Ser and Thr residues on cytosolic and nuclear proteins by O-GlcNAc transferases develops with elevated glucose levels [30]. The appearance of this PTM is increasingly recognized as a mechanism for pathological signaling in striated muscle [30]. Increases in O-GlcNAc protein modifications appear to protect against cell stress, and multiple contractile proteins are targeted for this modification in both skeletal and cardiac muscle [29, 69, 190]. Functionally, the presence of this modification in contractile proteins is associated with a reversible decrease in myofilament Ca^{2+} sensitivity [69, 190]. Interestingly, cTnI Ser151 in cardiac tissue is identified as the target site, which suggests O-GlcNAc may either prevent phosphorylation at this site and/or counteract the functional influence of cTnISer151 phosphorylation by kinases such as AMPK and PAK. While other contractile proteins contain multiple O-GlcNAc modifications, current predictive algorithms indicate TnT and TnC isoforms are unlikely to be targets for this modification [30].

Acetylation. Acetylation appears to be an additional PTM capable of modifying sarcomere function. Amino-terminal acetylation of Tn subunits was established with the identification of the amino acid sequence for each Tn subunit. The amino-terminal methionine is absent from each of Tn subunit isoform and the new N-terminal residue (Ala, Ser, or Gly) is acetylated and/or blocked and assumed to be acetylated [58, 115, 174, 267, 268]. Recent studies also verified the presence of this N-terminal acetylation in TnI [284]. While the purpose or function of Tn subunit acetylation has not been directly addressed in the intact sarcomere, N-terminal acetylation of another thin filament protein, Tm, improves actin and troponin binding [72, 269]. In general, N-terminal acetylation also influences protein stability, as it can act as a degradation signal in some proteins or protect a protein against ubiquitination and degradation [3, 78].

Earlier biochemical studies also established that significant modifications of contractile function develop in response to acetylation of lysine residues within Tn [56, 71, 103]. *In vitro* Lys acetylation of TnC or TnI increased myofilament Ca^{2+} sensitivity, but questions remained about whether Lys acetylation of contractile proteins is a significant pathway in intact muscle and heart. Work by Gupta and colleagues addressed this question and showed a histone acetyltransferase (HATs) as well as a histone deacetylase (HDACs) are localized to the z-band of sarcomeres, and this association requires muscle LIM protein (MLP) [60]. In agreement with earlier biochemical work, conditions designed to increase Lys acetylation in contractile proteins also increased myofilament Ca^{2+} sensitivity. In recent work, proteomic approaches indicated multiple Lys residues may be acetylated in skeletal or cardiac tissue from rat and human [130]. It is curious that rat ssTnI is only acetylated on Lys95, while multiple sites are phosphorylated in human tissue as well as in cTnI and fsTnI. In addition, up to 5 residues are acetylated in TnC (Lys6, Lys17, Lys43, Lys90, Lys92; rat sequence) and 4-10 residues are acetylated in TnT isoforms. Based on these results, future studies are needed to identify Lys acetylation profiles under physiological and pathophysiological conditions in addition to investigating the potential functional role these internal Lys acetylations play in combination with phosphorylation.

SUMOylation. While most studies on troponin PTMs examined their presence and/or impact on sarcomeric function, new areas of investigation also could emerge from studies on troponin PTMs. For example, there are tantalizing clues Tn also plays a role in the nucleus, especially during development. Starting over a decade ago, sporadic reports began to be published showing troponin expression in non-myocytes [52]. More recently, investigators showed TnI expression in unfertilized Drosophila eggs [204] along with TnC and TnT associated with the nucleus in adult human cardiac myocytes [12, 88]. Now it is recognized that a small but significant fraction of actin, Tm, and TnI are localized within the nucleus [4, 142, 204, 288]. A novel role for Tn became clear when the mutation of loci containing the TnI gene not only reduced nuclear TnI expression but also caused changes in cell polarity and abnormal nuclear division throughout Drosophila embryos [204]. This work demonstrated nuclear TnI and Tm are necessary to maintain nuclear integrity. Further work is needed to understand the mechanism(s), and to test ideas about the role troponin plays in this process.

The work in *Drosophila* also demonstrated Tn PTMs played an important role in the nuclear translocation of TnI. Translocation of Tn could result from the presence of a nuclear localization signal (NLS), and a consensus, bipartite nuclear localization sequence (R/K-R/K-

$(X)_{10\text{-}12}$-R/$K_{3\text{-}5}$ (with R/$K_{3\text{-}5}$ indicating 3R or K residues within the 5 C-terminal residues) is present in fsTnT, ssTnT and cTnT (rat cTnT 160-178 and 197-213), and cTnI (rat cTnI 193-208). These sequences may contribute to TnT and TnI nuclear localization, although a similar sequence is not present in mammalian TnC isoforms. In addition, nuclear TnI is SUMOylated in *Drosophila* experiments, and mutation of this endogenous SUMOylation site (ϕ-K-X-E/D) resulted in the absence of nuclear TnI and abnormal nuclear division [204]. Each isoform of TnI and TnC contain one of these sites (rat cTnI 85-88; cTnC145-148). Interestingly, the phosphorylated Thr181 (human sequence) site detected in failing human hearts is located adjacent to the cTnI SUMOylation site. In addition, cTnISer198 and cTnT Ser199 and Thr204 also are located within cTnI and cTnT NLS sequences. The close proximity of phosphorylated, oxidized and SUMOylated modifications within TnI and TnT suggests future studies to understand the role of kinases targeting these residues, their impact on PTMs, and interactions among PTMs may provide insight into the trafficking and/or function of nuclear Tn residues.

Acknowledgements

This work was supported by NIH R01 grant HL067254 and R21 HL089093. The author thanks Sarah Lang for critical review of the manuscript.

References

[1] Abello N, Kerstjens HA, Postma DS, Bischoff R. Protein tyrosine nitration: selectivity, physicochemical and biological consequences, denitration, and proteomics methods for the identification of tyrosine nitrated proteins. *J. Proteome Res.* 8:3222-3238,2009.

[2] Abraham SJ, Kobayashi T, Solaro RJ and Gaponenko V. Differences in lysine pKa values may be used to improve NMR signal dispersion in reductively methylated proteins. *J. Biomol. NMR* 43:239-246,2009.

[3] Arnesen T. Towards a functional understanding of protein N-terminal acetylation. *PLOS Biology* 9:e1001074,2011.

[4] Asumda FZ, Chase PB. Nuclear cardiac troponin and tropomyosin are expressed early in cardiac differentiation of rat mesenchymal stem cells. *Differentiation* 83:105-115,2012.

[5] Avkiran M, Rowland AJ, Cuello F, Haworth RS. Protein kinase D in the cardiovascular system: Emerging roles in health and disease. *Circ. Res.* 102:157-163,2008.

[6] Avner BS, Shioura KM, Scruggs SB, Grachoff M, et al., Myocardial infarction in mice alters sarcomere function via post-translational protein modification. *Mol. Cell Biochem.* 363:203-215,2012.

[7] Bain J, Plater L, Elliott M, Shapiro N, et al., The selectivity of protein kinase inhibitors: A further update. *Biochem. J.* 408:297-315,2007.

[8] Balasubramanian S, Kupuswamy D.RGD-containing peptides activate S6K through β_3Integrin in adult cardiac muscle cells. *J. Biol. Chem.* 278:42214-42224,2003.

[9] Bardswell SC, Cuello F, Rowland AJ, Sadayappan S, et al., Distinct sarcomeric substrates are responsible for protein kinase D-mediated regulation of cardiac myofilament Ca^{2+} sensitivity and cross-bridge cycling. *J. Biol. Chem.* 285:5674-5682,2010.

[10] Bartel S, Stein B, Eschenhagen T, Mende U, et al., Protein phosphorylation in isolated trabeculae from non-failing and failing human hearts. *Mol. Cell Biochem.* 157: 171-179,1996.

[11] Baryshnikova OK, Li MX, Sykes BD. Modulation of cardiac troponin C function by the cardiac-specific N-termnus of troponin I. *J. Mol. Biol.* 375:735-751,2008.

[12] Bergmann O, Bhardwaj RD, Bernard S, Zdunek S, et al., Evidence for cardiomyocyte renewal in humans.*Science*324:98-102,2009.

[13] Blom N, Sicheritz-Pontent T, Gupta R, Gammeltoft S, et al., Prediction of post-translational glycoslyation and phosphorylation of proteins from the amino acid sequence. *Proteomics* 4:1633-1649,2004.

[14] Blumenthal DK, Stull JT, Gill GN. Phosphorylation of cardiac troponin by guanosine 3':5'-monophosphate-dependent protein kinase.*J. Biol. Chem.*253:334-336,1978.

[15] Bodor G, Oakeley A, Allen P, Crimmins D, et al., Troponin I phosphorylation in the normal and failing adult human heart *Circulation* 96:1495-1500,1997.

[16] Bogoyevitch MA, Parker PJ, Sugden PH. Characterization of protein kinase C isotype expression in adult rat heart. *Circ. Res.* 72:757-767,1993.

[17] Boknik P, Khorchidi S, Bodor GS, Huke S, Knapp J, Linck B, Luss H, Muller FU, Schmitz W, Neumann J. Role of protein phosphatases in regulation of cardiac inotropy and relaxation. *Am. J. Physiol.*280:H786-H794,2001.

[18] Bossuyt J, Chang C-W, Helmstadter K, Kunkel MT, et al., Spatiotemporally distinct protein kinase D activation in adult cardiomyocytes in response to phenylaphrine and endothelin. *J. Biol. Chem.* 286:33390-33400, 2011.

[19] Braz JC, Gregory K, Pathak A, Zhao W, et al., PKC-alpha regulates cardiac contractility and propensity toward heart failure. *Nat. Med.* 10:248-254,2004.

[20] Browne GJ, Dickens M. Reversible translocation of ASK1 to a triton-X100 insoluble cytoplasmic compartment during cardiac myocyte cell stress. *J Recept Signal Transduct Res* 31:40-415, 2011.

[21] Brusq J-M, Mayoux E, Guigui L, Kirilovsky J. Effect of C-type natriuretic peptide on rat cardiac contractility.*Br. J. Pharmacol.* 128:206-212,1999.

[22] Bupha-Intr T, Holmes JW, Janssen PML. Induction of hypertrophy *in vitro* by mechanical loading in adult myocardium. *Am. J. Physiol.*293:H3759-H3767,2007.

[23] Burkart EM, Sumandea MP, Kobayashi T, Nili M, et al., Phosphorylation or glutamic acid substitution at protein kinase C sites on cardiac troponin I differentially depress myofilament tension and shortening velocity. *J. Biol. Chem.* 278:11265-11272,2003.

[24] Buscemi N, Foster DB, Neverova I, VanEyk JE.p21-Activated kinase increases the calcium sensitivity of rat triton-skinned cardiac muscle fiber unless via mechanism potentially involving novel phosphorylation of troponin I. *Circ. Res.* 91:509-516,2002.

[25] Canton M, Menazza, Sheeran FL, deLaureto PP, et al., Oxidation of myofibrillar proteins in human heart failure. *J. Am. Coll Cardiol.*57:300-309,2011.

[26] Chaudhri VK, Kumar D, Misra M, Dua R, et al., Integration of a phosphatase cascade with the mitogen-activated protein kinase pathway provides for a novel signal processing function. *J. Biol. Chem.*285:1296-1310,2010.

[27] Chen F, Kan H, Hobbs G, Finkel MS.p38 MAP kinase inhibitor reveres stress-induced myocardial dysfunction *in vivo*. *J. Appl. Physiol.*106:1132-1141,2009.

[28] Chen, G, Zhou X, Nicolaou P, Rodriguez P, et al., A human polymorphism of protein phosphatase-1 inhibitor-1 is associated with attenuated contractile response of crdiomyocytes to b-adrenergic stimulation. *FASEB J.* 22:179-1796,2007.

[29] Cieniewski-Bernard C, Bastide B, Lefebvre T, Lemoine J, et al., Identification of O-linked N-acetylglucosamine proteins in rat skeletal muscle using two-dimensional gel electrophoresis and mass spectrometry. *Mol. Cell Proteomics*3:577-585,2004.

[30] Cieniewski-Bernard C, Montel V, Stevens L, Bastide B.O-GlcNAcylation, an original modulator of contractile activity in striated muscle. *J. Musc. Res. Cell Motil.* 30:281-287,2009.

[31] Cohen P. The structure and regulation of protein phosphatases. *Ann. Rev. Biochem.* 58:453-508,1989.

[32] Cohen PTW. Protein phosphatase 1 – targeted in many directions. *J. Cell Sci.* 115:241-256,2002.

[33] Cuello F, Bardswell SC, Haworth RS, Ehler E, et al., Novel role for p90 ribosomal S6 kinase in the regulation of cardiac myofilament phosphorylation.*J. Biol. Chem.*286:5300-5310,2011.

[34] Cuello F, Bardswell SC, Haworth RS, Yin X, et al., Protein kinase D selective targets cardiac troponin I and regulates myofilament Ca^{2+} sensitivity in ventiruclar myocytes, *Circ. Res.* 100:864-873,2007.

[35] D'Cruz LG, Baboonian C, Phillimore HE, Taylor R, et al., Cytosine methylation confers instability on the cardic troponin T gene in hypertrophic cardiomyopathy. *J. Med. Genet.* 37:E18,2000.

[36] Davies SP, Reddy H, Caivano M, Cohen P. Specificity and mechanism of action of some commonly used protein kinase inhibitors. *Biochem. J.* 351:95-105,2000.

[37] DeGrande ST, Little SC, Nixon DJ, Wright P, et al., Molecular mechanisms underlying cardiac protein phosphatase 2A regulation in heart. *J. Biol. Chem.* 288:1032-1046,2013.

[38] Deshmukh PA, Blunt BC, Hofmann PA. Acute modulation of PP2A and troponin I phosphorylation in ventricular myocoytes: Studies with a novel PP2A peptide inhibitor. *Am. J. Physiol.* 292:H792-HJ799,2007.

[39] Dong X, Sumandea CA, Chen Y-C, Garcia-Cazarin ML, et al., Augmented phosphorylation of cardiac troponin I in hypertensive heart failure. *J. Biol. Chem.* 287:848-857,2012.

[40] Dubois E, Richard V, Mulder P, Lamblin N, et al., Decreased serine 207 phosphorylation of troponin T as a biomarker for left ventricular remodeling after myocardial infarction. *Eur. Heart J.* 32:115-123,2011.

[41] Duncan JG, Ravi R, Stull LB, Murphy AM. Chronic xanthine oxidase inhibition prevents myofibrillar protein oxidation and preserves cardiac function in a transgenic mouse model of cardiomyopathy. *Am. J. Physiol.*289:H1512-H1518,2005.

[42] Duquesnes N, Lezoualc'h G, Crozatier B.PKC-delta and PKC-epsilon: Foes of the same family or strangers? *J. Mol. Cell Cardiol.*51:665-673,2011.

[43] England PJ. Correlation between contraction and phosphorylation of the inhibitory subunit of troponin in perfused rat heart. *FEBS Lett.* 50:57-60,1975.

[44] Erikson E and Maller JL. Substrate specificity of ribosomal protein S6 kinase II from *Xenopus* eggs. *Second Messengers Phosphoproteins* 12:135-143,1988.

[45] Favre B, Zolnierowicz S, Turowski P, Hemmings BA. The catalytic subunit of protein phosphatase 2A is carboxyl-methylated *in vivo*.*J. Biol. Chem.*269:16311-16317,1994.

[46] Fedorova M, Todorovsky T, Kuleva N, Hoffmann R. Quantitative evaluation of tryptophan oxidation in actin and troponin I from skeletal muscles using a rat model of acute oxidative stress.*Proteomics*10:2692-2700,2010.

[47] Feng Y, Cao HQ, Liu Z, Ding JF, et al., Identification of the dual specificity and the funtional domains of the cardiac-specific protein kinase TNNI3K.*Gen. Physiol. Biophys.* 26:104-109,2007.

[48] Ferdinandy P, Danial H, Ambrus I, Rothery RA, et al.,Peroxynitrite is a major contributor to cytokine-induced myocardial contractile facilure. *Circ. Res.* 87:241-247,2000.

[49] Fielitz J, Kim MS, Shelton JM, Qi X, et al., Requirement of protein kinase D1 for pathological cardiac remodeling. *PNAS* 105:3059-3063,2008.

[50] Ford-Speelman DL, Roche JA, Bowman AL, Bloch RJ. The Rho-guanine nucleotide exchange factor domain of obscurin activates RhoA signaling in skeletal muscle. *Mol. Biol. Cell* 20:3905-3917,2009.

[51] Francis SH, Busch JL, Corbin JD. cGMP-Dependent protein kinase and cGMP phosphodiesterases in nitric oxide and cGMP action. *Pharmacol. Rev.* 62:525-563,2010.

[52] Gahlmann R, Wade R, Gunning P, Kedes L. Differential expression of slow and fast skeletal troponin C is expressed in human fibroblasts. *J. Mol. Biol.* 201:379-391,1988.

[53] Gergs U, Boknik P, Buchwalow I, Fabritz L, et al., Overexpression of the catalytic subunit of protein phosphatase 2A impairs cardiac function. *J. Biol. Chem.* 39:40827-40834,2004.

[54] Glantschnig H, Rodan GA, Reszka AA. Mapping of MST1 Kinase Sites of Phosphorylation *J. Biol. Chem.* 277:42987-42996,2002.

[55] Gorza L, Menabo R, Vitadello M, Bergamini CM, et al., Cardiomyocyte troponin T immunoreactivity is modified by cross-linking resulting from intracellular calcium overload. *Circulation* 93:1896-1904,1996.

[56] Grabarek Z, Mabuchi Y, Gergely J. TnC acetylation at lysine residues. *Biochemistry* 34:11872-11881,1995.

[57] Grabarek Z, Tan RY, Wang J, Tao T, et al., Inhibition of mutant troponin C activity by an intra-domain disulphide bond. *Nature* 345:132-135,1990.

[58] Grand RJA, Wilkinson JM.The amino acid sequence of rabbit cardiac troponin I. *Biochem. J.* 159:633-641,1976.

[59] Guo J, Gertsberg Z, Ozgen N, Sabri A, et al., Protein kinase D isoforms are activated in an agonist-specific manner in cardiomyocytes. *J. Biol. Chem.* 286:6500-6509,2011.

[60] Gupta MP, Samant SA, Smith SH, Shroff SG.HDAC4 and PCAF bind to cardiac sarcomeres and play a role in regulating myofilament contractility. *J. Biol. Chem.* 283:10135-10146,2008.

[61] Gupta RC, Neumann J, Watanabe AM, Lesch M, et al., Evidence for the presence and hormonal regulation of protein phosphatase inhibitor-1 in ventricular cardiomyocyte. *Am. J. Physiol.* 270:H1159-H1164,1996.

[62] Gusev NB, Dobrovolskii AB, Severin SE. Isolation and some properties of troponin T kinase from rabbit skeletal muscle. *Biochem. J.* 189:219-226,1980.

[63] Hamdani N, Bishu KG, von Frieling-Salewsky M, Redfield MM, et al., Deranged myofilament phosphorylation and function in experimental heart failure with preserved ejection fraction. *Cardiovasc. Res.* 97:464-71,2013.
[64] Hannon JD, Chase PB, Martyn DA, Huntsman LL, et al., Calcium-independent activation of skeletal muscle fibers by a modified form of cardiac troponin C. *Biophys. J.* 64:1632-1637,1993.
[65] Haworth RS, Cuello F, Herron TJ, Franzen G, et al., Protein kinase D is a novel mediator of cardiac troponin I phosphorylation and regulates myofilament function. *Circ. Res.* 95:1091-1099,2004.
[66] Haworth RS, Goss MW, Rozengurt E, Avkiran M. Expression and activity of protein kinase D/Protein kinase C in myocardium. *J. Mol. Cell Cardiol.*32:1013-1023,2000.
[67] Haworth RS, Roberts NA, Cuello F, Avkiran M. Regulation of protein kinase D activity in adult myocardium. *J. Mol. Cell Cardiol.* 43:686-695,2007.
[68] He X Liu Y Sharma V, Dirksen RT Waugh R, et al., ASK1 associates with troponin T and induces troponin T phosphorylation and contractile dysfunction in cardiomyocytes. *Am. J. Pathol.* 163:243-251,2003.
[69] Hedou J, Cieniewski-Bernard C, Leroy Y, Michalski JC, et al., O-linked N-acetylglucosaminylation is involved in the Ca^{2+} activation properties of rat skeletal muscle. *J. Biol. Chem.* 282:10360-10369,2007.
[70] Herron TJ, Korte S, McDonald KS. Power output is increased after phosphorylation of myofibrillar protein sin rat skinned cardiac myocytes. *Circ. Res.* 89:1184-1190,2001.
[71] Hitchcock-Degregori SE. Study of the structure of troponin-I by measuring the relative reactivities of lysines with acetic anhydride. *J. Biol. Chem.*257.7372-7380,1982.
[72] Hitchcock-DeGregori SE, Heald RW. Altered actin and troponin binding of amino-erminal variants of chicken striated muscle α-tropomyosin expressed in *Escherichia coli*. *J. Biol. Chem.* 262:9730-9735,1987.
[73] Hong SJ, Gokulrangan G, Schoneich C. Proteomic analysis of age dependent nitration of rat cardiac proteins by solution isoelectric focusing coupled to nano-HPLC tandem mass spectrometry. *Exp. Gerontol.* 42:639-651,2007.
[74] Hu J, Rho HS, Newman RH, Hwang W, et al., Global analysis is phosphorylation networks in humans. *Biochim. Biophys. Acta*pII: S1570-9639(13)00122-2. doi.org/10.1016/j.bbapap.2013.03.009. [Epub ahead of print], 2013.
[75] Huang L, Wolska BM, Montgomery DE, Burkart EM, et al., Increased contractility and altered Ca^{2+} transients of mouse heart myocytes conditionally expressing PKCβ.*Am. J. Physiol.* 280:C1114-C1120,2001.
[76] Huang X, Walker JW. Myofilament anchoring of protein kinase C-epsilon in cardiac myocoytes. *J. Cell Sci.* 117: 1971-1978,2004.
[77] Hunlich M, Begin KJ, Gorga JA, Fishbaugher DE, et al., Protein kinase A mediated modulation of acto-myosin kinetics. *J. Mol. Cell Cardiol.*38:119-125,2005.
[78] Hwang C-S, Shemorry A, Varshavsky A.N-terminal acetylation of cellular proteins creates specific degradation signals. *Science*327:973-977,2010.
[79] Hwang H, Robinson D, Stevenson TK, Rogers JB, et al., Agonist activated $PKC\beta_2$ translocation and modulation of cardiac myocyte contractile function. *Sci. Rep.*3:1971,2013.
[80] Hwang H, Robinson DA, Stevenson TK, Wu HC, et al.,$PKC\beta_{II}$ modulation of myocyte contractile performance. *J. Mol. Cell Cardiol.*53:176-186,2012.

[81] Iskratsch T, Lange S, Dwyer J, Kho AL, et al., Formin follows function: A muscle-specific isoform of FHOD3 is regulated by CK2 phoshorylation and promotes myofibril maintenance. *J. Cell Biol.* 191:1159-1172,2010.

[82] Itoh S, Ding B, Bains CP, Wang N, et al., Role of p90 ribosomal S6 kinase (p90RSK) in reactive oxygen species and protein kinase C β (PKC-β)-mediated cardiac troponin I phosphorylation. *J. Biol. Chem.* 280:24135-24142,2005.

[83] Iwasa T, Inoue N, Fukunaga K, Isobe T, et al., Purification and characterization of a multifunctional calmodulin-dependent protein kinase from canine myocardial cytosol. *Arch. Biochem. Biophys.* 248:21-29,1986.

[84] Jaffer ZM, Chernoff J.p21-Activated kinases: Three more join the Pak. *Intl. J. Biochem. Cell Biol.* 34:713-717,2002.

[85] Jaquet K, Fukunaga K, Miyamoto E, Meyer HE.A site phosphorylated in bovine cardiac troponin T by CaMKII. *Biochim. Biophys. Acta* 1248:193-195,1995.

[86] Jideama NM, Crawford BH, Hussain AKM, Raynor RL. Dephosphorylation specificities of protein phosphatase for cardiac troponin I, troponin T, and sites within troponin T. *Int. J. Biol. Sci.* 2:1-9,2006.

[87] Jideama NM, Noland TA Jr, Raynor RL, Blobe GC, et al., Phosphorylation specificities of protein kinase C isozymes for bovine cardiac troponin I and troponin T and sites within these proteins and regulation of myofilament properties. *J. Biol. Chem.* 271:23277-23283,1996.

[88] Kajstura J, Urbanek K, Perl S, Hosoda T, et al., Cardiomyogenesis in the adult human heart. *Circ. Res.* 107:305-315,2010.

[89] Kang M, Walker JW. Protein kinase C δ and ε mediate positive inotropy in adult ventricular myocytes. *J. Mol. Cell Cardiol.* 38:753-764,2005.

[90] Kanski J, Behring A, Pelling J, Shoneich C. Proteomic identification of 3-nitrotyrosine-containing rat cardiac proteins: effects of biological aging. *Am. J. Physiol.* 288:H371-H381,2005.

[91] Kanski J, Hong SJ, Schoneich C. Proteomic analysis of protein nitration in aging skeletal muscle and identification of nitrotyrosine-containing sequences *in vivo* by nanoelectrospray ionization tandem mass spectrometry. *J. Biol. Chem.* 280:24261-24266,2005.

[92] Katoh N, Wise BC, Kuo JF. Phosphorylation of cardiac troponin inhibitory subunit (troponin I) and tropomyosin-binding subunit (troponin T) by cardiac phospholipid-sensitive Ca^{2+}-dependent protein kinase. *Biochem. J.*209:189-195,1983.

[93] Kaye DM, Wivott SD, Kelly RA. Activation of nitric oxide synthase (NOS3) by mechancal activity alters contractile activity in a Ca^{2+}-independent manner in cariac myocytes. *Biochem. Biophys. Res. Comm.* 256:398-403,1999.

[94] Ke Y, Sheehan KA, Eroume K, Egom A, et al., Novel bradykinin signaling in adult rat cardiac myocytes through activation of p21-activated kinase. *Am. J. Physiol.* 298:H1283-H1289,2010.

[95] Ke Y, Wang L, Pyle WG, de Tombe PP, et al., Intracellular localization and functional effects of p21-activated kinase-1 (Pak1) in cardiac myocytes. *Circ. Res.* 94:194-200,2004.

[96] Keane NE, Quirk PG, Gao Y, Patchell VB, et al., The ordered phosphorylation of cardiac troponin I by the cAMP-dependent protein kinase. *Eur. J. Biochem.* 248: 329-337,1997.

[97] Kedar V, McDonough H, Arya R, Li H-H, et al., Muscle-specific RING finger 1 is a bona fide ubiquitin ligase that degrades cardiac troponin I.*PNAS* 101:18135-18140,2004.

[98] Kelly ML and Chernoff J. Getting smart about p21-activated kinases. *Mol. Cell Biol.* 31:386-387,2011.

[99] Kelly ML, Chernoff J. Mouse models of PAK function. *Cellular Logistics*22:84-88,2012

[100] Kemp BE, Pearson RB. Protein kinase recognition sequence motifs.*TIBS*15:342-346,1990.

[101] Kentish JC, McCloskey DT, Layland J, Palmer S, et al., Phosphorylation of troponin I by protein kinase A accelerates relaxation and crossbridge cycle kinetics in mouse ventricular muscle. *Circ. Res.* 88:1059-1065,2001.

[102] Kirk JA, MacGowan GA, Evans C, Smith SH, et al., Left ventricular and myocardial function in mice expressing constitutively pseudophosphorylated cardiac troponin I. *Circ. Res.* 105:1232-1239,2009.

[103] Kobayashi T, Grabarek Z, Gerbely J, Collins JH. Extensive interactions between troponins C and I. *Biochemistry* 34:10946-10952,1995.

[104] Kobayashi T, Yang X, Walker LA, VanBreemen RB, et al., A non-equilibrium isoelectric focusing method to determine states of phosphorylation of cardiac troponin I.*J. Mol. Cell Cardiol.* 38:213-218,2005.

[105] Kohr MJ, Aponte A, Sun J, Gucek M, et al., Measurement of S-nitrosylation occupancy in the myocardium with cysteine-reactive tandem mass tags. *Circ. Res.* 111:1308-1312,2012.

[106] Kumar D, Dua R, Srikanth R, Jayaswal S, et al., Cellular phosphatases facilitate combinatorial processing of receptor-activated signals. *BMC Res. Notes* 1:81-92,2008.

[107] Lang SE, Robinson DA, Wu HC, Herron TJ, et al., Myofilament incorporation and contractile function after gene transfer of cardiac troponin-ISer43/45Ala. *Arch. Biochem. Biophys.*535:49-55,2013.

[108] Larrea MD, Hong F, Wander SA, da Silva TG, et al., RSK1 drives p27Kip1 phosphorylation at T198 to promote RhoA inhibition and increase cell motility.*PNAS*106:9268-9263,2009.

[109] Layland J, Li J-M, Shah AM. Role of cyclic GMP-dependent protein kinase in the contractile response to exogenous nitric oxide in rat cardiac myocytes. *J. Physiol.*540:457-467,2002.

[110] Layland J, Solaro RJ, Shah AM. Regulation of cardiac contractile function by troponin I phosphorylation. *Cardiovasc. Res.* 66:12-21,2005.

[111] Lee DI, Vahebi S, Tocchetti CG, Barouch LA, et al., PDE5A suppression of acute β-adrenergic activation requires modulation of myocyte β_3 signaling coupled to PKG-mediated troponin I phosphorylation. *Basic Res. Cadiol.* 105:337-347,2010.

[112] Lee J, Stock J. Protein phosphatase 2A catalytic subunit is methyl-esterified at its carboxyl terminus by a novel methyltransferase. *J. Biol. Chem.* 268:19192-19195,1993.

[113] Lee T-Y, Chen Y-J, Lu C-T, Ching C-T, et al., dbSNO:a database of cysteine S-nitrosylation. *Bioinformatics* 28:2293-2295,2012.

[114] Lessey EC, Guilluy C, Burridge K. From mechanical force to RhoA activation. *Biochemistry* 51:7420-7432,2012.

[115] Leszyk J, Dumaswala R, Potter JD, Gusev NB, et al., Bovine cardiac troponin T:Amino acid sequences of the two isoforms. *Biochemistry* 26:7035-7042,1987.

[116] Li J, Kritzer MD, Michel JJ, Le A, et al., Anchored p90 ribosomal S6 kinase 3 is required for cardiac myocyte hypertrophy. *Circ. Res.* 112:128-139,2013.

[117] Li P-F, Li J, Muller E-C, Otto A, et al., Phosphorylation oy protein kinase CK2:A signaling switch for the caspase-inhibiting protein ARC. *Molecular Cell* 10:247-258,2002.

[118] Liao R, Wang C-K and Cheung HC. Coupling of calcium to the interaction of troponin I with troponin C from cardiac muscle. *Biochemistry* 33:12729-12734,1994.

[119] Lima B, Forrester MT, Hess DT, Stamler JS.S-Nitrosylation in cardiovascular signaling. *Circ. Res.* 106:633-646,2010.

[120] Lincoln TM, Corbin JD. Purified cyclic GMP-dependent protein kinase catalyzes the phosphorylation of cardiac troponin inhibitor subunit. *J. Biol. Chem.*253:337-339,1978.

[121] Ling S, Sun Q, Li Y, Zhang L, et al., CKIP-1 inhibits cardiac hypertrophy by regulating class II histone deacetylase phosphorylation through recruiting PP2A. *Circulation* 126:3028-2040,2012.

[122] Liu Q, Hofmann PA. Modulation of protein phosphatase 2A by adenosine A1 receptors in cardiomyocytes: Role for p38 MAPK. *Am. J. Physiol.* 285:H97-H103,2003.

[123] Liu Q, Molkentin JD. Protein kinase Cα as a heart failure therapeutic target. *J. Mol. Cell Cardiol.*51:474-478,2011.

[124] Liu Q, Chen X, MacDonnell SM, Kranias EG, et al., Protein kinase-Cα, but not PKCβ or PKCγ, regulates contractility and heart failure susceptibility. *Circ. Res.* 105:194-200,2009.

[125] Liu Q, Sargent MA, York AJ, Molkentin JD.ASK1 regulates cardiomyocyte death but not hypertrophy in transgenic mice. *Circ. Res.* 105:1110-1117,2009.

[126] Liu X, Ramjiganesh T, Chen Y-H, Chung SW, et al., Disruption of striated peferentially expressed gene locus leads to dilated cardiomyopathy in mice. *Circulation* 119:261-268,2009.

[127] Lu Q-W, Hinken AC, Patrick SE, Solaro RJ, et al., Phosphorylation of cardiac troponin I at protein kinase C site threonine144 depresses cooperative activation of thin filaments. *J. Biol. Chem.* 285:11810-11817,2010.

[128] Luckey SW, Walker LA, Smyth T, Mansoori J, et al., The role of Akt/GSK-3β signaling in familial hypertrophic cardiomyopathy. *J. Mol. Cell Cardiol.* 46:739-747,2009.

[129] Luft FC. Hearts of this ILK rely on TNNI3K, a MAPKKK that regulated TNNI3.*J. Mol. Med.* 81:279-280,2003.

[130] Lundby A, Lage K, Welnert BT, Bekker-Jensen DB, et al., Proteomic analysis of lysine acetylation sites in rat tissues reveals organ specificity and subcellular patterns. *Cell Reports* 4:419-431,2012.

[131] Luo Y, Li B, Yang G, Gergely J, Tao T. Cross-linking between the regulatory regions of troponin-I and troponin-C abolishes the inhibitory function of troponin. *Biochemistry* 41:12891-12898,2002.

[132] Luss H, Klein-Wiele O, Boknik P, Herzig S, et al., Regional expression of protein phosphatase type 1 and 2A catalytic subunit isoforms in the human heart. *J. Mol. Cell Cardiol.* 32:2349-2359,2000.

[133] MacGowan GA, Du C, Cowan DB, Stamm C, et al., Ischemic dysfuntion in transgenic mice expressing troponin I lacking protein kinase C phosphorylation sites. *Am. J. Physiol.* 280:H835-H843,2001.

[134] MacGowan GA, Evans C, Hu TC-C, Debrah D, et al., Troponin I protein kinase C phosphorylation sites and ventricular function. *Cardiovasc. Res.* 63:245-255,2004.

[135] Manning G, Whyte DB, Martinez R, Hunter T, et al., The protein kinase complement of the human genome. *Science* 298:1912-1934,2002.

[136] Manser E, Lim L. Roles of PAK family kinases. *Prog. Mol. Subcell Biol.* 22:115-133,1999.

[137] Marshall M, Anilkumar N, Layland J, Walker SJ, et al., Protein phosphatase 2A contributes to the cardiac dysfunction induced by endotoxemia. *Cardiovasc. Res.* 82:67-76,2009.

[138] Martin DDO, Ahpin CY, Heit RJ, Perinpanayagam MA, et al., Tandem reporter assay for myristoylated proteins post-translationally (TRAMPP) identifies novel substrates for post-translational myristoylation: PKCε, a case study. *FASEB J.* 26:13-28,2012.

[139] Martin M, Kettmann R, Dequiedt F. Recent insights into protein phosphatase 2A structure and regulation. *Biotechnol Agron Soc. Environ.* 14:243-252,2010.

[140] Mazzei GJ, Kuo JF. Phosphorylation of skeletal-muscle troponin I and troponin T by phospholipid-sensitive Ca^{2+}-dependent protein kinase and its inhibition by troponin C and tropomyosin. *Biochem. J.* 218:361-369,1984.

[141] McConnell BK, Moravek CS, Bond M. Troponin I phosphorylation and myofilament calcium sensitivity during decompensated cardiac hypertrophy. *Am. J. Physiol.*274:H385-H396,1998.

[142] McDonald D, Carrero G, Andrin C, de Vries G, et al., Nucleo-plasmic β-actin exists in a dynamic equilibrium between low-mobility polymeric species and rapidly diffusing species. *J. Cell Biol.* 172:541-552,2006.

[143] McDonough JL, VanEyk JE. Developing the next generation of cardiac markers. *Prog. Cardiovasc. Dis.* 47:207-216,2004.

[144] McDonough JL, Arrell DK, VanEyk JE. Troponin I degradation and covalent complex formation accompanies myocardial ischemia/reperfusion injury. *Circ. Res.* 84:9-20,1999.

[145] Meggio F, Pinna LA. One-thousand-and-one substrates of protein kinase CK2?*FASEB J.* 17:349-368,2003.

[146] Merida I, Avila-Flores A, Merino E. Diacylglycerol kinases: at the hub of cell signaling. *Biochem. J.* 409:1-18,2008.

[147] Meurs KM and Mealey KL. Evaluation of the flanking nucleotide sequences of sarcomeric hypertophic cardiomyopathy substitution mutations. *Mutation Res.* 642:86-89,2008.

[148] Mihm MJ, Yu F, Reiser PJ, Bauer JA. Effects of peroxynitrite on isolated cardiac trabeculae: selective impact on myofibrillar energetic controllers. *Biochimie* 85:587-596,2003.

[149] Miller ML, Jensen LJ, Diella F, Jorgensen C, et al., Linear motif atlas for phosphorylation-dependent signaling. *Sci. Signal* 1(35):ra2,2008.
[150] Miranda L, Carpentier S, Platek A, Hussain N, et al., AMP-activated protein kinase induces actin cytoskeleton reorganization in epithelial cells. *Biochem. Biophys. Res. Comm.* 396:656-661,2010.
[151] Mittmann K, Jaquet K, and Heilmeyer LMG. Ordered phosphorylation of a duplicated minimal recognition motif for cAMP-dependent protein kinase present in cardiac troponin I.*FEBS Lett.* 302:133-137,1992.
[152] Moens AL, Leyton-Mange JS, Niu X, Yang R, et al., Adverse ventricular remodeling and exacerbated NOS uncoupling from pressure-overload in mice lacking the beta3-receptor. *J. Mol. Cell Cardiol.*47:576-585,2009.
[153] Moir AJ, Cole HA and Perry SV. The phosphorylation sites of troponin T from white skeletal muscle and the effects of interaction with troponin C on their phosphorylation by phosphorylase kinase. *Biochem. J.* 161:371-382,1977.
[154] Moir AJG, Perry SV. Phosphorylation of rabbit cardiac-muscle troponin I by phosphorylase kinase .*Biochem. J.* 191:547-554,1980.
[155] Mollica JP, Dutka TL, Merry TL, Lamboley CR, et al., S-glutathionylation of troponin I (fast) increases contractile apparatus Ca^{2+} sensitivity in fast-twitch muscle fibres of rats and humans. *J. Physiol.* 590:1443-63,2012.
[156] Monasky MM, Taglieri DM, Jacobson AK, Haizlip KM, et al., Post-translational modification of myofilament proteins involved in length-dependent prolongation of relaxation in rabbit right ventricular myocardium. *Arch. Biochem. Biophys.*535:22-29, 2013.
[157] Monasky MM, Taglieri DM, Patel BG, Chernoff J, et al., p21-activated kinase improves cardiac contractility during ischemia-reperfusion concomitant with changes in troponin-T and myosin light chain 2 phosphorylation. *Am. J. Physiol.* 302:H224-H230,2012.
[158] Moniotte S, Kobzik L, Feron O, Trochu JN, et al., Upregulation of β_3-adrenoceptors and altered contractile response to inotropic amines in human failing myocardium. *Circulation* 103:1649-1655,2001.
[159] Montgomery DE, Chandra M, Huang Q-Q, Jin J-P, et al., Transgenic incorporation of skeletal TnT into cardiac myofilaments blunts PKC-mediated depression of force. *Am. J. Physiol.* 280:H1011-H1018,2001.
[160] Montgomery DE, Wolska BM, Pyle WG, Roman BB, et al., α-Adrenergic response and myofilament activity in mouse hearts lacking PKC phosphorylation site on cardiac TnI. *Am. J. Physiol.*282:H2397-H2405,2002.
[161] Neulen, A, Blaudeck N, Zittrich S, Metzler D, et al., Mn^{2+}-dependent protein phosphatase 1 enhances protein kinase A-induced Ca^{2+} desensitization in skinned murine myocardium. *Cardiovasc. Res.*74:124-132,2007.
[162] Newton AC. Protein kinase C:Poised to signal. *Am. J. Physiol.*298:E395-402,2010.
[163] Nixon BR, Thawornkaiwong A, Jin J, Brundage EA, et al., AMP-activated protein kinase phosphorylates cardiac troponin I at Ser-150 to increase myofilament calcium sensitivity and blunt PKA-dependent function. *J. Biol. Chem.* 287:19136-19147,2012.
[164] Noguchi T, Hunlich M, Camp PC, Begin KJ, et al., Thin-filament based modulation of contractile performance in human heart failure. *Circulation* 110:982-987,2004.

[165] Noland TA Jr., Kuo JF. Protein kinase C phosphorylation of cardiac troponin I and troponin T inhibits Ca^{2+}-stimulated MgATPase activity in reconstituted actomyosin and isolated myofibrils, and decreases actin-myosin interactions. *J. Mol. Cell Cardiol.*25:53-65,1993.

[166] Noland TA JR, Guo X, Raynor RL, Jideama NM, Averyhart-Fullard V, Solaro RJ, Kuo JF. Cardiac troponin I mutants. *J. Biol. Chem.* 270:25445-25454,1995.

[167] Noland TA Jr., Kuo JF. Protein kinase C phosphorylation of cardiac troponion I or troponin T inhibits Ca^{2+}-stimulated actomyosin MgATPase activity. *J. Biol. Chem.*266:4974-4978,1991.

[168] Noland TA Jr., Raynor RL, Jideama NM, Guo X, et al., Differential regulation of caridac actomyosin S-1 MgATPase by protein kinase C isozyme-specific phosphorylation of specifisies in cardiac troponin I and its phosphorylation site mutants. *Biochemistry* 35:14923-14931,1996.

[169] Noland TA Jr., Raynor RL, Kuo JF. Identification of sites phosphorylated in bovine cardiac troponin I and troponin T by protein kinase C and comparative substrate activity of synthetic peptides containing the phosphorylation sites. *J. Biol. Chem.* 2264:20778-20785,1989.

[170] Obenauer JC, Cantley LC, Yaffe MB. Scansite 2.0:Proteome-wide prediction of cell signaling interactions using short sequence motifs. *Nucleic Acids Res.* 31:3635-3641,2003.

[171] Oliveira SM, Zhang Y-H, Solis RS, Isackson H, et al., AMP-activated protein kinase phosphorylates cardiac troponin I and alters contractility of murine ventricular myocytes. *Circ. Res.* 110:1192-1201,2012.

[172] Ouyang Y, Mamidi R, Jayasundar JJ, Changra M, et al., Structural and kinetic effects of PAK3 phosphorylation mimic of cTnI (S151E) on the cTnC-cTnI interaction in the cardiac thin filament. *J. Mol. Biol.* 400:1036-1045,2010.

[173] Park HS, Gong BJ Tao T.A disulfide crosslink between Cys98 of troponin C and Cys133 of troponin-I abolishes the activity of rabbit skeletal troponin. *Biophys. J.* 66:2062-2065,1994.

[174] Pearlstone JR, Carpenter MR, Johnson P, Smillie LB. Amino acid sequence of tropomyosin-binding component of rabbit skeletal muscle troponin. *Proc. Nat. Acad. Sci.*73:1902-1906,1976.

[175] Pearson RB, Kemp BE. Protein kinase phosphorylation site sequences and consensus specificity motifs. *Meth. Enzymol.* 200:62-81,1991.

[176] Perry SV, Cole HA. Phosphorylation of troponin and the effects on interactions between the components of the complex. *Biochem. J.* 141:733-743,1974.

[177] Pfleiderer P, Sumandea MP, Rybin VO, Wang C, et al., Raf-1: A novel cardiac troponin T kinase. *J. Muscle Res. Cell Motil* 30:67-72,2009.

[178] Pi YQ, Kemnitz KR, Zhang D, Kranias EG, et al., Phosphorylation of troponin I controls cardiac twitch dynamic: Evidence from phosphorylation site mutants expressed on a troponin I-null background in mice. *Circ. Res.* 90:649-656,2002.

[179] Pi YQ, Zhang D, Kemnitz, Wang H, et al., Protein kinse C and A sites on troponin I regulate myofilament Ca^{2+} sensitivity and ATPase activity in the mouse myocardium. *J. Physiol.*552:845-857,2003.

[180] Pierkes M, Gambaryan S, Boknik P, Lohmann SM, et al., Increased effects of C-type natriuretic peptide on cardiac ventricular contractility and relaxation in guanylyl cyclase A-deficient mice. *Cardiovasc. Res.*53:852-861,2002.

[181] Pinna LA, Allende JE. Protein kinase CK2: an ugly duckling in the kinome pond. *Cell Mol. Life Sci.* 66:1797-1799,2009.

[182] Pinto JR, de Sousa VP, Sorenson MM. Redox state of troponin C cysteine in the D/E helix alters the C-domain affinity for the thin filament of vertebrate striated muscle. *Biochim. Biophys. Acta* 1810:391-397,2011.

[183] Popola A, Autore G, Pinto A, Marzocco S. Oxidative stress in cardiovascular disease and chronic renal failure. *Free Radical Res.* In press, 2013.

[184] Puceat M, Hilal-Dandan R, Strulovici B, Brunton LL, et al., Differential regulation of protein kinase C isoforms in isolated neonatal and adult rat cardiomyocytes. *J. Biol. Chem.* 269:16938-16944,1994.

[185] Putkey JA, Dotson DG, Mouawad P. Formation of inter- and intra-molecular disulfide bonds can activate cardiac troponin C. *J. Biol. Chem.*268:6827-6830,1993.

[186] Pyle WG, Hart MC, Cooper JA, Sumandea MP, et al., Actin capping protein: An essential element in protein kinase signaling to the myofilaments. *Circ. Res.* 90:1299-1306,2002.

[187] Pyle WG, Sumandea MP, Solaro RJ, deTombe PP. Troponin I serines 43/45 and regulation of cardiac myofilament function. *Am. J. Physiol.*283:H12151214,2002.

[188] Raeker MO, Bieniek AN, Ryan AS, Tsai HJ, et al., Targeted deletion of the zebra fish obscurin A RhoGEF domain affects heart, skeletal muscle and brain development. *Dev. Biol.*337:432-43,2010.

[189] Raggi A, Grand RJA, Moir AJG, Perry SV. Structure-function relationships in cardiac troponin T. *Biochim. Biophys. Acta* 997:135-143,1989.

[190] Ramirez-Correa GA, Jin W, Wang Z, Zhong X, et al., O-linked GlcNAc modification of cardiac myofilament proteins: a novel regulation of myocardial function. *Circ. Res.* 103:1354-1358,2008.

[191] Reddy YS, Ballard D, Giri NY, Schwartz A. Phosphorylation of cardiac native tropomyosin and troponin. *J. Mol. Cell Cardiol.* 5:461-471,1973.

[192] Reiser PJ, Westfall MV, Schiaffino S, Solaro RJ. Tension production and thin-filament protein isoforms in developing rat myocardium. *Am. J. Physiol.* 267:H1589-1596,1994.

[193] Rennefahrt UE, Deacon SW, Parker SA, Devarajan K, et al., Specificity profiling of Pak kinases allows identification of novel phosphorylation sites. *J. Biol. Chem.* 282:15667-15678,2007.

[194] Reporter M. Protein synthesis in cultured muscle cells: methylation of nascent proteins. *Arch. Biochem. Biophys.* 158:577-585,1973.

[195] Rhee SG. Regulation of phosphoinositide-specific phospholipase C. *Ann. Rev. Biochem.* 70:281-312,2001.

[196] Rincon E, Gharbi SI, Santos-Mendoza T, Merida I. Diacylglycerol kinase ζ: at the crossroads of lipid signaling and protein complex organization. *Progress Lipid Res* 51:1-10,2012.

[197] Risnik VV, Dobrovolskii AB, Gusev NB, Severin SE. Phosphorylase kinase phosphorylation of skeletal-muscle troponin T. *Biochem. J.* 191:851-854,1980.

[198] Risnik VV, Gusev NB. Some properties of the nucleotide-binding site of troponin T kinase-casein kinase type II from skeletal muscle. *Biochim. Biophys. Acta* 790:108-116,1984.

[199] Robertson SP, Johnson JD, Holroyde MJ, Kranias EG, et al., The effect of troponin I phosphorylation on the Ca^{2+}-binding properties of the Ca^{2+}-regulatory site of bovine cardiac troponin. *J. Biol. Chem.* 257:260-263,1982.

[200] Robia SL, Ghanta J, Robu VG, Walker JW. Localization and kinetics of protein kinase C-ε anchoring in cardiac myocytes. *Biophys J* 80:2140-2151,2001.

[201] Robia SL, Kang M, Walker JW. Novel determinant of PKC-ε anchoring at cardiac z-lines. *Am. J. Physiol.*289:H1941-H1950,2005.

[202] Romeo Y, Zhang X, Roux PP. Regulation and function of the RSK family of protein kinases. *Biochem. J.* 441:553-569,2012.

[203] Ruse CI, Willard B, Jin JP, Haas T, et al., Quantitative dynamics of site-specific protein phosphorylation determined using liquid chromatography electrospray ionization mass spectrometry. *Anal. Chem.* 74:1658-1664,2002.

[204] Sahota VK, Grau BF, Mansilia A, Ferrus A.Troponin I and tropomyosin regulate chromosomal stability and cell polarity. *J. Cell Sci.* 122:2623-2631,2009.

[205] Sakthivel S, Finley NL, Rosevear PR, Lorenz JN, et al., *In vivo* and *in vitro* analysis of cardiac troponin I phsphorylation. *J. Biol. Chem.* 280:703-714,2005.

[206] Satoh S, Makino N. Intracellular mechanisms of cGMP-mediated regulation of myocardial contraction. *Basic Res. Cardiol.* 96:652-658,2001.

[207] Satoh S, Ueda Y, Koyanagi M, Kadokami T, et al., Chronic inhibition of Rho kinase blunts the process of left ventricular hypertrophy leading to cardiac contractile dysfunction in hypertension-induced heart failure. *J. Mol. Cell Cardiol.* 35:59-70,2003.

[208] Schild A, Ittner LM, Gotz J. Altered phosphorylation of cytoskeletal proteins in mutant protein phosphatase 2A transgenic mice. *Biochem. Biophys. Res. Comm.* 343:1171-1178,2006.

[209] Schriemer DC, Yalcin T, Li L.MALDI mass spectrometry combined with avidin-biotin chemistry for analysis of protein modifications. *Anal. Chem.* 70:1569-1575,1998.

[210] Sciduna BP, Tanck MW, Remme CA, Beekman L, et al., Quantitative trait loci for electrocardio-graphic parameters and arrhythmia in the mouse. *J. Mol. Cell Cardiol.*50:380-389,2011.

[211] Scruggs SB, Walker LA, Lyu T, Geenen DL, et al., Partial replacement of cardiac troponin I with a non-phosphorylatable mutant at servines 43/45 attenuates the contractile dysfunction associated with PKCε phosphorylation. *J. Mol. Cell Cardiol.*40:465-473,2006.

[212] Sfichi-Duke L, Garcia-Cazarin ML, Sumandea CA, Sievert GA, et al., Cardiomyopathy-causing ΔK210 in cardiac troponin-T alters phosphorylation propensity of sarcomeric proteins. *J. Mol. Cell Cardiol.* 48:934-942,2010.

[213] Sheehan KA, Ke Y, Solaro RJ.p21-Activated kinase-1 and its role in integrated regulation of cardiac contractility. *Am. J. Physiol.*293:R963-R973,2007.

[214] Sheehan KA, Ke Y, Wolska BM, Solaro RJ. Expression of active p21-activated kinase-1 induces Ca2+ flux modification with altered regulatory protein phosphorylation in cardiac myocytes. *Am. J. Physiol.* 296:C47-C58,2009.

[215] Shenouda SK, Lord KC, McIlwain E, Lucchesi PA, et al., Ecstasy produces left ventricular dysfunction and oxidative stress in rats. *Cardiovasc. Res.* 79:662-670,2008.

[216] Shi Y. Serine/Threonine Phosphatases: Mechanism through structure. *Cell* 139:468-485,2009.

[217] Silverstein AM, Davis AJ, Bielinski VA, Esplin ED, et al., Protein phosphatase A.*Handbook Cell Signal* 2:1353-1365,2010.

[218] Singh TJ, Akatsuka A, Blake KR, Huang KP.Phosphorylation of troponin and myosin light chain by cAMP-independent casein kinase-2 from rabbit skeletal muscle. *Arch. Biochem. Biophys.* 220:615-622,1983.

[219] Smolen P, Baxter DA, Byrne JH. Bistable MAP kinase activity. *Am. J. Physiol.* 294:C503-C515,2007.

[220] Smrcka AV, Brown JH, Holz GG. Role of phospholipase Cε in physiological phosphoinositide signaling networks. *Cell Signal*24:1333-1343,2012.

[221] Snook JH, Li J, Helmke BP, Guilford WH. Peroxynitrite inhibits myofibrillar protein function in an *in vitro* assay of motility. *Free Radic Biol Med* 44:14-23,2008.

[222] Solaro RJ, Kobayashi T. Protein phosphorylation and signal transduction in cardiac thin filaments. *J. Biol. Chem.*286:9935-9940,2011.

[223] Solaro RJ, Moir AJG, Perry SV. Phosphorylation of troponin I and the inotropic effect of adrenaline in the perfused rabbit heart.*Nature*262:615-617,1976.

[224] Solis RS, Ge Y, Walker JW. A preferred AMP phosphorylation site adjacent to the inhibitory loop of cardiac and skeletal troponin I. *Protein Sci.* 20:894-907,2011.

[225] Solis RS, Ge Y, Walker JW. Single amino acid sequence polymorphisms in rat cardic troponin revealed by top-down tandem mass spectrometry. *J. Muscle Res. Cell Motil.*29:203-212,2008.

[226] Squerzanti M, Cervellati C, Ura B, Mischiati C, et al., The side chain of glutamine 13 in the acyl-donor amino acid modified by type 2 transglutaminase in subunit T of the native rabbit skeletal muscle troponin complex. *Amino Acids* 44:227-234,2013.

[227] Steinberg SF. Cardiac actions of protein kinase C isoforms. *Physiology* 27:130-139,2012.

[228] Steinberg SF. Distinctive activation mechanisms and functions for protein kinase Cδ. *Biohem. J.* 384:449-459,2004.

[229] Steinberg SF. Oxidative stress and sarcomeric proteins. *Circ. Res.* 112:393-405,2013.

[230] Steinberg SF. Structural basis of protein kinase C isoform function. *Physiol. Rev.*88:1341-1378,2008.

[231] Stelzer JE, Patel JR, Walker JW, Moss RL. Differential roles of cardiac myosin-binding protein C and cardiac troponin I in the myofibrillar force responses to protein kinase A phosphorylation. *Circ. Res.* 101:503-511,2007.

[232] Stull JT, Buss JE. Phosphorylation of cardiac troponin by cyclic adenosine 3':5'-monophosphoate-dependent protein kinase. *J. Biol. Chem.* 252:851-857,1977.

[233] Stull JT, High CW. Phosphorylation of skeletal muscle contractile proteins *in vivo*. *Biochem. Biophys. Res. Comm.*77:1078-83,1977.

[234] Suematsu N, Satoh S, Kinugawa S, Tsutsui H, et al., α_1-adrenoceptor-G1-RhoA signaling is upregulated to increase myofibrillar Ca^{2+} sensitivity in failing hearts. *Am. J. Physiol.* 281:H367-H646,2001.

[235] Sugamura K, Keaney JF Jr. Reaction oxygen species in cardiovascular disease *Free Rad. Biol. Med.* 51:978-992,2011.

[236] Sulakhe PV, Vo XT, Phan TD, Morris TE. Phosphorylation of inhibitory subunit of troponin and phospholamban in rat cardiomyocytes: Modulation by exposure of cardiomyocytes to hydroxyl radicals and sulfhydryl group reagents. *Mol. Cell Biochem.* 175:99-107,1997.

[237] Sumandea MP, Burkart EM, Kobayashi T, deTombe PP, et al., Molecular and integrated biology of thin filament protein phosphorylation in heart muscle. *Ann. NY Acad. Sci.* 1015:39-52,2004.

[238] Sumandea MP, Pyle WG, Kobayashi T, deTombe PP, et al. ,Identification of a functionally critical protein kinase C phosphorylation residue of cardiac troponin T.*J. Biol. Chem.* 278:35135-35144,2003.

[239] Sumandea MP, Rybin VO, Hinken AC, Wang C, et al., Tyrosine phosphorylation modifies protein kinase C -dependent phosphorylation of cardiac troponin I.*J. Biol. Chem.*283:22680-22689,2008.

[240] Sussman MA, Welch S, Walker A, Klevitsky R, et al., Altered focal adhesion regulation correlates with cardiomyopathy in mice expressing constitutively active rac1.*J. Clin. Invest* 105:875-886,2000.

[241] Sussman MA. Developing hearts need their SPEG. *Circulation* 119:213-214,2009.

[242] Swiderek K, Jaquet K, Mayer HE, Heilmeyer LMC Jr. Cardiac troponin I, isolated from bovine heart, contains two adjacent phosphoserines. A first example of phosphoserine determination by derivatization to S-ethylcysteine. *Eur. J. Biochem.* 176:335-342,1988.

[243] Swiderek K, Jaquet K, Meyer HE, Schachtele C, et al., Sites phosphorylated in bovine cardiac troponin T and troponin I.*Eur. J. Biochem.*190:575-582,1990.

[244] Takeishi Y, Chu G, Kirkpatrick DM, Li Z, et al., *In vivo* phosphorylation of cardiac troponin I by protein kinse Cβ_2 decreases cardiomyocyte calcium responsiveness and contractility in transgenic mouse hearts. *J. Clin. Invest* 102:72-78,1998.

[245] Takimoto E, Soergel DG, Janssen PML, Stull LB, et al., Frequency- and afterload-dependent cardiac modulation *in vivo* by troponin I with constitutively active protein kinase A phosphoryation sites. *Circ. Res.* 94:496-504,2004.

[246] Tang H, Xiao K, Mao L, Rockman HA, et al., Overexpression of TNNI3K, a cardiac-specific MAPKKK, promotes cardiac dysfunction. *J. Mol. Cell Cardiol.*54:101-111,2013.

[247] Thannickal VJ, Fanburg BL. Reactive oxygen species in cell signaling. *Am. J. Physiol.* 279:L1005-L1028,2000.

[248] Tolstykh T, Lee J, Vafai S, Stock JB. Caroxyl methylation regulates phosphoprotein phosphatase 2A by controlling the association of regulatory B subunits. *EMBO J.* 19:5682-5691,2000.

[249] Toyo-oka T, Ross J Jr.Ca^{2+} sensitivity change and troponin loss in cardiac natural actomyosin after coronary occlusion. *Am. J. Physiol.*240:H704-708,1981.

[250] Ubersax JA, Ferrell JE Jr.Mechanisms of specificity in protein phosphorylation. *Nat. Rev.Mol. Cell Biol.* 8:30-541,2007.

[251] Utsumi T, Nakano K, Funakoshi T, Kayano Y, et al., Vertical-scanning mutagenesis of amino acids in a moel N-myristoylation motif reveals the major amnio-terminal sequence requirements for protein N-myristoylation. *Eur. J. Biochem.* 271:863-874,2004.

[252] Vahebi S, Kobayashi T, Warren CM, deTombe PP, et al., Functional effects of Rho-kinase-dependent phosphorylation of specific sites on cardiac troponin. *Circ. Res.* 96:740-747,2005.

[253] Vahebi S, Ota A, Li, M, Warren CM, et al.,p38-MAPK induced dephosphorylation of alpha-tropomyosin is associated with depression of myocardial sarcomeric tension and ATPase activity. *Circ. Res.* 100:408-415 2007.

[254] Venardos KM, Zatta AJ, Marshall T, Ritchie R, et al., Reduced L-arginine transport contributes to the pathogenesis of myocardial ischemia-reperfusion. *J. Cell Biochem.* 108:156-168,2009.

[255] Venema RC, Kuo JF. Protein kinase C-mediated phosphorylation of troponin I and C-protein in isolated myocardial cells is associated with inhibition of myofibrillar actomyosin MgATPase. *J. Biol. Chem.* 268:2705-2711,1993.

[256] Villar-Palasi C, Kumon A. Purification and properties of dog cardiac troponin T kinase. *J. Biol. Chem.* 256:7409-7415,1981.

[257] Walker LA, Medway AM, Walker JS, Cleveland JC Jr., et al., Tissue procurement strategies affect the protein biochemistry of human heart samples. *J. Muscle Res. Cell Motil.* 31:309-314,2011.

[258] Walker LA, Walker JS, Ambler SK, Buttrick PM. Stage-specific changes in myofilament protein phosphorylation following myocardial infarction in mice. *J. Mol. Cell Cardiol.*48:1180-1186,2010.

[259] Wang H, Chen C, Song X, Chen J, et al., Mef2c is an essential regulatory element required for unique expression of the cardiac-specific CARK gene. *J. Cel Mol. Med.* 12:304-315,2008.

[260] Wang H, Grant JE, Doede CM, Sadayappan S, et al., PKC-β_{II} sensitizes cardiac myofilaments to Ca^{2+} by phosphorylating troponin I on threonine-144.*J. Mol. Cell Cardiol.* 41:823-833,2006.

[261] Wang H, Wang L, Song L, Zhang Y-W, et al., TNNI3K is a novel mediator of myofilament function and phosphorylates cardiac troponin I. *Braz J. Med. Biol. Res.*46:128-137, 2013.

[262] Ward DG, Ashton PR, Trayer HR, Trayer IP. Additional PKA phosphorylation sites in human cardiac troponin I. *Eur. J. Biochem.*268:179-185,2001.

[263] Westfall MV, Borton AR. Role of troponin I phosphorylation in protein kinase C-mediated enhanced contractile performance of rat myocytes. *J. Biol. Chem.*278:33694-33700,2003.

[264] Westfall MV, Lee AM, Robinson DA. Differential contribution of troponin I phosphorylation sites to the endothelin-modulated contractile response. *J. Biol. Chem.* 280:41324-41331,2005.

[265] Westfall MV, Rust EM, Metzger JM. Slow skeletal troponin-I gene transfer, expression and myofilament incorporation enhances adult cardiac myocyte contractile function.*PNAS*94:5444-5449,1997.

[266] Wijnker PJM, Boknik P, Gergs U, Muller FU, et al., Protein phosphatase 2A affects myofilament contractility in non-failing but not in failing human myocardium. *J. Muscle Res. CellMotil.*32:221-233,2011.

[267] Wilkinson JM. Troponin C from rabbit slow skeletal and cardiac muscle is the product of a single gene. *Eur. J. Biochem.* 103:179-188,1980.

[268] Wilkinson JM, Grand RJA. The amino acid sequence of troponin I from rabbit skeletal muscle. *Biochem. J.* 149:493-496,1975.

[269] Willadsen KA, Butters CA, Hill LE, Tobacman LS. Effects of amino-terminal regions of tropomyosin and troponin T on thin filament assembly. *J. Biol. Chem.* 267:23746-23752,1992.

[270] Wiltshire SA, Leiva-Torres GA, Vidal SM. Quantitative trait locus analysis, pathway analysis, and consomic mapping show genetic variants of TNNI3K, FPGT, or H28 control susceptibility to viral myocarditis. *J. Immunol.* 186:6398-6405,2011.

[271] Wise BC, Kuo JF. Phosphorylation of cardiac troponin inhibitory subunit (troponin I) and tropomyosin-binding subunit (troponin T) by cardiac phospholipid-sensitive Ca^{2+}-dependent protein kinase. *Biochem. J.* 209:189-195,1983.

[272] Wolff MR, Buck SH, Stoker SW, Greaser ML, Mentzer RM. Myofibrillar calcium sensitivity of isometric tension is increased in human dilated cardiomyopathies. *J. Clin. Invest* 98:167-176,1996.

[273] Wollert KC, Yurukova S, Kilic A, Begrow F, et al., Increased effects of C-type natriuretic peptide on contractility and calcium regulation in murine hearts overexpressing cycling GMP-dependent protein kinase I.*Br. J. Pharmacol.*140:1227-1236,2003.

[274] Wu S, Solaro RJ. Protein kinase Cζ: A novel regulator of both phosphorylation and de-phosphorylation of cardiac sarcomeric proteins. *J. Biol. Chem.*282:30691-30698,2007.

[275] Xing Y, Chen Y, Stock JB, Jeffrey PD, Shi Y. Structural mechanism of demethylation and inactivation of protein phosphatase 2A. *Cell* 133:154-163,2008.

[276] Yasuda S, Coutu P, Sadayappan S, Robbins J, et al., Cardiac transgenic and gene transfer strategies converge to support an important role for troponin I in regulating relaxation in cardiac myocytes. *Circ. Res.* 101:377-386,2007.

[277] Yin X, Cuello F, Mayr U, Hao Z, et al., Proteomics analysis of the cardiac myofilament subproteome reveals dynamic alterations in phosphatase subunit distribution. *Mol. Cell Prot.* 9:497-509,2010.

[278] You B, Huang S, Qin Q, Yi B, et al., Glyceraldehyde-3-phosphate dehydrogenase interacts with proapoptotic kinase Mst1 to promote cardiomyocyte apoptosis. *PLOS One*8: e58697,2013.

[279] You B, Yan G, Zhang Z, Yan L, et al., Phosphorylation of cardiac troponin I by mammalian sterile 20-like kinase 1.*Biochem. J.*418:93-101,2009.

[280] Zabrouskov V, Ge Y, Schwartz J, Walker JW. Unraveling molecular complexity of phosphorylated human cardiac troponin I by top down electron capture dissociation/electron transfer dissociation mass spectrometry. *Mol. Cell Proteom.* 7:1838-1849,2008.

[281] Zaha VG and Young LH.AMP-activated protein kinase regulation and biological actions in the heart. *Circ. Res.* 111:800-814,2012.

[282] Zakhary DR, Moravec CS, Stewart RW, Bond M. Protein kinase A (PKA)-dependent troponin I phosphorylation and PKA regulatory subunits are decreased in human dilated cardiomyopathy.*Circulation*99:505-510,1999.

[283] Zemskova MA, Shur SA, Skolysheva LK, Vul'fson PL. Interaction of phosphorylase kinase with thin filament proteins of rabbit skeletal muscles. *Biokhimiia* 56:100-108,1991.

[284] Zhang J, Dong X, Hacker TA, Ge Y. Deciphering modifications in swine cardiac troponin I by top-down high-resolution tandem mass spectrometry. *J. Am. Soc. Mass Spectrom*21:940-948,2010.

[285] Zhang J, Zhang H, Ayaz-Guner S, Chen Y-C, et al., Phosphorylation, but not alterative splicing or proteolytic degradation, is conserved in human and mouse cardiac troponin T.*Biochemistry*50:6081-6092,2011.

[286] Zhang T, Birbrair A, Wang Z-M, Taylor J, et al., Troponin T nuclear localization and its role in aging skeletal muscle. *AGE*35:353-370, 2013.

[287] Zhang P, Kirk JA, Ji W, dos Remedios CG, et al., Multiple reaction monitoring to identify site-specific troponin I phosphorylated residues in the failing human heart. *Circulation* 126:1828-1837,2012.

[288] Zhang R, Zhao J, Potter JD. Phosphorylation of both serine residues in cardiac troponin I is required to decrease the Ca^{2+} affinity of cardiac troponin-C. *J. Biol. Chem.* 270:30773-30780,1995.

[289] Zhao Y, Meng XM, Wei YJ, Zhao XW, Liu DQ, Cao HQ, Liew CC, Ding JF. Cloning and characterization of a novel cardiac-specific kinase that interacts specifically with cardiac troponin I.*J. Mol. Med.* 81:297-304,2003.

In: Troponin
Editor: J.-P. Jin

ISBN: 978-1-62948-320-7

Chapter IX

Role of Troponin in Length-Tension Relation

Pieter P. de Tombe, PhD
Department of Cell and Molecular Physiology
Loyola University Chicago, Stritch School of Medicine
Maywood IL, US

Abstract

The Frank-Starling law of the heart describes the interrelationship between ventricular volume and the strength of the heart to pump blood independent of other, mostly hormonal, influences that affect the intrinsic strength of the heart (inotropic state). A change in ventricular volume is accompanied by an immediate change in contraction strength (termed the "Frank-Starling" mechanism), followed by an additional change in contraction strength that develops over several minutes following the change in volume (termed the "Anrep" or "homeometric auto regulation" mechanism); this review is concerned with the former mechanism, the immediate beat-to-beat modulation of contractile strength by ventricular volume. The main cellular mechanism that underlies this phenomenon is an increase in the responsiveness of cardiac myofilaments to activating Ca^{2+} ions at a longer sarcomere length, commonly referred to as myofilament length dependent activation. This review focuses on what molecular mechanisms may underlie myofilament length dependency. Specifically, the roles of inter-filament spacing, thick and thin filament based regulation, as well as sarcomeric regulatory proteins are discussed. Although the "Frank-Starling law of the heart" constitutes a fundamental cardiac property that has been appreciated for well over a century, it is still not known in muscle how the contractile apparatus transduces the information concerning sarcomere length to modulate ventricular pressure development.

Keywords: Sarcomere, Frank-Starling Relation, Contractile Proteins, Myofilament Length Dependent Activation

Grants: Supported by grants from the National Institutes of Health and the American Heart Association

1. Frank-Starling's of the Heart

Over a century ago, Otto Frank in Germany and Ernest Starling in England reported on the relationship between the extent of ventricular filling and pump function of the heart, a phenomenon collectively referred to as Frank-Starling's Law of the Heart. A modern view of this phenomenon is illustrated in Figure 1 [1].

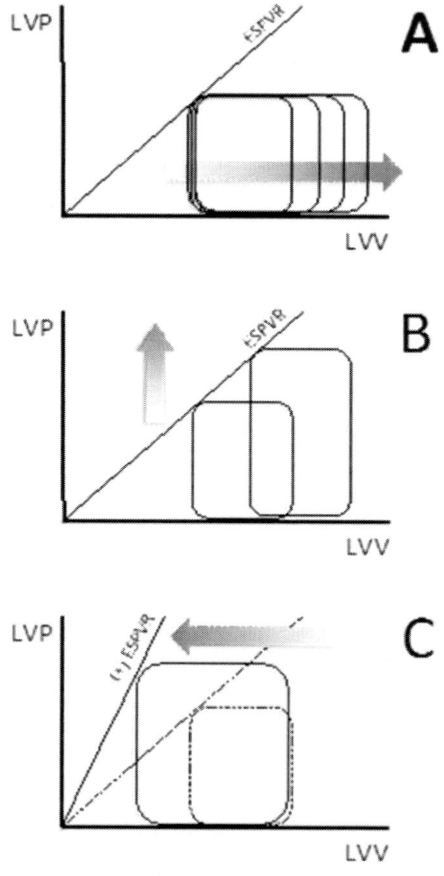

Figure 1. In the heart, a unique relationship exists between end-systolic pressure and end-systolic volume (ESPVR). The ESPVR allows for increased cardiac stroke volume (and, thus, cardiac output in response to increased cardiac filling (increased preload; Panel A), maintenance of stroke volume in the face of increased systolic pressure (increased afterload; Panel B). The ESPVR is an index of cardiac contractility. Panel C illustrates a typical response to exercise where increased cardiac contractility allows for increased systolic pressure *and* stroke volume for a given amount of ventricular filling (i. e. constant preload).

In general, there is a unique relationship between end-systolic volume and end-systolic pressure in the heart, the steepness of which is modulated by contractile state. For a given contractile state, ventricular stroke volume is *i)* proportional to diastolic filling (i. e., preload; c. f. panel A), and *ii)* stroke volume can be maintained in the face of increased aortic pressures (i. e., afterload; c. f. panel B) simply by increasing preload. Contractile state, within this framework, can be viewed as any factor that alters end-systolic pressure *independently* of end-systolic volume and can conveniently be estimated semi-quantitatively by the slope of the end-systolic pressure-volume relationship (ESPVR). The slope of the ESPVR is a very useful, albeit empirical, index of cardiac contractility that can be measured in situ by various methods; a convenient and popular approach is the use of the pressure-volume conductance catheter [2]. The cellular mechanisms that underlie the ESPVR are discussed in the following section. An increase in contractile state, for example such as occurs during exercise, allows for increases stroke volume even in the face of increased afterload (illustrated in panel C). Note that cardiac output is determined by stroke volume *and* heart rate; it has been suggested that the increased cardiac output during exercise is modulated more prominently by heart rate in young individuals, while the Frank-Starling mechanism is of greater significance in the elderly [3]. Finally, the Frank-Starling mechanism in conjunction with pericardial constraint serves to balance the output of the left and right ventricle on a beat-to-beat basis [4].

2. Myofilament Length Dependent Activation (LDA)

Pump function of the heart is determined by force generation, active shortening, and relaxation of the cardiac sarcomeres [1,5]. The relationship between biomechanical properties of the cardiac sarcomere and mechanical behavior of the heart's chamber is complex. It is determined not only by the orientation and density of the cardiac muscle fibers, but also by the timing of cardiac muscle fiber activation and relaxation. Nevertheless, there is ample evidence to support the notion that the biochemical properties of the cardiac cell, and indeed the cardiac sarcomere itself, are directly responsible for many, if not most, of the mechanical properties of the heart [1, 5, 6]. Consistent with this notion, twitch force in isolated cardiac muscle is directly proportional to systolic sarcomere length; furthermore, the shape of the force-sarcomere length relationship is modulated by contractile state such that more force is generated at a given sarcomere length when contractile activation is elevated (e. g. by raising extracellular [Ca^{2+}]) [7]. These phenomena are schematically illustrated in Figure 2: increases or decreases in contractile state of the heart are accompanied by increases or decreases in the slope of the end-systolic pressure-volume relationship (ESPVR; panel A), while at the muscle fiber level modulation of contractile state is accompanied by increases or decreases in slope of the force-sarcomere length relationship (F-SL; panel B), albeit in a non-linear fashion. Due to the non-linear transformations from twitch force and sarcomere length to ventricular pressure and ventricular volume, respectively, a non-linear F-SL relationship forms the basis for the close to linear ESPVR as is seen in the whole heart.

Activation of the contractile apparatus is initiated upon a transient increase in the cytosolic calcium concentration [8]. Under normal physiological conditions, calcium entry during the plateau phase of the cardiac action potential is not sufficient to directly activate the myofilaments, but instead serves as a trigger to release calcium from the sarcoplasmic reticulum. It is important to note that, in general, there is a direct relationship between the magnitude of the calcium transient and the contractile state of the cardiac cell and, therefore, the ventricle [8, 9].

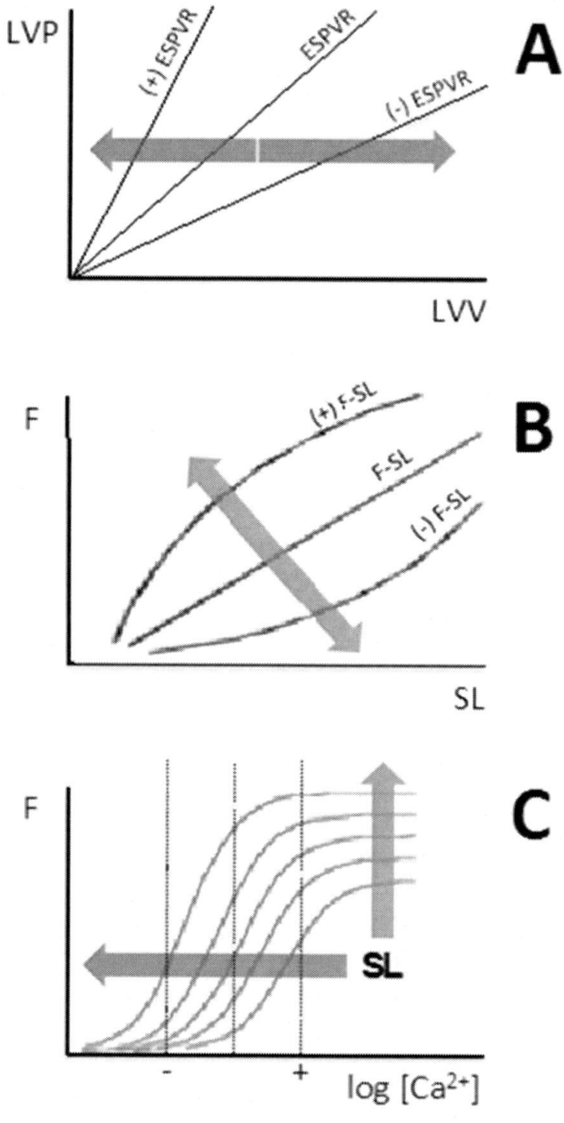

Figure 2. Close to linear end-systolic pressure-volume relationships (ESPVR; panel A) are observed in the whole heart; the slope of the ESPVR is a sensitive index of contractile state. At the muscle fiber level, twitch force (F) is determined by sarcomere length (SL); the shape of the F-SL relationship is modulated by inotropic state. The non-linear mathematical transformation between F/SL and LVP/LVV results in the linear ESPVRs at the whole heart level. The sigmoidal relationship between activator Ca^{2+} ions and myofilament force development is shifted to the left and upward in response to an increase in SL (panel C); this phenomenon is termed "myofilament length dependent activation" and forms the basis of modulation of the F-SL (panel B) and ESPVR (panel A) with inotropic state (c. f. vertical dashed lines in Panel C).

However, the level of myofilament activation is by no means a simple proportional function of cytosolic calcium concentration. Rather, it constitutes a complex and dynamic signal transduction process that itself is also subject to regulation and modulation by both intrinsic (mechanical loading, sarcomere length) and extrinsic (neuro-hormonal) factors. Early experiments by Fabiato suggested that the amount of this activator calcium released varies with sarcomere length, but these results have not since been confirmed [10]. Instead, more recent experiments [11-13] clearly demonstrated that it is the level of activation of the cardiac contractile apparatus itself that is sensitive to changes in sarcomere length. These early experiments were performed on chemically permeabilized (skinned) isolated cardiac muscle, a preparation that allows direct access to the contractile apparatus such that steady state force can be measured as function of activator calcium concentration and sarcomere length; more recent experiments on intact twitching isolated myocardium employing fluorescent Ca^{2+} probes have confirmed these results [14]. Therefore, there is a direct proportionality between sarcomere length and the sensitivity of the cardiac sarcomere to Ca^{2+} ions, such that more force is generated at a given concentration of activator Ca^{2+} as sarcomere length is increased (the curves are both shifted to the left on the activator $[Ca^{2+}]$ axis and up to higher forces at the higher sarcomere length. Hence, it can be said that the myofilaments possess a length dependency property and this is termed "myofilament length dependent activation". This phenomenon is illustrated in the bottom panel of Figure 2: because of the sarcomere length modulation of sarcomeric properties, myofilament force development for a given level of activator $[Ca^{2+}]$ during the cardiac cycle (Panel C, dashed vertical lines) is also modulated and this, in turn, results in modulation of twitch force (Panel B) and ventricular pressure (Panel A) development at end-systole. Thus, the whole heart Frank-Starling property has, as its basis, the myofilament length dependent activation property of the cardiac sarcomere [15]. As mentioned above, intact twitching cardiac muscle as well as the ventricle responds to a change in length (volume) in two distinct phases: an immediate change in twitch force (pressure), and a slower phase that develops over the course of several minutes. Experimental evidence suggests that the latter phase is due to a change in sarcoplasmic reticulum calcium release secondary to altered calcium loading of the cell [16,17], while the immediate response is due to the change in calcium sensitivity of the cardiac sarcomere described above; this review is focused on the immediate response to a change in sarcomere length. Despite the importance for the Frank-Starling mechanism, the molecular mechanisms that may underlie the immediate response of myofilament force generation upon a change in sarcomere length have not been entirely elucidated.

3. Ca^{2+} Ion Activation of Contractile Proteins

The change in myofilament force upon variation of sarcomere length is due to a change in the number of active cycling, force producing cross-bridges in the cardiac sarcomere [18, 19]. The major proteins that make up the striated muscle sarcomere are actin, troponin, and tropomyosin (in the thin filament), and myosin, myosin light chains, and myosin binding protein C (in the thick filament). These proteins and details regarding structure-function relations can be found in [20] and the various chapters in this book. Dimers of the asymmetric molecule myosin are located in the thick filament; their tail portions form the backbone of this filament, while the globular heads protrude from the thick filament reaching out towards actin. These globular heads, or cross-bridges, cyclically attach to the actin filament during contraction. It is now generally accepted that cyclic interactions of the actin-myosin cross-bridge form the fundamental basis for the generation of active force in muscle [21]. When strongly bound, the cross-bridges produce a pulling force on actin and, through the thin filament, on the proteins that make up the Z-band. The Z-bands are mechanically coupled via a multitude of proteins that include the integrin receptors at the cell surface to the extracellular collagen network thereby transmitting actin-myosin produced force to the tendons in skeletal muscle or towards the heart's cavity so as to produce pressure. It should be noted that an additional sarcomeric protein has been identified, titin, a high molecular weight giant protein that spans the entire half sarcomere from the Z-band to the M-line and that binds tightly with the myosin thick filament [22-25]. As will be discussed below, recent evidence implicates titin in myofilament length dependent activation. Striated muscle contraction is regulated by troponin and the tropomyosin (Tm) filament, both located on the thin filament and these proteins are essential for the regulation of myofilament contractile activation [20]. The heteromeric protein complex troponin (Tn) is composed of troponin-C (TnC), the calcium receptor; troponin-I (TnI), an inhibitor of contraction that switches between being bound to actin and cTnC; and troponin-T (TnT), a Tm binding protein that relays the Ca^{2+} binding signal from TnC to the thin filament by interacting with TnI and Tm. All striated muscle (skeletal and cardiac) are fundamentally identical in terms of the type of proteins that make up the sarcomere, but differ in terms of the specific isoforms expression of these proteins (which usually derive from different gene products in different muscle types). In diastole, cross-bridges are either blocked from interacting with actin or in a weak, rapid binding-unbinding state with actin without generating force [26-28]. In systole, activation of the thin filament allows for a strong-binding cross-bridge state, associated with force generation and high rate of ATP hydrolysis. Ca^{2+} is the physiological activator of the contractile machinery in both skeletal and cardiac muscle. Skeletal muscle TnC contains two high affinity and two low affinity binding sites [29]. The cardiac isoform (cTnC) also has two high affinity sites, but only one low affinity binding site [30]. The nature of Ca^{2+} regulation is such that activation of the sarcomere occurs by binding to the low affinity site(s). Ca^{2+} always occupies the high affinity sites of TnC under physiological conditions [31]; hence, these sites may serve only a structural function allowing for the anchoring of TnC to the troponin complex. In either cardiac muscle or skeletal muscle, the relationship between free Ca^{2+} and isometric tension is very steep, much greater than would be expected on the basis of equilibrium binding of Ca^{2+} to one or two binding sites on TnC [13, 32-35]. Hence, the transduction processes induced by Ca^{2+} binding that lead to activation of muscle contraction clearly involve cooperative interactions. In the absence of Ca^{2+}, TnI exerts an inhibitory effect, preventing actin-myosin interaction. Ca^{2+} binding induces a conformational change in TnC such that a hydrophobic region becomes exposed thereby allowing interaction with TnI [36, 37]. However, the affinity of skeletal TnC for TnI in the presence of Ca^{2+} is much greater than the affinity of cardiac TnC for TnI in the presence of Ca^{2+} [38, 39] which may be related to the diminished ability of Ca^{2+}+ to fully activate cardiac thin filaments at low (or initial) levels of cross-bridge binding [20, 40-43]. This phenomenon may also explain the greater myofilament length dependent activation observed in cardiac muscle compared to skeletal muscle [44]. Prior to Ca^{2+} activation, Tn constrains Tm in a position that sterically hinders myosin-S1 binding [38, 45]. The movement of the troponin complex induces a conformational change in Tm, increasing its binding to actin, thereby revealing myosin-S1 binding sites on actin. Furthermore, upon forming strongly-bound cross-bridges, the binding of additional cross-bridges is greatly enhanced in a cooperative manner, presumably by initiating the movement of tropomyosin to a third, more favorable position [20, 27, 38, 45].

4. Role of Inter-Filament Spacing in LDA

A unifying theory that has gained acceptance proposes that the impact of sarcomere length on myofilament Ca^{2+} sensitivity is due to changes in the spacing between the thick and thin filaments [40, 46-48]. This theory is schematically illustrated in Figure 3 (panel A). Because myofibrils maintain close to constant volume [49-52], elongation of the sarcomere is expected to lead to a reduction of the distance between the thick and thin filaments. A closer approximation of the myosin heads to actin may be expected to increase the probability of strong cross-bridge formation at a given concentration of activator calcium. Thus, one would predict an increase in myofilament Ca^{2+} sensitivity at the higher sarcomere length. Several experiments have provided support for this theory. Osmotic compression of the myofilaments by high molecular weight molecules that cannot enter the myofilament lattice structure, such as dextran, induces an increase in myofilament Ca^{2+} sensitivity concomitant with a reduction of muscle diameter [46-48]. A reduction in muscle diameter occurs in parallel to a reduction in myofilament lattice spacing in muscle, hence, the increase in myofilament Ca^{2+} sensitivity following dextran treatment without a change in sarcomere length is consistent with the myofilament lattice spacing theory. More direct support was provided by Fuchs et al, who showed that myofilament Ca^{2+} sensitivity could be rendered length *independent* when muscle diameter was kept constant by applying an appropriate amount dextran at each sarcomere length [48]. In addition, selective removal of titin by mild treatment with trypsin has been shown to affect both myofilament Ca^{2+} sensitivity and inter-filament spacing [53]; however, whether these observations are causally linked could not be determined in that study (see also below). To shed further light on these issues, we have reexamined the inter-filament spacing hypothesis, this time employing x-ray diffraction (c. f. Figure 3B) to directly measure myofilament lattice spacing, in a comprehensive series of experiments [49, 51, 54-58]. First, we found an inverse relation between inter-filament spacing and sarcomere length in isolated cardiac [51] as well as fast and slow skeletal muscle preparations [56], a finding that is overall consistent with, although not proof of, the inter-filament spacing hypothesis. Second, we found that muscle diameter was not a consistent indicator of inter-filament spacing [57]. Third, in a study comparing skinned myocardium obtained from a transgenic murine model in which cTnI was replaced by ssTnI to wild-type myocardium, we found profound and opposing changes in inter-filament spacing upon PKA treatment that were not correlated to myofilament Ca^{2+} sensitivity [55]. Fourth, a detailed comparison between fast and slow skeletal muscle of the rat revealed no differences in inter-filament spacing as function of sarcomere length, despite large differences in myofilament length dependency [56]. Finally, in experiments illustrated in Figure 3 (panel C), we used osmotic compression to reduce myofilament inter-filament spacing at a short sarcomere length to match inter-filament spacing recorded at a high sarcomere length (i. e. reduced lattice spacing despite the fact that sarcomere length had not changed) and asked the question whether myofilament Ca^{2+} sensitivity at the short sarcomere length would match that recorded at the high sarcomere length [57].

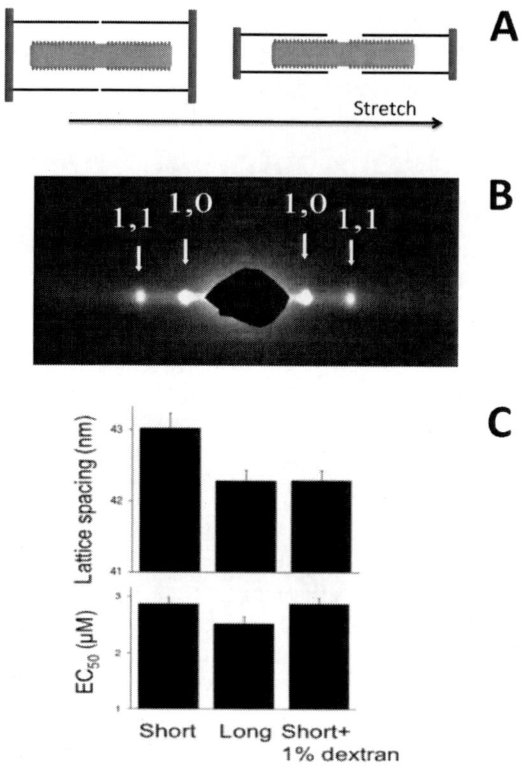

Figure 3. Because the sarcomere maintains close to constant volume, stretch of the muscle is predicted to result in compression of the filament lattice structure, moving actin and myosin closer together that may result in increased myofilament Ca^{2+} sensitivity (panel A). The only way to measure inter filament spacing in non-fixed tissue is by small angle x-ray diffraction. Panel B displays a 2D x-ray diffraction obtained from an isolated rat skinned cardiac muscle; equatorial reflections contain strong first order reflections (1,0 & 1,1) that allow for accurate measurement of inter filament spacing. Sarcomere length is measured either by FFT analysis of the microscope video image [160] or by laser diffraction techniques [161]. Panel C shows reduction of inter filament spacing by stretch from short to long SL, or by compression at short SL with 1% Dextran to the same inter filament spacing as measured at the long SL. Stretch, but not 1% Dextran compression induced increased myofilament Ca^{2+} sensitivity (as indexed by EC_{50}), demonstrating that inter filament may not determine myofilament Ca^{2+} sensitivity and, thus, is unlikely the molecular mechanism underlying myofilament length dependent activation (modified from [57]).

A similar study, based on cell diameter *not* inter-filament spacing, had previously been reported by McDonald et al. [47], and their results were in support of the inter-filament spacing hypothesis. As illustrated by the bar graphs in Figure 3C, despite the matching of inter-filament spacing, myofilament Ca^{2+} sensitivity was only affected by sarcomere length and not by inter-filament spacing *per se*. The apparent discrepancy between those experiments and the previous experiments may be explained by *i)* the difference impact of osmotic compression on inter-filament spacing compared to muscle diameter; to mimic sarcomere length induced changes in inter-filament spacing much less osmotic compression is required than to mimic the changes in muscle diameter [57], *ii)* dextran induced osmotic compression causes increased myofilament Ca^{2+} sensitivity, but this effect appears to be due to a direct effect on weakly bound cross-bridges rather than inter-filament spacing per se [54, is associated with a marked blunting of myofilament length dependent activation in cardiac (Figure 7B and [157]) as well as skeletal muscle [149]. Access to this mutant rat model as well as murine animal models in which titin strain is modulated by mutations [158, 159] will greatly enhance future research into the role of this giant elastic protein in myofilament length dependent activation.

9. Overview, Molecular Models, and Remaining Questions

The "Frank-Starling law of the heart" constitutes a fundamental cardiac property that has been appreciated for well over a century. At its basis, this ventricular property is the result of a modulation of myofilament Ca^{2+} sensitivity upon a change in sarcomere length. Despite intense investigations, the molecular mechanisms that underlie this phenomenon are still not known. As discussed in this review, controversy remains whether inter-filament spacing constitutes the molecular strain sensing mechanism of the sarcomere. Rather, evidence points to a length modulated regulation of thin filament activation state via regulation of thin filament activation state. How the thin filament is regulated by sarcomere length is not known; some of the potential mechanisms are illustrated in figure 8 that include cooperative feedback between force generating cross-bridges, Ca^{2+} binding affinity, as well as the giant cytoskeletal protein titin as the mechanical linker protein that transmits information regarding sarcomere length to the contractile apparatus.

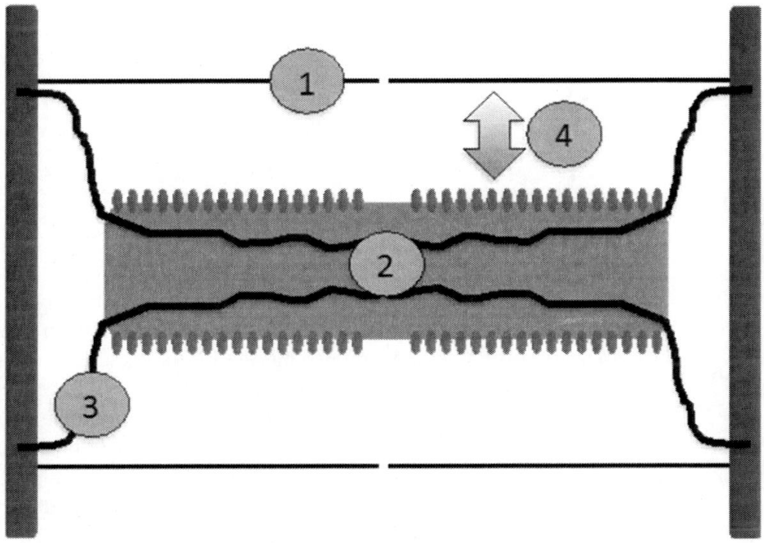

Figure 8. Various potential molecular mechanisms for myofilament length dependent activation. Stretch of the sarcomere may: 1) affect thin filament cooperative activation, 2) alter the structure of weakly bound cross-bridges, 3) these mechanisms may be induced directly by strain transmission via the giant elastic protein titin, 4) or impinge on sarcomere dynamics via modulation of the cooperative feedback processes between the thin- and thick-filament.

of thin filament activation. A different method to reduce strongly bound cross-bridges during Ca^{2+} activation is application of Blebbistatin, a compound that blocks the weak to strong acto-myosin transition [68, 69].

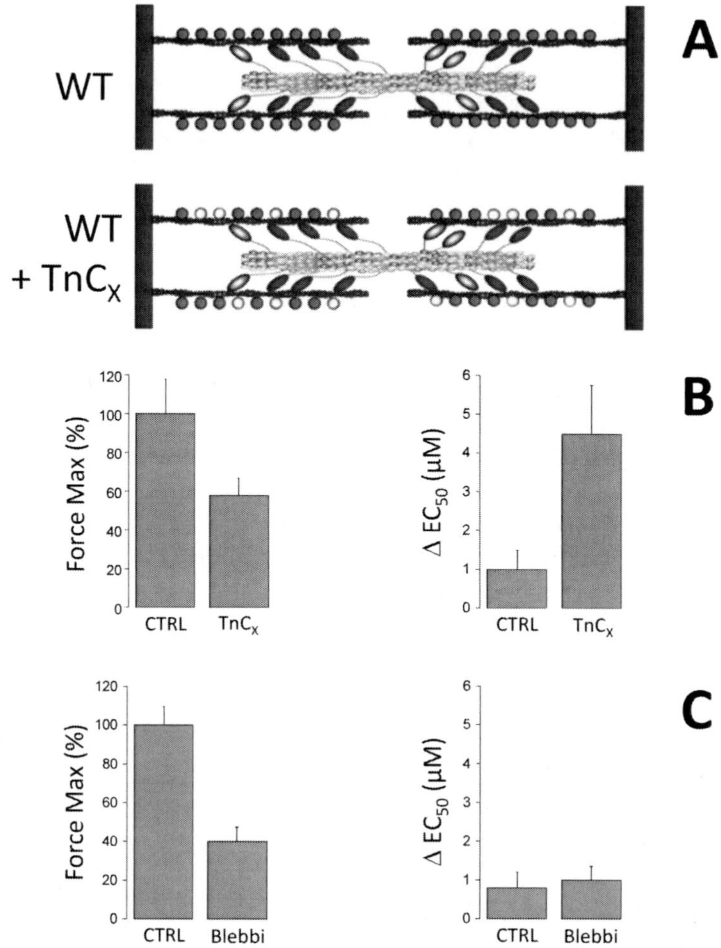

Figure 4. Panel A, schematic diagram of the cardiac sarcomere indicating thin filaments attached to Z-disks (vertical black bars) and a thick filament situated sandwiched between two thin filaments. Exchange with recombinant mutant troponin C incapable of binding Ca^{2+} at the regulatory site (cTnC-Glu40Ala,Glu76Ala; TnC_X) is illustrated in the bottom sarcomere; Ca^{2+} "dead" troponin regulatory units are indicated by the open troponin circles. Panel B, incorporation of TnC_X results in ~40% reduction in maximum Ca^{2+} saturated force development (Force Max) and a fourfold increase in myofilament length dependent activation (indexed by ΔEC_{50}, the difference in myofilament Ca^{2+} sensitivity between short and long SL). Panel C, application of Blebbistatin (1 μM) reduced maximum force by ~60%, similar to TnC_X exchange, but did not impact myofilament length dependent activation (modified from [67]).

Figure 4C illustrates the impact of Blebbistatin application (1 μM) on maximum Ca^{2+} activated force development and myofilament length dependent activation. Inhibition of force was similar to that obtained with recombinant TnC_X exchange (panel B), but in contrast length dependency was not significantly affected. We interpret these findings to imply, at least in the cardiac sarcomere that, first, strong binding cross-bridges *per se* do not participate significantly in the processes that affect myofilament Ca^{2+} sensitivity or myofilament length dependency. This conclusion is consistent with a recent report regarding the impact of strong binding cross-bridges on troponin structure employing site directed environmental fluorescent probes [70]. Second, the finding that interruption of cooperative activation impacts myofilament Ca^{2+} sensitivity most at short sarcomere length indicates that the length sensing mechanism involves a long range interaction along the thin filament that is strengthened at longer sarcomere length. In other words, the "range" of cooperativety along the thin filament is variable and depends on sarcomere length, such that at high sarcomere length Ca^{2+} activation of one troponin regulatory unit will be sensed by troponin units far removed from this locus, effectively "skipping" sites containing TnC_X. The nature of the molecular moiety responsible for this mechanism is currently unknown, but may very well be the giant molecule titin (see below). Such a mechanism would also imply that the length sensing process is based on simple mechanically induced protein conformational changes induced by strain. This notion is supported by our recent finding that myofilament length dependent activation develops very rapidly, within 5 ms upon either stretch or release of the cardiac sarcomere [71].

An alternative mechanism underling length dependent activation may not be related to cross-bridges entering the strongly bound state, but rather the structure of the weakly bound state [40,41,72]. Indeed, data from isolated skinned myocardium suggests that the number of cross-bridges in the weakly bound state increases when sarcomere length is increased, or, when relaxed skinned muscle is treated with dextran. Enhanced weakly bound cross-bridges were detected both mechanically by high-speed stiffness and by x-ray diffraction [54, 72, 73]. Since cardiac muscle thin filament activation appears to be sub-maximal under normal conditions [20,40,42,43,72,74], expansion of the pool of weakly bound cross-bridges at the longer sarcomere length that can contribute to thin-filament activation via feedback could give rise to a robust length dependent activation. Consistent with this notion, dextran application induces increases in the intensity ratios of the 1,1 and 1,0 equatorial reflections in x-ray diffraction measurements, indicative of alterations in myosin structure upon dextran compression [54,72]. We obtained a similar result recently in intact isolated rat myocardium under conditions of osmotic compression with 1% mannitol [58]. Both equatorial and meridional myosin x-ray reflections were enhanced at long sarcomere length (in relaxed muscle), regardless of inter filament spacing, indicating that increased sarcomere length enhances the structural order of weakly bound cross-bridges prior to Ca^{2+} activation of twitch contraction. Incidentally, in our studies relaxed cross-bridge disposition correlates with myofilament Ca^{2+} sensitivity, while inter-filament spacing does not [54,58], again indicating that inter filament spacing may play but a limited role in the mechanisms underlying myofilament length dependent activation.

Myofilament length dependency is significantly blunted in slow skeletal muscle[44] (Figure 5B). The protein composition of the cardiac sarcomere shares some similarities with slow skeletal muscle, such as the cardiac isoform of TnC and β-myosin. In contrast, in the heart the slow skeletal isoform of TnI is replaced by a unique cardiac isoform of Tn-I shortly after birth in most species [75]. We have previously demonstrated that the blunted length dependency seen in slow skeletal muscle is recapitulated when cardiac TnI is replaced by ssTnI in the hearts of a transgenic mouse model [55], indicating that cardiac TnI plays a pivotal role in the length signal transduction process within the sarcomere.

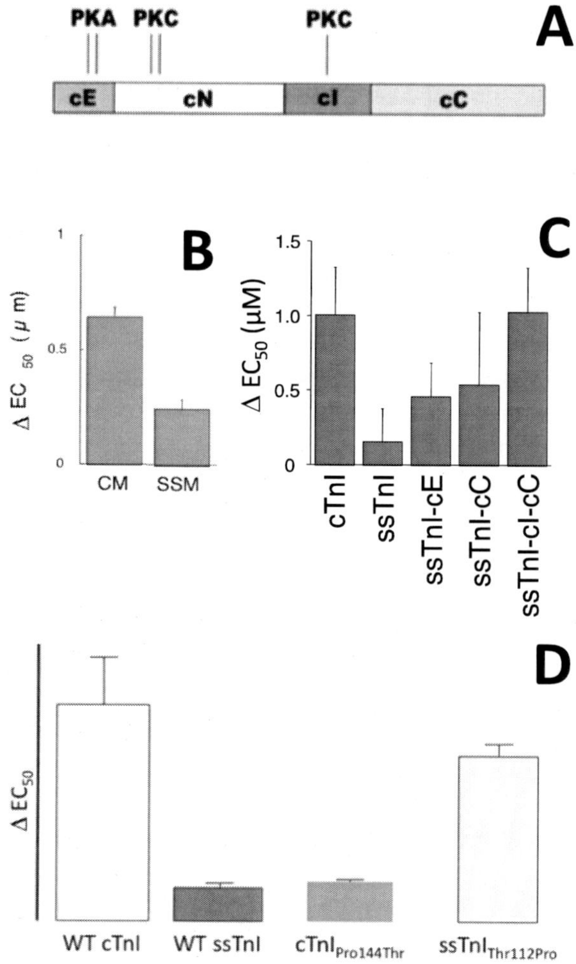

Figure 5. Panel A shows a schematic of the structure of cardiac troponin I (cTnI). In contrast to the skeletal isoforms of this protein, cTnI contains a unique N-terminal extension (33 AA; cE) that contains two serine residues (S22, S23 in the mouse) that is target for protein kinase A (PKA). The N-terminus (cN) contains two target sites for protein kinase C (PKC). In addition, the inhibitory domain (cI) contains a threonine at position 144 that is also a PKC target; the inhibitory domain is identical between slow skeletal TnI (ssTnI) with the exception of Thr144 which is a Proline in ssTnI. Panel B illustrates the markedly blunted myofilament length dependency (LDA) in rat slow skeletal muscle compared to rat cardiac muscle [44]. Panel C shows the impact of cTnI/ssTnI domain swapping employing recombinant protein exchange in isolated rat skinned trabeculae; neither addition of the unique cardiac N-terminus extension (cE) nor the cardiac C-terminus (cC) to ssTnI restored LDA to that found upon cTnI WT troponin exchange [82]. Panel D, substitution of cardiac Thr144 by ssTnI Pro in cTnI virtually eliminates LDA, while introduction of a Pro at the equivalent (Pro122) site in ssTnI greatly enhances LDA [83]; these data indicate the pivotal role of Thr144 in cTnI in myofilament length dependency.

The cardiac isoform of TnI differs from slow skeletal TnI in several significant ways as illustrated in Figure 5A [38,75-79]: a unique 33 amino-acid N-terminus extension (cE) that contains two serines that are the target for protein kinase A (PKA). In addition, two protein-kinase C (PKC) target sites that are located in the cardiac N-terminus domain (cN), which are lacking/non-functional in the equivalent domain in slow skeletal TnI. Finally, a threonine at position 144 within the cardiac inhibitory domain (cTnI) that is also a PKC target, is replaced by a proline residue in the equivalent domain of slow skeletal TnI (ssTnI); of interest, this is the only difference between the cardiac and slow skeletal inhibitory domain of TnI. Using a protein exchange technique in chemically permeabilized isolated myocardium it is possible to replace the endogenous troponin complex with a recombinant derived troponin complex [80, 81]. Application of this technique in which domains within ssTnI were replaced with the equivalent cTnI domains resulted in the data displayed in Figure 5C [82]. First, exchange for wild type troponin, containing cTnI, resulted in maintenance of robust myofilament length dependency, while exchange for troponin cardiac T/C + ssTnI virtually eliminated length dependency, as is the case in slow skeletal muscle (panel B). Neither addition of the unique cardiac N-terminal extension (cE) nor replacement of the slow skeletal C-terminus domain (cC) with the cardiac equivalent domain restored myofilament length dependency; only when the inhibitory domain (cI) is also included does robust cardiac-like length dependency immerse. These data implicate a critical role for the cardiac inhibitory domain in the length sensing mechanism. This hypothesis was tested in a follow up study, the results of which are summarized in Figure 5D [83]. Again, exchange with recombinant troponin complex containing wild-type cardiac troponin resulted in robust length dependency, while exchange with troponin containing wild-type slow skeletal TnI virtually eliminated length dependency. Substitution of the cardiac threonine with proline residue at the 144 position in cardiac TnI virtually eliminated length dependency, while introduction of a threonine at the equivalent (122) position in slow skeletal TnI markedly increased length dependency, almost to the level that is seen in cardiac muscle. Thus, the inhibitory region of cTnI plays a prominent role in the length transducing process of the sarcomere and, furthermore, this property is governed by the presence of the threonine residue at position 144 in cardiac TnI. The molecular mechanisms underlying this phenomenon, however, are presently unknown. Of note, we recently have generated a murine mouse model in which cTnI-Thr144Pro replaces endogenous cTnI; early preliminary data show blunted myofilament length dependent activation in skinned isolated myocardium derived from the hearts of these mice. Although cTnI-Thr144 is clearly important for myofilament length dependent activation, the regional differences of this parameter observed within the heart [84, 85] indicate that other factors must also play a significant role in modulating length sensitivity. Of interest, a gradient of myosin light chain phosphorylation has long been recognized to exist in the heart [86], and phosphorylation of this contractile protein is known to be involved in stretch dependent regulation of cardiac contraction [86-88] (see below).

The role of cardiac Troponin T, the protein that anchors the troponin complex to tropomyosin, in myofilament length dependent activation is less clear. Different isoforms of this protein are expressed in cardiac and skeletal muscle. Recombinant protein exchange in skinned cardiac isolated muscle with cardiac troponin complex containing either cTnT or fast skeletal TnT (fsTnT) revealed a blunting of length dependency in cardiac sarcomeres containing fsTnT [89]. Moreover, a recent study by this group revealed two domains in the highly acidic N-terminus of cTnT, where the first 44 residues were found to be important for maximum force generation and the cardiac specific residues 45-74 were found to be important for cooperativety and myofilament length dependent activation [90]. Thus, cTnT may also play an important role in myofilament length dependent activation, but full elucidation of the role of this troponin awaits further study. One complication that hampers study of this protein is the wide spectrum of cardiac isoform expression, which appears to be modulated both developmentally and by disease [91-94].

6. Role of Contractile Protein Phosphorylation in LDA

As discussed above, cardiac pump function is determined not only by *intrinsic* factors, but also by *extrinsic* factors such as neuro-hormonal stimulation [79, 95]. The underlying mechanisms at the level of the sarcomere involve post-translational modification of contractile proteins by phosphorylation [38, 78, 79, 95]. For example, stimulation of ß-adrenergic receptors by catecholamines in the heart has a profound effect on myocardial function, such as increased peak force development, decreased time to peak force, and enhanced rate of force relaxation. ß-adrenergic stimulation causes activation of a cyclic-AMP dependent protein kinase (PKA) that catalyzes phosphorylation of several key proteins that are intimately involved in calcium handling by the cardiac myocyte [8]. The contractile protein targets of PKA are the cardiac troponin-I isoform (cTnI) and cardiac myosin binding protein C that are uniquely expressed in heart muscle [76, 78, 79, 96], as well as the giant elastic protein titin [25, 97, 98]. PKA mediated TnI phosphorylation induces reduced myofilament Ca^{2+} sensitivity that is likely beneficial for enhanced cardiac relaxation kinetics in the face of increased heart rate [96], as is increased cross-bridge cycling rate [99, 100] and myofilament power output [101]. Moreover, data suggest that both cTnI and cardiac myosin binding protein C need to be phosphorylated by PKA, or least be present in the sarcomere, for these effects to manifest [55, 102 103, 104]. The decreased myofilament calcium sensitivity may due to PKA induced enhanced Ca^{2+}-dissociation from cardiac troponin-C [105]. Indeed, rapid removal of activator calcium via flash photolysis has provided direct evidence to support this hypothesis [100, 106]. Of note, a similar impact was observed upon removal of the N-terminus of TnI [107], consistent with the notion that S22, 23 phosphorylation induces a conformational change in cTnI that weakens the interaction between the N-lobes of cTnC and cTnI [99,108]. Several proteins in the cardiac sarcomere are also substrates of other protein kinases, notably protein kinase C (PKC) [38, 79, 95]. This kinase is comprised of a large family of proteins of at least 12 members, and is activated by several sarcolemmal receptor signal transduction systems that include the α-adrenergic, endothelin, and the angiotensin receptors [109]. It has been suggested that phosphorylation of contractile proteins by protein kinase C may have a profound impact on cross-bridge cycling and maximum force development [38, 79, 95, 110-112]. However, the precise impact of PKC mediated phosphorylation is not clear at present which is likely due, in part, to the fact that there are many protein targets sites for these kinases, often with apparent opposite effects on cross-bridge cycling and calcium sensitivity [79, 113-116]. Of note, we recently found in heart failure that PKC-α mediated phosphorylation caused marked depression of force development and myofilament Ca^{2+} sensitivity [117, 118]. In addition, PKC specific phosphorylation sites have been identified on other cardiac regulatory proteins, such as troponin-T and myosin binding protein C [79], although the functional consequence of phosphorylation on these PKC sites is less clear. We have examined in detail the functional impact of PKC mediated phosphorylation of 4 sites on cTnT (Thr197, Ser201, Thr206, and

Thr287) by site directed mutagenesis and found Thr206 to be specifically important for modulating Ca^{2+} sensitivity and cross-bridge cycling rate [119]. Phosphorylation of regulatory myosin light chain (MLC_2) has emerged as important modulator of myofilament function [86, 88, 120-122], consistent with the recent identification of a cardiac specific isoform of MLC_2 kinase [120, 123, 124]. Finally, recently serine 150 in cTnI has been identified as target for phosphorylation by AMP kinase, inducing increased myofilament Ca^{2+} sensitivity [125].

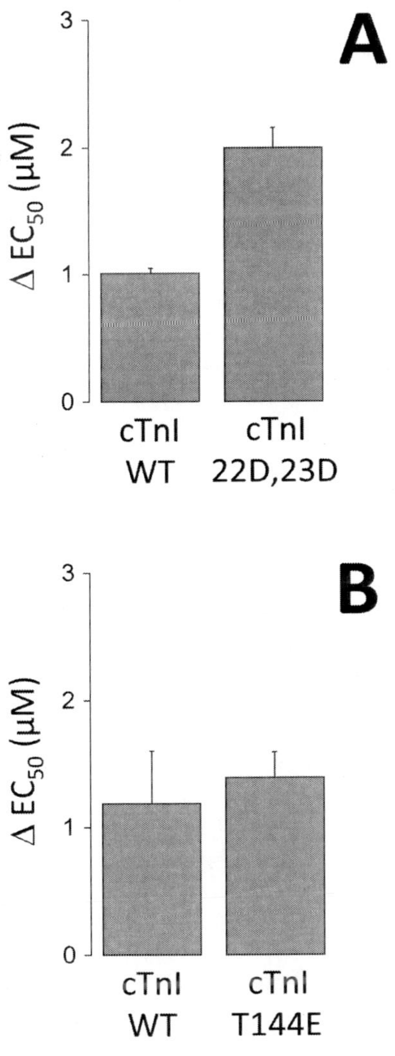

Figure 6. Panel A, replacement of the PKA target sites (S22, S23) within cTnI by charged amino acid residues in recombinant protein. Exchange of PKA phosphomimic cardiac troponin complex into skinned isolated rat myocardium results in increased myofilament length dependent activation (as indexed by ΔEC_{50}). Panel B, in contrast, introduction of a charged amino acid mutation at the target site T144 did not affect length dependency.

Phosphorylation of contractile proteins mediated by PKA has been shown to increase myofilament length dependent activation, both in rodent models [55, 126, 127] and more recently in human myocardium [128], albeit not uniformly [129, 130]. As illustrated in Figure 6A, this phenomenon can be recapitulated by substitution of endogenous cTnI by mutant cTnI in which Ser22, 23 are replaced by phosphomimic charged residues. In contrast, S150 AMPK mediated phosphorylation was recently shown to induce a decrease in myofilament length dependent activation [125]. Although cTnI-T144 phosphorylation induces reduced myofilament Ca^{2+} sensitivity, there is no impact on length dependency, as is illustrated in Figure 6B. The impact of other troponin phosphorylation targets [79], notably cTnI-S43, 45 [131] and cTnT-T206 [119], has been established in terms of Ca^{2+} sensitivity and cross-bridge cycling kinetics. However, the impact on LDA has not been studied in detail. With regards to the other sarcomeric proteins targeted by kinases, phosphorylation of myosin binding protein C (MyoBPC) [132, 133] has been shown to accelerate stretch activation [134], a property that is likely related to LDA [135], and absence of Myosin binding protein C in the cardiac sarcomere has been shown to reduce stretch activation [135] and LDA [103]. Likewise, as is the case with myosin binding protein C, myosin light chain phosphorylation has been shown to accelerate stretch activation [88], but whether LDA is also impacted has yet to be determined. Recently, we found that transgenic ablation of myosin light chain phosphorylation results in reduced maximum force *in-vitro* and delayed ejection dynamics *in-vivo*; moreover, we found that absence of this protein's phosphorylation blunts the impact of PKA mediated phosphorylation on myofilament function [136]. Finally, recent data indicates that the elasticity of titin is modulated by phosphorylation [25, 137], but whether this directly affects LDA is not known (see also below). This is due, in part, to the inherent difficulty that phosphorylation of other contractile proteins usually accompanies titin phosphorylation and because it's large size renders this protein resistant to transgenic or recombinant protein exchange.

In conclusion, contractile protein phosphorylation prominently impacts myofilament function in terms of Ca^{2+} responsiveness and cross-bridge cycle kinetics. Evidence shows that cTnI phosphorylation at PKA sites S22, 23 enhances, AMPK site S150 reduces, while PKC site T144 is without impact on LDA. Phosphorylation of myosin binding protein C and regulatory light chain enhances stretch activation kinetics, but its impact on LDA has not been established, as has the impact of other contractile protein phosphorylation targets.

7. Role of Contractile Protein Mutations in LDA

Mutations in contractile proteins are now recognized as major cause for the development of non-ischemic hypertrophic (HCM) and dilated (DM) cardiomyopathy [132, 133, 138-140]. The affected loci appear to be concentrated in myosin and myosin binding protein C and to a lesser, but still significant level, in other contractile proteins such as the troponins, myosin light chains, titin, etc. The impact of specific disease related mutations on myofilament function has been extensively studied over the past decade: revealing impacts on Ca^{2+} responsiveness and cross-bridge cycle kinetics in biophysical studies employing reconstituted proteins in solution, the motility assay, and intact and skinned isolated muscle. Despite these data, a clear insight into the relation between the presence of a mutation and its impact on myofilament function and the clinical presentation of HCM or DCM is still lacking [140]. Moreover, some disease related mutations appear to be without significant impact on myofilament function in isolation, but rather appear to interfere with kinase/phosphatase mediated cellular signaling. For example, we found no impact of the DCM related TnC-G159D mutation on Ca^{2+} sensitivity in isolation [99]; however, presence of this mutation virtually eliminated the myofilament Ca^{2+} desensitization that normally accompanies PKA mediated cTnI phosphorylation. In contrast, the impact of PKA phosphorylation on cross-bridge cycle kinetics was not affected by the TnC mutation. We recently found a similar "cross-talk" between ablated MLC_2 phosphorylation and PKA cTnI phosphorylation [136], and the Marston group has suggested that this may be a general phenomenon [141]. The impact of disease related mutations on LDA, in general, have received much less attention. Ford et al. reported blunted LDA associated with the FHC cTnT-R92L mutation, albeit only within a sarcomere composed of □-myosin [142]. Likewise, Regan et al. reported in a preliminary study reduced LDA in skinned human myocytes harboring the cTnI-R145W mutation [143]. Finally, a blunting of length dependency has recently been reported in human hypertrophic cardiomyopathy, provided that PKA phosphorylation state was normalized across the different groups studied [128]. With regard to myosin binding protein C, in most cases, mutated protein fails to incorporate into the sarcomere leading to haploinsufficiency [132, 138, 139] (as mentioned above, ablation of myosin binding protein C results in reduced length dependency). An exception is myosin binding protein-ΔC10, a mutation predominantly found in South East Asia [144, 145]; the impact of this mutation on myofilament function is not completely known. We have obtained preliminary data that indicate blunting of LDA in skinned myocardium upon treatment with a N-terminal proteolytic fragment of MyoBPC, which also depresses myocyte contractility [146]. Finally, it should be noted that interactions between presences of contractile protein mutations, phosphorylation, and LDA have not been studied to date and this should be an interesting area for research in the future.

In conclusion, HCM and DCM associated sarcomeric protein mutations have been shown to impact myofilament function, albeit not universally. Besides a few recent studies, the impact of mutations on LDA is incompletely understudied. Finally, mutations have been reported to "uncouple" PKA induced myofilament Ca^{2+} desensitization; whether mutations induce "cross-talk" between contractile protein phosphorylation and LDA has not been studied to date.

8. Role of Titin in LDA

It has been suggested that titin may be involved in myofilament length dependent activation [24, 25, 147-150]. It is perceivable that titin strain influences the number of weakly bound cross-bridges since titin is known to possess both actin and myosin binding domains [24, 25, 151]. Indeed, release of the passive tension borne by the titin molecule, either by mild treatment with trypsin [53] or by brief over-stretching of the sarcomere [147], markedly reduces myofilament length dependency. Moreover, regional differences between the epi- and endo-cardial layer of the heart have been identified in passive stiffness [84, 152]. The larger endo-myocardial stiffness was found to correlate with enhanced myofilament length dependent activation in several species [84, 152]. Of interest, these regional differences are diminished in heart failure, but can be restored by exercise [153]. The mechanisms that underlie this regional phenomenon are not fully understood. Differences in titin isoform distribution were but almost certainly excluded; rather, marked regional differences in contractile protein phosphorylation were suggested as the possible mechanism that underlies regional variation in length dependent activation [84]; which contractile protein(s), and whether titin itself is among the affected contractile protein remains to be investigated. Finally, the group of Greaser [154, 155] has recently identified a strain of rats in which the developmentally regulated splice variation of the titin gene is disrupted in the heart. As a result, the normal reduction in the titin persistence length upon aging does not occur in the hearts of these animals, leading to low passive stiffness and unusually long resting sarcomere length, as is illustrated in Figure 7A.

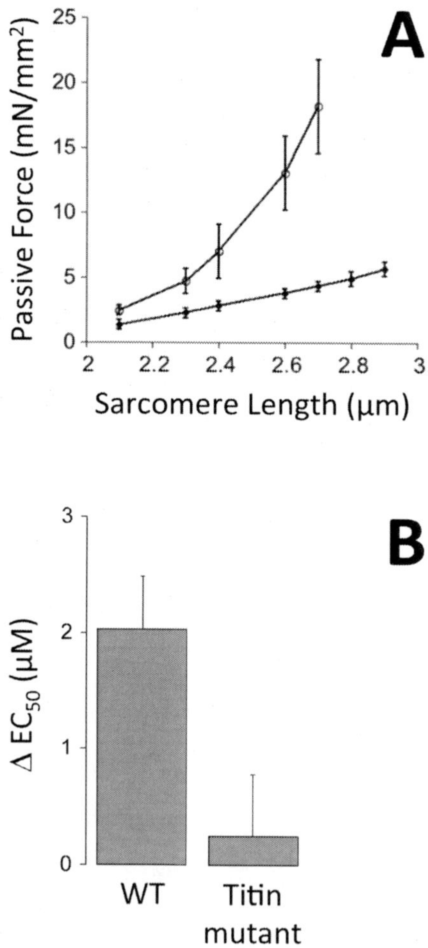

Figure 7. Panel A, passive force development as function of sarcomere length in isolated rat skinned myocyte fragments in wild-type (open circles) and giant titin mutant (closed circles); presence of the mutation[154] leading to an unusually high molecular weight titin resulted in marked reduced passive muscle stiffness. Panel B, myocardium isolated from the titin mutant rat heart shows a significantly blunted myofilament length dependent activation property (as indexed by EC_{50}).

Subsequently, a transcription factor mutation was identified that, in humans, correlates with the development of cardiac disease [156]. Moreover, presence of the giant titin isoform is associated with a marked blunting of myofilament length dependent activation in cardiac (Figure 7B and [157]) as well as skeletal muscle [149]. Access to this mutant rat model as well as murine animal models in which titin strain is modulated by mutations [158, 159] will greatly enhance future research into the role of this giant elastic protein in myofilament length dependent activation.

9. Overview, Molecular Models, and Remaining Questions

The "Frank-Starling law of the heart" constitutes a fundamental cardiac property that has been appreciated for well over a century. At its basis, this ventricular property is the result of a modulation of myofilament Ca^{2+} sensitivity upon a change in sarcomere length. Despite intense investigations, the molecular mechanisms that underlie this phenomenon are still not known. As discussed in this review, controversy remains whether inter-filament spacing constitutes the molecular strain sensing mechanism of the sarcomere. Rather, evidence points to a length modulated regulation of thin filament activation state via regulation of thin filament activation state. How the thin filament is regulated by sarcomere length is not known; some of the potential mechanisms are illustrated in figure 8 that include cooperative feedback between force generating cross-bridges, Ca^{2+} binding affinity, as well as the giant cytoskeletal protein titin as the mechanical linker protein that transmits information regarding sarcomere length to the contractile apparatus.

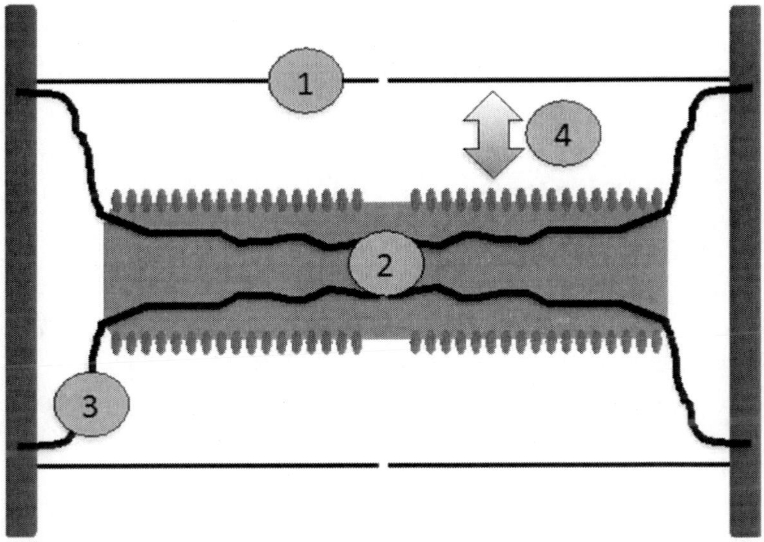

Figure 8. Various potential molecular mechanisms for myofilament length dependent activation. Stretch of the sarcomere may: 1) affect thin filament cooperative activation, 2) alter the structure of weakly bound cross-bridges, 3) these mechanisms may be induced directly by strain transmission via the giant elastic protein titin, 4) or impinge on sarcomere dynamics via modulation of the cooperative feedback processes between the thin- and thick-filament.

Clearly, the question as to as to how myofilament Ca^{2+} sensitivity is regulated by sarcomere length is far from being solved. Future biophysical, biochemical, and detailed ultra-structural investigations are clearly required to solve this intriguing cellular molecular physiology puzzle.

References

[1] Sagawa K, Maughan L, Suga H, Sunagawa K. Cardiac contraction and the pressure-volume relationship. New York, Oxford: Oxford University Press; 1988.

[2] Pacher P, Nagayama T, Mukhopadhyay P, Batkai S, Kass DA. Measurement of cardiac function using pressure-volume conductance catheter technique in mice and rats. *Nat. Protoc.* 2008; 3:1422–1434.

[3] Lakatta EG. Cardiovascular regulatory mechanisms in advanced age. *Physiological reviews.* 1993; 73:413–467.

[4] Kroeker CAG, Shrive NG, Belenkie I, Tyberg JV. Pericardium modulates left and right ventricular stroke volumes to compensate for sudden changes in atrial volume. *Am. J. Physiol. Heart Circ. Physiol.* 2003; 284:H2247–54.

[5] de Tombe PP. Cardiac myofilaments: mechanics and regulation. *J Biomech.* 2003; 36:721–730.

[6] Keurs ter HEDJ, Tyberg JV. Mechanics of the Circulation. Dordrecht: Martinus Nijhoff; 1986.

[7] Keurs ter HEDJ, Rijnsburger WH, van Heuningen R, Nagelsmit MJ. Tension development and sarcomere length in rat cardiac trabeculae: Evidence of length-dependent activation. *Circ. Res.* 1980; 46:703–714.

[8] Bers DM. Excitation-Contraction Coupling and Cardiac Contractile Force. 2nd ed. Dordrecht, The Netherlands: Kluwer Academic Press; 2001.

[9] Backx PH, Gao WD, Azan-Backx MD, Marban E. The relationship between contractile force and intracellular [Ca2+] in intact rat cardiac trabeculae. *J. Gen. Physiol.* 1995; 105:1–19.

[10] Fabiato A, Fabiato F. Dependence of the contractile activation of skinned cardiac cells on the sarcomere length. *Nature.* 1975; 256:54–56.

[11] Hibberd MG, Jewell BR. Calcium- and length-dependent force production in rat ventricular muscle. *J. Physiol. (Lond.).* 1982; 329:527–540.

[12] Kentish JC, Keurs ter HE, Ricciardi L, Bucx JJ, Noble MI. Comparison between the sarcomere length-force relations of intact and skinned trabeculae from rat right ventricle. Influence of calcium concentrations on these relations. *Circ. Res.* 1986; 58:755–768.

[13] Dobesh DP, Konhilas JP, de Tombe PP. Cooperative activation in cardiac muscle: impact of sarcomere length. *Am. J. Physiol. Heart Circ. Physiol.* 2002; 282:H1055–62.

[14] Claflin DR, Morgan DL, Julian FJ. The effect of length on the relationship between tension and intracellular [Ca2+] in intact frog skeletal muscle fibres. *The Journal of Physiology.* 1998; 508 (Pt 1):179–186.

[15] Nowak G, Pena JR, Urboniene D, Geenen DL, Solaro RJ, Wolska BM. Correlations between alterations in length-dependent Ca2+ activation of cardiac myofilaments and

the end-systolic pressure-volume relation. *J. Muscle Res. Cell. Motil.* 2007; 28:415–419.

[16] Cingolani OH, Kirk JA, Seo K, Koitabashi N, Lee D-I, Ramirez-Correa G, Bedja D, Barth AS, Moens AL, Kass DA. Thrombospondin-4 is required for stretch-mediated contractility augmentation in cardiac muscle. *Circ. Res.* 2011; 109:1410–1414.

[17] Cingolani HE, Pérez NG, Cingolani OH, Ennis IL. The Anrep effect: 100 years later. *AJP: Heart and Circulatory Physiology*. 2013; 304:H175–82.

[18] Wannenburg T, Heijne GH, Geerdink JH, Van Den Dool HW, Janssen PM, de Tombe PP. Cross-bridge kinetics in rat myocardium: effect of sarcomere length and calcium activation. *Am. J. Physiol Heart Circ. Physiol.* 2000; 279:H779–90.

[19] Wannenburg T, Janssen PM, Fan D, de Tombe PP. The Frank-Starling mechanism is not mediated by changes in rate of cross-bridge detachment. *Am. J. Physiol.* 1997; 273:H2428–35.

[20] Gordon AM, Homsher E, Regnier M. Regulation of contraction in striated muscle. *Physiological reviews*. 2000; 80:853–924.

[21] Huxley HE. The mechanism of muscular contraction. *Science*. 1969; 164:1356–1365.

[22] Wang K, McClure J, Tu A. Titin: major myofibrillar components of striated muscle. *Proc. Natl. Acad. Sci. U. S. A.* 1979; 76:3698–3702.

[23] Maruyama K, Kimura S, Ohashi K, Kuwano Y. Connectin, an elastic protein of muscle. Identification of "titin" with connectin. *J. Biochem. (Tokyo)*. 1981; 89:701–709.

[24] Granzier HL, Labeit S. The giant protein titin: a major player in myocardial mechanics, signaling, and disease. *Circ. Res.* 2004; 94:284–295.

[25] Krüger M, Linke WA. The giant protein titin: a regulatory node that integrates myocyte signaling pathways. *Journal of Biological Chemistry*. 2011; 286:9905–9912.

[26] McKillop DF, Geeves MA. Regulation of the interaction between actin and myosin subfragment 1: evidence for three states of the thin filament. *Biophys. J.* 1993; 65:693–701.

[27] Lehrer SS, Geeves MA. The muscle thin filament as a classical cooperative/allosteric regulatory system. *Journal of molecular biology*. 1998; 277:1081–9.

[28] Maytum R, Lehrer SS, Geeves MA. Cooperativity and switching within the three-state model of muscle regulation. *Biochemistry*. 1999; 38:1102–1110.

[29] Potter JD, Gergely J. The calcium and magnesium binding sites on troponin and their role in the regulation of myofribrillar adenosine triphosphatase. *Journal of Biological Chemistry*. 1975; 250:4628–4633.

[30] Holroyde MJ, Robertson SP, Johnson JD, Solaro RJ, Potter JD. The calcium and magnesium binding sites on cardiac troponin and their role in the regulation of myofibrillar adenosine triphosphatase. *J. Biol. Chem.* 1980; 255:11688–11693.

[31] Pan BS, Solaro RJ. Calcium-binding properties of troponin C in detergent-skinned heart muscle fibers. *J. Biol. Chem.* 1987; 262:7839–7849.

[32] Rice JJ, Stolovitzky G, Tu Y, de Tombe PP. Ising model of cardiac thin filament activation with nearest-neighbor cooperative interactions. *Biophys. J.* 2003; 84:897–909.

[33] Rice JJ, Wang F, Bers DM, de Tombe PP. Approximate model of cooperative activation and crossbridge cycling in cardiac muscle using ordinary differential equations. *Biophys. J.* 2008; 95:2368–2390.

[34] Hill TL. Cooperativity theory in Biochemistry: steady-state and equilibrium systems. New York: Springer-Verlag; 1985.

[35] Sun Y-B, Irving M. The molecular basis of the steep force-calcium relation in heart muscle. *J. Mol. Cell Cardiol.* 2010; 48:859–865.

[36] Slupsky CM, Sykes BD. NMR solution structure of calcium-saturated skeletal muscle troponin C. *Biochemistry*. 1995; 34:15953–15964.

[37] Gagne SM, Tsuda S, Li MX, Smillie LB, Sykes BD. Structures of the troponin C regulatory domains in the apo and calcium-saturated states. *Nat. Struct. Biol.* 1995; 2:784–789.

[38] Kobayashi T, Jin L, de Tombe PP. Cardiac thin filament regulation. *Pflugers Arch.* 2008; 457:37–46.

[39] Kobayashi T, Solaro RJ. Calcium, thin filaments, and the integrative biology of cardiac contractility. *Annual review of physiology*. 2005; 67:39–67.

[40] Smith L, Tainter C, Regnier M, Martyn DA. Cooperative cross-bridge activation of thin filaments contributes to the Frank-Starling mechanism in cardiac muscle. *Biophys. J.* 2009; 96:3692–3702.

[41] Adhikari BB, Regnier M, Rivera AJ, Kreutziger KL, Martyn DA. Cardiac length dependence of force and force redevelopment kinetics with altered cross-bridge cycling. *Biophys. J.* 2004; 87:1784–1794.

[42] Clemmens EW, Entezari M, Martyn DA, Regnier M. Different effects of cardiac versus skeletal muscle regulatory proteins on in vitro measures of actin filament speed and force. *The Journal of Physiology*. 2005; 566:737–746.

[43] Regnier M, Martin H, Barsotti RJ, Rivera AJ, Martyn DA, Clemmens E. Cross-bridge versus thin filament contributions to the level and rate of force development in cardiac muscle. *Biophys. J.* 2004; 87:1815–1824.

[44] Konhilas JP, Irving TC, Smith SH, de Tombe PP. Length-dependent activation compared in three striated muscle types of the rat. *Biophys. J.* 2001; 80:1053.

[45] Poole KJ, Lorenz M, Evans G, Rosenbaum G, Pirani A, Craig R, Tobacman LS, Lehman W, Holmes KC. A comparison of muscle thin filament models obtained from electron microscopy reconstructions and low-angle X-ray fibre diagrams from non-overlap muscle. *Journal of structural biology*. 2006; 155:273–284.

[46] Godt RE, Maughan DW. Influence of osmotic compression on calcium activation and tension in skinned muscle fibers of the rabbit. *Pflugers Arch.* 1981; 391:334–7.

[47] McDonald KS, Moss RL. Osmotic compression of single cardiac myocytes eliminates the reduction in Ca2+ sensitivity of tension at short sarcomere length. *Circ. Res.* 1995; 77:199–205.

[48] Fuchs F, Smith SH. Calcium, cross-bridges, and the frank-starling relationship. *News Physiol Sci*. 2001; 16:5–10.

[49] Farman GP, Allen EJ, Gore D, Irving TC, de Tombe PP. Interfilament spacing is preserved during sarcomere length isometric contractions in rat cardiac trabeculae. *Biophys. J.* 2007; 92:L73–5.

[50] Rome E. X-ray diffraction studies of the filament lattice of striated muscle in various bathing media. *Journal of molecular biology*. 1968; 37:331–44.

[51] Irving TC, Konhilas J, Perry D, Fischetti R, de Tombe PP. Myofilament lattice spacing as a function of sarcomere length in isolated rat myocardium. *Am. J. Physiol Heart Circ Physiol*. 2000; 279:H2568–73.

[52] Matsubara I, Umazume Y, Yagi N. Lateral filamentary spacing in chemically skinned murine muscles during contraction. *J. Physiol. (Lond.)*. 1985; 360:135–148.

[53] Fukuda N, Wu Y, Farman G, Irving TC, Granzier H. Titin-based modulation of active tension and interfilament lattice spacing in skinned rat cardiac muscle. *Pflugers Arch.* 2005; 449:449–457.

[54] Farman GP, Walker JS, de Tombe PP, Irving TC. Impact of osmotic compression on sarcomere structure and myofilament calcium sensitivity of isolated rat myocardium. *Am. J. Physiol Heart Circ. Physiol*. 2006; 291:H1847–55.

[55] Konhilas JP, Irving TC, Wolska BM, Jweied EE, Martin AF, Solaro RJ, Tombe PP de. Troponin I in the murine myocardium: influence on length-dependent activation and interfilament spacing. *The Journal of Physiology*. 2003; 547:951–961.

[56] Konhilas JP, Irving TC, de Tombe PP. Length-dependent activation in three striated muscle types of the rat. *The Journal of Physiology*. 2002; 544:225–236.

[57] Konhilas JP, Irving TC, de Tombe PP. Myofilament calcium sensitivity in skinned rat cardiac trabeculae: role of inter-filament spacing. *Circ. Res.* 2002; 90:59–65.

[58] Farman GP, Gore D, Allen E, Schoenfelt K, Irving TC, de Tombe PP. Myosin head orientation: a structural determinant for the Frank-Starling relationship. *AJP: Heart and Circulatory Physiology*. 2011; 300:H2155–60.

[59] Moss RL, Fitzsimons DP. Frank-Starling relationship: long on importance, short on mechanism. *Circ. Res.* 2002; 90:11–13.

[60] Fuchs F, Martyn DA. Length-dependent Ca(2+) activation in cardiac muscle: some remaining questions. *J. Muscle Res. Cell. Motil.* 2005; 26:199–212.

[61] Fitzsimons DP, Moss RL. Strong binding of myosin modulates length-dependent Ca2+ activation of rat ventricular myocytes. *Circ. Res.* 1998; 83:602–607.

[62] Fukuda N, Kajiwara H, Ishiwata S, Kurihara S. Effects of MgADP on length dependence of tension generation in skinned rat cardiac muscle. *Circ. Res.* 2000; 86:E1–6.

[63] Smith SH, Fuchs F. Effect of ionic strength on length-dependent Ca(2+) activation in skinned cardiac muscle. *J. Mol. Cell Cardiol*. 1999; 31:2115–25.

[64] Arteaga GM, Kobayashi T, Solaro RJ. Molecular actions of drugs that sensitize cardiac myofilaments to Ca2. *Ann Med*. 2002; 34:248–258.

[65] Stelzer JE, Larsson L, Fitzsimons DP, Moss RL. Activation dependence of stretch activation in mouse skinned myocardium: implications for ventricular function. *J. Gen. Physiol*. 2006; 127:95–107.

[66] Fukuda N, J OU, Sasaki D, Kajiwara H, Ishiwata S, Kurihara S. Acidosis or inorganic phosphate enhances the length dependence of tension in rat skinned cardiac muscle. *J. Physiol. (Lond.)*. 2001; 536:153–160.

[67] Farman GP, Allen EJ, Schoenfelt KQ, Backx PH, de Tombe PP. The role of thin filament cooperativity in cardiac length-dependent calcium activation. *Biophys. J.* 2010; 99:2978–2986.

[68] Kovacs M, Toth J, Hetenyi C, Malnasi-Csizmadia A, Sellers JR. Mechanism of blebbistatin inhibition of myosin II. *J. Biol. Chem*. 2004; 279:35557–35563.

[69] Farman GP, Tachampa K, Mateja R, Cazorla O, Lacampagne A, de Tombe PP. Blebbistatin: use as inhibitor of muscle contraction. *Pflugers Arch.* 2008; 455:995–1005.

[70] Sun YB, Lou F, Irving M. Calcium- and myosin-dependent changes in troponin structure during activation of heart muscle. *J. Physiol. (Lond.).* 2009; 587:155–163.

[71] Mateja RD, de Tombe PP. Myofilament Length-Dependent Activation Develops within 5 ms in Guinea-Pig Myocardium. *Biophys. J.* 2012; 103:L13–5.

[72] Martyn DA, Adhikari BB, Regnier M, Gu J, Xu S, Yu LC. Response of equatorial x-ray reflections and stiffness to altered sarcomere length and myofilament lattice spacing in relaxed skinned cardiac muscle. *Biophys. J.* 2004; 86:1002–1011.

[73] Xu S, Martyn D, Zaman J, Yu LC. X-ray diffraction studies of the thick filament in permeabilized myocardium from rabbit. *Biophys. J.* 2006; 91:3768–3775.

[74] Regnier M, Rivera AJ, Chen Y, Chase PB. 2-deoxy-ATP enhances contractility of rat cardiac muscle. *Circ. Res.* 2000; 86:1211–1217.

[75] Kruger M, Kohl T, Linke WA. Developmental changes in passive stiffness and myofilament Ca2+ sensitivity due to titin and troponin-I isoform switching are not critically triggered by birth. *Am. J. Physiol Heart Circ. Physiol.* 2006; 291:H496–506.

[76] Solaro RJ, Rarick HM. Troponin and tropomyosin: proteins that switch on and tune in the activity of cardiac myofilaments. *Circ. Res.* 1998; 83:471–480.

[77] Solaro RJ, Montgomery DM, Wang L, Burkart EM, Ke Y, Vahebi S, Buttrick P. Integration of pathways that signal cardiac growth with modulation of myofilament activity. *J. Nucl. Cardiol.* 2002; 9:523–533.

[78] Solaro RJ. Protein phosphorylation in heart muscle. Boca Raton: CRC Press; 1986.

[79] Solaro RJ, Stull JT. Thematic minireview series on signaling in cardiac sarcomeres in health and disease. *Journal of Biological Chemistry.* 2011; 286:9895.

[80] Brenner B, Chalovich JM. Kinetics of thin filament activation probed by fluorescence of N-((2-(Iodoacetoxy)ethyl)-N-methyl)amino-7-nitrobenz-2-oxa-1, 3-diazole-labeled troponin I incorporated into skinned fibers of rabbit psoas muscle: implications for regulation of muscle contraction. *Biophys. J.* 1999; 77:2692–2708.

[81] Chandra M, Kim JJ, Solaro RJ. An improved method for exchanging troponin subunits in detergent skinned rat cardiac fiber bundles. *Biochem. Biophys. Res. Commun.* 1999; 263:219–223.

[82] Smith SH, Solaro RJ, Martin AF, de Tombe PP. Chimeric Troponin I and the Sarcomere Length Dependence of Calcium Sensitivity in Skinned Rat Trabeculae. *Biophys. J.* 2003; 84:251a.

[83] Tachampa K, Wang H, Farman GP, de Tombe PP. Cardiac troponin I threonine 144: role in myofilament length dependent activation. *Circ. Res.* 2007; 101:1081–1083.

[84] Ait Mou Y, le Guennec J-Y, Mosca E, de Tombe PP, Cazorla O. Differential contribution of cardiac sarcomeric proteins in the myofibrillar force response to stretch. *Pflugers Arch.* 2008; 457:25–36.

[85] Cazorla O, Le Guennec J-Y, White E. Length-tension relationships of sub-epicardial and sub-endocardial single ventricular myocytes from rat and ferret hearts. *J. Mol. Cell Cardiol.* 2000; 32:735–744.

[86] Davis JS, Hassanzadeh S, Winitsky S, Lin H, Satorius C, Vemuri R, Aletras AH, Wen H, Epstein ND. The overall pattern of cardiac contraction depends on a spatial gradient of myosin regulatory light chain phosphorylation. *Cell.* 2001; 107:631–641.

[87] Scruggs SB, Walker LA, Lyu T, Geenen DL, Solaro RJ, Buttrick PM, Goldspink PH. Partial replacement of cardiac troponin I with a non-phosphorylatable mutant at serines

43/45 attenuates the contractile dysfunction associated with PKCepsilon phosphorylation. *J. Mol. Cell Cardiol.* 2006; 40:465–473.

[88] Stelzer JE, Patel JR, Moss RL. Acceleration of stretch activation in murine myocardium due to phosphorylation of myosin regulatory light chain. *J. Gen. Physiol.* 2006; 128:261–272.

[89] Chandra M, Tschirgi ML, Rajapakse I, Campbell KB. Troponin T modulates sarcomere length-dependent recruitment of cross-bridges in cardiac muscle. *Biophys. J.* 2006; 90:2867–2876.

[90] Mamidi R, Mallampalli SL, Wieczorek DF, Chandra M. Identification of two new regions in the N-terminus of cardiac troponin T that have divergent effects on cardiac contractile function. *J. Physiol. (Lond.).* 2013; 591:1217–1234.

[91] Biesiadecki BJ, Chong SM, Nosek TM, Jin JP. Troponin T core structure and the regulatory NH2-terminal variable region. *Biochemistry.* 2007; 46:1368–1379.

[92] Biesiadecki BJ, Jin JP. Exon skipping in cardiac troponin T of turkeys with inherited dilated cardiomyopathy. *J. Biol. Chem.* 2002; 277:18459–18468.

[93] Greig A, Hirschberg Y, Anderson PAW, Hainsworth C, Malouf NN, Oakeley AE, Kay BK. Molecular basis of cardiac troponin T isoform heterogeneity in rabbit heart. *Circ. Res.* 1994; 74:41–47.

[94] Anderson PAW, Greig A, Mark TM, Malouf NN, Oakeley AE, Ungerleider RM, Allen PD, Kay BK. Molecular basis of human cardiac troponin T isoforms expressed in the developing, adult, and failing heart. *Circ. Res.* 1995; 76:681–686.

[95] Solaro RJ, de Tombe PP. Review focus series: sarcomeric proteins as key elements in integrated control of cardiac function. *Cardiovasc. Res.* 2008; 77:616–618.

[96] Solaro RJ, Moir AJ, Perry SV. Phosphorylation of troponin I and the inotropic effect of adrenaline in the perfused rabbit heart. *Nature.* 1976; 262:615–617.

[97] Yamasaki R, Wu Y, McNabb M, Greaser M, Labeit S, Granzier H. Protein kinase A phosphorylates titin's cardiac-specific N2B domain and reduces passive tension in rat cardiac myocytes. *Circ. Res.* 2002; 90:1181–1188.

[98] Kruger M, Linke WA. Protein kinase-A phosphorylates titin in human heart muscle and reduces myofibrillar passive tension. *J. Muscle Res. Cell. Motil.* 2006; 27:435–444.

[99] Biesiadecki BJ, Kobayashi T, Walker JS, John Solaro R, de Tombe PP. The troponin C G159D mutation blunts myofilament desensitization induced by troponin I Ser23/24 phosphorylation. *Circ. Res.* 2007; 100:1486–1493.

[100] Kentish JC, McCloskey DT, Layland J, Palmer S, Leiden JM, Martin AF, Solaro RJ. Phosphorylation of troponin I by protein kinase A accelerates relaxation and crossbridge cycle kinetics in mouse ventricular muscle. *Circ. Res.* 2001; 88:1059–1065.

[101] McDonald KS. The interdependence of Ca(2+) activation, sarcomere length, and power output in the heart. *Pflugers Arch.* 2011; 462:61–67.

[102] Fentzke RC, Buck SH, Patel JR, Lin H, Wolska BM, Stojanovic MO, Martin AF, Solaro RJ, Moss RL, Leiden JM. Impaired cardiomyocyte relaxation and diastolic function in transgenic mice expressing slow skeletal troponin I in the heart. *The Journal of Physiology.* 1999; 517 (Pt 1):143–157.

[103] Cazorla O, Szilagyi S, Vignier N, Salazar G, Kramer E, Vassort G, Carrier L, Lacampagne A. Length and protein kinase A modulations of myocytes in cardiac myosin binding protein C-deficient mice. *Cardiovasc. Res.* 2006; 69:370–380.

[104] Stelzer JE, Patel JR, Walker JW, Moss RL. Differential roles of cardiac myosin-binding protein C and cardiac troponin I in the myofibrillar force responses to protein kinase A phosphorylation. *Circ. Res.* 2007; 101:503–511.

[105] Robertson SP, Johnson JD, Holroyde MJ, Kranias EG, Potter JD, Solaro RJ. The effect of troponin I phosphorylation on the Ca2+-binding properties of the Ca2+-regulatory site of bovine cardiac troponin. *J. Biol. Chem.* 1982; 257:260–263.

[106] Zhang R, Zhao J, Mandveno A, Potter JD. Cardiac troponin I phosphorylation increases the rate of cardiac muscle relaxation. *Circ. Res.* 1995; 76:1028–1035.

[107] Biesiadecki BJ, Tachampa K, Yuan C, Jin J-P, de Tombe PP, Solaro RJ. Removal of the cardiac troponin I N-terminal extension improves cardiac function in aged mice. *J. Biol. Chem.* 2010; 285:19688–19698.

[108] Solaro RJ, Rosevear P, Kobayashi T. The unique functions of cardiac troponin I in the control of cardiac muscle contraction and relaxation. *Biochem. Biophys. Res. Commun.* 2008; 369:82–87.

[109] Steinberg SF. Structural basis of protein kinase C isoform function. *Physiological reviews.* 2008; 88:1341–1378.

[110] Strang KT, Moss RL. a1-adrenergic receptor stimulation decreases maximum shortening velocity of skinned single ventricular myocytes from rats. *Circ. Res.* 1995; 77:114–120.

[111] Lester JW, Hofmann PA. Role for PKC in the adenosine-induced decrease in shortening velocity of rat ventricular myocytes. *Am. J. Physiol. Heart Circ. Physiol.* 2000; 279:H2685–93.

[112] Sumandea MP, Rybin VO, Hinken AC, Wang C, Kobayashi T, Harleton E, Sievert G, Balke CW, Feinmark SJ, Solaro RJ, Steinberg SF. Tyrosine phosphorylation modifies protein kinase C delta-dependent phosphorylation of cardiac troponin I. *J. Biol. Chem.* 2008; 283:22680–22689.

[113] Noland TAJ, Guo XD, Raynor RL, Jideama NM, Averyhart-Fullard V, Solaro RJ, Kuo JF. Cardiac troponin I mutants - Phosphorylation by protein kinases C and A and regulation of Ca2+-stimulated MgATPase of reconstituted actomyosin S-1. *Journal of Biological Chemistry.* 1995; 270:25445–25454.

[114] Damron DS, Darvish A, Murphy L, Sweet W, Moravec CS, Bond M. Arachidonic acid-dependent phosphorylation of troponin I and myosin light chain 2 in cardiac myocytes. *Circ. Res.* 1995; 76:1011–1019.

[115] McClellan G, Weisberg A, Winegrad S. Effect of endothelin-1 on actomyosin ATPase activity. Implications for the efficiency of contraction. *Circ. Res.* 1996; 78:1044–1050.

[116] Venema RC, Kuo JF. Protein kinase C-mediated phosphorylation of troponin I and C-protein in isolated myocardial cells is associated with inhibition of myofibrillar actomyosin MgATPase. *J. Biol. Chem.* 1993; 268:2705–2711.

[117] Belin RJ, Sumandea MP, Allen EJ, Schoenfelt K, Wang H, Solaro RJ, de Tombe PP. Augmented protein kinase C-alpha-induced myofilament protein phosphorylation contributes to myofilament dysfunction in experimental congestive heart failure. *Circ. Res.* 2007; 101:195–204.

[118] Belin RJ, Sumandea MP, Sievert GA, Harvey LA, Geenen DL, Solaro RJ, de Tombe PP. Interventricular differences in myofilament function in experimental congestive heart failure. *Pflugers Arch.* 2011; 462:795–809.

[119] Sumandea MP, Pyle WG, Kobayashi T, de Tombe PP, Solaro RJ. Identification of a functionally critical protein kinase C phosphorylation residue of cardiac troponin T. *J. Biol. Chem.* 2003; 278:35135–35144.

[120] Ding P, Huang J, Battiprolu PK, Hill JA, Kamm KE, Stull JT. Cardiac myosin light chain kinase is necessary for myosin regulatory light chain phosphorylation and cardiac performance in vivo. *Journal of Biological Chemistry*. 2010; 285:40819–40829.

[121] Scruggs SB, Solaro RJ. The significance of regulatory light chain phosphorylation in cardiac physiology. *Arch. Biochem. Biophys.* 2011;

[122] Sheikh F, Ouyang K, Campbell SG, Lyon RC, Chuang J, Fitzsimons D, Tangney J, Hidalgo CG, Chung CS, Cheng H, Dalton ND, Gu Y, Kasahara H, Ghassemian M, Omens JH, Peterson KL, Granzier HL, Moss RL, McCulloch AD, Chen J. Mouse and computational models link Mlc2v dephosphorylation to altered myosin kinetics in early cardiac disease. *The Journal of clinical investigation*. 2012;:–.

[123] Chan JY, Takeda M, Briggs LE, Graham ML, Lu JT, Horikoshi N, Weinberg EO, Aoki H, Sato N, Chien KR, Kasahara H. Identification of cardiac-specific myosin light chain kinase. *Circ. Res.* 2008; 102:571–580.

[124] Seguchi O, Takashima S, Yamazaki S, Asakura M, Asano Y, Shintani Y, Wakeno M, Minamino T, Kondo H, Furukawa H, Nakamaru K, Naito A, Takahashi T, Ohtsuka T, Kawakami K, Isomura T, Kitamura S, Tomoike H, Mochizuki N, Kitakaze M. A cardiac myosin light chain kinase regulates sarcomere assembly in the vertebrate heart. *The Journal of clinical investigation*. 2007; 117:2812–2824.

[125] Nixon BR, Thawornkaiwong A, Jin J, Brundage EA, Little SC, Davis JP, Solaro RJ, Biesiadecki BJ. AMP-activated Protein Kinase Phosphorylates Cardiac Troponin I at Ser-150 to Increase Myofilament Calcium Sensitivity and Blunt PKA-dependent Function. *Journal of Biological Chemistry*. 2012; 287:19136–19147.

[126] Hanft LM, McDonald KS. Sarcomere length dependence of power output is increased after PKA treatment in rat cardiac myocytes. *Am. J. Physiol. Heart Circ. Physiol.* 2009; 296:H1524–31.

[127] McDonald KS, Hanft LM, Domeier TL, Emter CA. Length and PKA Dependence of Force Generation and Loaded Shortening in Porcine Cardiac Myocytes. *Biochem. Res. Int.* 2012; 2012:371415.

[128] Sequeira V, Wijnker PJM, Nijenkamp LLAM, Kuster DWD, Najafi A, Witjas-Paalberends R, Regan JA, Boontje N, Cate Ten F, Germans T, Carrier L, Sadayappan S, van Slegtenhorst M, Zaremba R, Foster DB, Murphy A, Poggesi C, Remedios Dos CG, Stienen GJM, Ho CY, Michels M, van der Velden J. Perturbed Length-Dependent Activation in Human Hypertrophic Cardiomyopathy with Missense Sarcomeric Gene Mutations. *Circ. Res.* 2013;

[129] van der Velden J, de Jong JW, Owen VJ, Burton PB, Stienen GJ. Effect of protein kinase A on calcium sensitivity of force and its sarcomere length dependence in human cardiomyocytes. *Cardiovasc. Res.* 2000; 46:487–495.

[130] Rao VS, Korte FS, Razumova MV, Feest ER, Hsu H, Irving TC, Regnier M, Martyn DA. N-terminal phosphorylation of cardiac troponin-I reduces length-dependent calcium sensitivity of contraction in cardiac muscle. *The Journal of Physiology*. 2013; 591:475–490.

[131] Pyle WG, Sumandea MP, Solaro RJ, de Tombe PP. Troponin I serines 43/45 and regulation of cardiac myofilament function. *Am. J. Physiol Heart Circ Physiol.* 2002; 283:H1215–24.

[132] Sadayappan S, de Tombe PP. Cardiac myosin binding protein-C: redefining its structure and function. *Biophys Rev.* 2012; 4:93–106.

[133] Barefield D, Sadayappan S. Phosphorylation and function of cardiac myosin binding protein-C in health and disease. *J. Mol. Cell Cardiol.* 2010; 48:866–875.

[134] Stelzer JE, Patel JR, Moss RL. Protein kinase A-mediated acceleration of the stretch activation response in murine skinned myocardium is eliminated by ablation of cMyBP-C. *Circ. Res.* 2006; 99:884–890.

[135] Stelzer JE, Moss RL. Contributions of stretch activation to length-dependent contraction in murine myocardium. *J. Gen. Physiol.* 2006; 128:461–471.

[136] Scruggs SB, Hinken AC, Thawornkaiwong A, Robbins J, Walker LA, de Tombe PP, Geenen DL, Buttrick PM, Solaro RJ. Ablation of ventricular myosin regulatory light chain phosphorylation in mice causes cardiac dysfunction in situ and affects neighboring myofilament protein phosphorylation. *J. Biol. Chem.* 2009; 284:5097–5106.

[137] Anderson BR, Granzier HL. Titin-based tension in the cardiac sarcomere: Molecular origin and physiological adaptations. *Prog. Biophys. Mol. Biol.* 2012;

[138] Harris SP, Lyons RG, Bezold KL. In the thick of it: HCM-causing mutations in myosin binding proteins of the thick filament. *Circ. Res.* 2011; 108:751–764.

[139] Seidman CE, Seidman JG. Identifying sarcomere gene mutations in hypertrophic cardiomyopathy: a personal history. *Circ. Res.* 2011; 108:743–750.

[140] Tardiff JC. Thin filament mutations: developing an integrative approach to a complex disorder. *Circ. Res.* 2011; 108:765–782.

[141] Messer AE, Memo M, Bayliss CR, Leung MC, Marston SB. Does Uncoupling of Troponin I Phosphorylation from Changes in Myofibrillar Ca2+-Sensitivity Play a Role in the Pathogenesis of Cardiomyopathy? *Biophys. J.* 2012; 102:556a–556a.

[142] Ford SJ, Mamidi R, Jimenez J, Tardiff JC, Chandra M. Effects of R92 mutations in mouse cardiac troponin T are influenced by changes in myosin heavy chain isoform. *J. Mol. Cell Cardiol.* 2012; 53:542–551.

[143] Regan J, Sequeira V, Stienen GJM, Michels M, Cate ten FJ, van Slegtenhorst M, Remedios Dos CG, Konhilas JP, van der Velden J. Reduced length-dependent activation in human cardiomyocytes harboring the troponin-I mutation R145W. *Biophysical J.* 2012; 102:158a.

[144] Dhandapany PS, Sadayappan S, Xue Y, Powell GT, Rani DS, Nallari P, Rai TS, Khullar M, Soares P, Bahl A, Tharkan JM, Vaideeswar P, Rathinavel A, Narasimhan C, Ayapati DR, Ayub Q, Mehdi SQ, Oppenheimer S, Richards MB, Price AL, Patterson N, Reich D, Singh L, Tyler-Smith C, Thangaraj K. A common MYBPC3 (cardiac myosin binding protein C) variant associated with cardiomyopathies in South Asia. *Nature genetics.* 2009; 41:187–191.

[145] Knöll R. Myosin binding protein C: implications for signal-transduction. *J. Muscle Res. Cell. Motil.* 2012; 33:31–42.

[146] Govindan S, Sarkey J, Ji X, Sundaresan NR, Gupta MP, de Tombe PP, Sadayappan S. Pathogenic properties of the N-terminal region of cardiac myosin binding protein-C in vitro. *J. Muscle Res. Cell. Motil.* 2012;

[147] Cazorla O, Wu Y, Irving TC, Granzier H. Titin-based modulation of calcium sensitivity of active tension in mouse skinned cardiac myocytes. *Circ. Res.* 2001; 88:1028–1035.

[148] Fukuda N, Wu Y, Farman G, Irving TC, Granzier H. Titin isoform variance and length dependence of activation in skinned bovine cardiac muscle. *J. Physiol. (Lond.).* 2003; 553:147–154.

[149] Mateja RD, Greaser ML, de Tombe PP. Impact of titin isoform on length dependent activation and cross-bridge cycling kinetics in rat skeletal muscle. *Biochim. Biophys. Acta.* 2012;:–.

[150] Helmes M, Lim CC, Liao R, Bharti A, Cui L, Sawyer DB. Titin determines the Frank-Starling relation in early diastole. *J. Gen. Physiol.* 2003; 121:97–110.

[151] Soteriou A, Gamage M, Trinick J. A survey of interactions made by the giant protein titin. *J. Cell Sci.* 1993; 104 (Pt 1):119–123.

[152] Cazorla O, Szilagyi S, Le Guennec J-Y, Vassort G, Lacampagne A. Transmural stretch-dependent regulation of contractile properties in rat heart and its alteration after myocardial infarction. *FASEB J.* 2005; 19:88–90.

[153] Mou YA, Reboul C, Andre L, Lacampagne A, Cazorla O. Late exercise training improves non-uniformity of transmural myocardial function in rats with ischaemic heart failure. *Cardiovasc. Res.* 2009; 81:555–564.

[154] Greaser ML, Warren CM, Esbona K, Guo W, Duan Y, Parrish AM, Krzesinski PR, Norman HS, Dunning S, Fitzsimons DP, Moss RL. Mutation that dramatically alters rat titin isoform expression and cardiomyocyte passive tension. *J. Mol. Cell Cardiol.* 2008; 44:983–991.

[155] Cazorla O, de Tombe PP. Some rat: a very special rat with a rather special titin. *J. Mol. Cell Cardiol.* 2008; 44:976–978.

[156] Guo W, Schafer S, Greaser ML, Radke MH, Liss M, Govindarajan T, Maatz H, Schulz H, Li S, Parrish AM, Dauksaite V, Vakeel P, Klaassen S, Gerull B, Thierfelder L, Regitz-Zagrosek V, Hacker TA, Saupe KW, Dec GW, Ellinor PT, MacRae CA, Spallek B, Fischer R, Perrot A, Özcelik C, Saar K, Hubner N, Gotthardt M. RBM20, a gene for hereditary cardiomyopathy, regulates titin splicing. *Nat. Med.* 2012; 18:766–773.

[157] Patel JR, Pleitner JM, Moss RL, Greaser ML. The magnitude of length-dependent changes in contractile properties varies with titin isoform in rat ventricles. *AJP: Heart and Circulatory Physiology.* 2011;

[158] Granzier HL, Radke MH, Peng J, Westermann D, Nelson OL, Rost K, King NMP, Yu Q, Tschöpe C, McNabb M, Larson DF, Labeit S, Gotthardt M. Truncation of titin's elastic PEVK region leads to cardiomyopathy with diastolic dysfunction. *Circ. Res.* 2009; 105:557–564.

[159] Nedrud J, Labeit S, Gotthardt M, Granzier H. Mechanics on myocardium deficient in the N2B region of titin: the cardiac-unique spring element improves efficiency of the cardiac cycle. *Biophys. J.* 2011; 101:1385–1392.

[160] Fan D, Wannenburg T, de Tombe PP. Decreased myocyte tension development and calcium responsiveness in rat right ventricular pressure overload. *Circulation.* 1997; 95:2312–2317.

[161] de Tombe PP, Keurs ter HE. Force and velocity of sarcomere shortening in trabeculae from rat heart. Effects of temperature. *Circ. Res.* 1990; 66:1239–1254.

In: Troponin
Editor: J.-P. Jin

ISBN: 978-1-62948-320-7

Chapter X

Troponin Abnormality and Systolic Function of the Heart

Brandon J. Biesiadecki, PhD and Jonathan P. Davis, PhD
Department of Physiology and Cell Biology, Davis Heart and Lung Research Institute, The Ohio State University, Columbus, Ohio, USA

Abstract

The function of the heart is to distribute blood, providing substance to and removing waste from the tissues of the body. The heart achieves this function through the contraction of its muscular walls generating pressure that propels blood throughout the body. The troponin complex (Tn) is a central regulatory component responsible for translating the calcium (Ca^{2+}) signal into contraction. It is clear that Tn does not behave in the heart as a simple Ca^{2+}-dependent switch, but that its regulatory properties are tuned by alteration of multiple protein-protein interactions within the myofilament. Abnormal Tn linked to disease can be employed as tools to delineate important Tn interactions involved in the Ca^{2+} regulation of cardiac contraction. In this chapter we discuss abnormal Tn modifications as models to understand how their Tn structural alteration affects the different protein-protein interactions involved in the Ca^{2+}-dependent regulation of cardiac pressure development.

Introduction

The heart is an incredible machine that continuously beats roughly 3 billion times in an average lifespan. With each beat, the heart must produce a pressure and then contract to expel the blood from the chambers. The troponin complex (Tn) is a central regulatory figure in this process linking the action potential generated calcium (Ca^{2+}) signal to the mechanical force needed to expel the blood from the heart and circulate it throughout the body. Tn is composed of three subunits that include troponin C (TnC) the Ca^{2+}-binding subunit, troponin T (TnT) the tropomyosin (Tm) binding subunit and troponin I (TnI) the inhibitory subunit. It is clear that Tn does not behave in the heart as a simple Ca^{2+}-dependent switch, but that its regulatory properties are tuned through alteration of its multiple protein-protein interactions within the complex, the thin filament where it resides and the greater myofilament. These interactions are further tuned by post-translational modification in response to altered cardiac demand and disease. In essence, the Tn complex acts as a central hub integrating information to control the active state of the thin filament as one of the major inputs regulating the force producing interaction of myosin and actin. In this chapter we will discuss the consequences of aberrantly altered Tn as models of regulatory interactions within the Tn complex. Through investigation of the protein, cellular and animal effects displayed by Tn point mutations and aberrant alternative splicing we highlight important Tn protein-protein structural alterations critical to the Ca^{2+}-dependent regulation of cardiac pressure development, cellular shortening and ultimately contractile function.

Cardiac Systole

Following completion of the diastolic filling phase of the cardiac cycle, an action potential is generated in the sinoatrial node of the heart that eventually propagates to the left ventricle where it initiates the systolic phase of ventricular contraction. Systole is composed of two phases; isovolumic contraction and ejection. Initially the ventricle contracts isovolumically against the chamber volume until the intra-ventricular pressure exceeds that in the aorta. At this point, the aortic valve opens and the ventricle begins to actively shorten against this aortic pressure ejecting blood down its pressure gradient from the ventricle out to the vasculature of the body (for a more thorough presentation of systolic function see [1]). Since there is some internal shortening of the muscle during the isovolumic contraction and the afterload (aortic pressure that the ventricle must work against to eject the blood) is not constant, the contraction of the heart is better described as auxotonic (simultaneous changes in force and length). Under normal conditions, the systolic ventricular pressure required to open the aortic valve is only twenty-five to fifty percent of what the ventricle can be called upon to maximally produce[1] [3]. The heart therefore contains a significant contractile reserve. Importantly, this reserve can be called upon by any one or a combination of events including an increased intracellular Ca^{2+} concentration, increased Tn Ca^{2+} sensitivity or an increased number of force producing cross-bridges. Thus, modulation of this ventricular contractile reserve serves as a mechanism for the heart to modulate ventricular performance and match cardiac function to systemic demand.

As part of the isovolumic and ejection phases of systole, both the extent of force production and the rate of ventricular pressure rise are significant muscle level parameters in the determination of ventricular contraction and overall cardiac function. Both the extent of isometric (or isovolumic) contraction and the rate of its rise are highly dependent on Ca^{2+} [4]. Since Ca^{2+} has no direct effect on the ability of myosin to produce force, the Ca^{2+} dependent regulation of force and velocity must be transmitted through Tn. While the Ca^{2+} dependent

[1] This sub maximal level of contraction appears to be tuned to the peak power output of the muscle (2).

extent of force development is a common measurement of function, the extent to which Ca^{2+} regulates velocity is less studied. This phenomenon is most easily observed in detergent skinned (the removal of membrane bound structures leaving behind the functional contractile apparatus) ventricular tissue or single myofibril[2] preparations. In these preparations force development can be activated by either a rapid rise in free Ca^{2+} (k_{Ca}; that does not rely on large, rate-limiting diffusive distances) or by a rapid shortening and re-lengthening procedure performed at a constant free Ca^{2+} concentration. The rapid adjustments in muscle length are thought to dislodge force producing cross-bridges from the thin filament, which subsequently rebind and reproduce force, at a rate termed force redevelopment (k_{tr}). The general consensus is that k_{Ca} and k_{tr} are of similar values for a similar final achieved force (for review see [5]). For cardiac muscle, the rate of force development typically increases (more or less linearly) with an increase in force (or Ca^{2+}). In other words, the more thin filament units that can be activated by Ca^{2+}, the faster and to a greater extent isometric force will be generated. The Ca^{2+} dependence of force/velocity is critical in early systole to heart performance. During this phase of the contraction cycle of the heart, where the intracellular Ca^{2+} levels are the highest, the heart may initially activate excess Tn to recruit more myosin than it actually needs to pump the stroke volume. This may be accomplished to maximize the speed at which the isovolumic pressure is reached before ejection begins. This in essence minimizes the time the muscle is expending energy to produce a force that is not generating any external work. Additionally, this "extra-activation" also increases the rate at which the muscle shortens isotonically, moving it further up the force-velocity relationship. Modulation of the Ca^{2+} mediated regulation of either the extent or rate of force production can therefore significantly alter cardiac contractility.

As discussed above, the magnitude of both the extent and rate of ventricular force generation is highly dependent on the free Ca^{2+} concentration bathing the contractile apparatus. Interestingly, the force produced at a given Ca^{2+} concentration of the intact cardiac muscle preparation (apparent Ca^{2+} sensitivity) is nearly an order of magnitude more sensitive to Ca^{2+} than following skinning [6]. There are any number of reasons that may explain this observation (swelling of the myofilament lattice, loss of cytosolic and membrane bound signaling systems, rather high free Mg^{2+}, etc…), but it is mentioned here to highlight that the apparent Ca^{2+} sensitivity of any cardiac parameter is determined by a number of specific interactions within the cardiac muscle that appears to be highly system dependent. In any regard, under resting conditions, the transient rise in Ca^{2+} during activation of the cardiac muscle is not sufficient to saturate the available regulatory Tn Ca^{2+} binding sites (~1.2 billion Tn in a single cardiac myocyte assuming a 30pL volume and 70uM Tn[3]) and fully activate the entire myofilament. This sub-maximal myofilament activation at resting conditions is consistent with the significant contractile reserve in the heart that is at least in part regulated by the ability of Ca^{2+} to activate the myofilament. Increases or decreases in the Tn regulated Ca^{2+} sensitivity of the muscle is therefore a significant mechanism in the modulation of both

[2] The heart is composed of electrically coupled rod-shaped muscle cells called cardiomyocytes (~100 μm long and 25 μm diameter). Packed within each myocyte are numerous rod-shaped myofibrils (~ 2 μm diameter) that contain the repeating contractile element of the muscle called the sarcomere. Each sarcomere contains interdigitating thin (actin/Tn/Tm) and thick (myosin) filaments.

[3] Tn is actually the largest intracellular Ca^{2+} buffer in the cardiomyocytes. There is growing evidence that changing the Ca^{2+} buffering potential of Tn by altering its Ca^{2+} sensitivity can feedback on Ca^{2+} handling and trigger arrhythmias (7).

the extent and rate of force development, both critical factors that influence systolic performance and contractile reserve.

Tn Regulation of Systolic Function

Unfortunately, there are no techniques available to directly observe the behavior of Tn within the heart. Its function in the heart can only be inferred based on its behavior in simplified and reduced physiological and biochemical systems. Having an excess or lack of any one individual Tn component can cause cardiac muscle to either produce force in the absence of Ca^{2+} (loss of TnI) or to not contract at all in the presence of Ca^{2+} (loss of TnC). Furthermore, naturally occurring or designed mutations in all three Tn subunits can also alter the response of force development across the gamut of gain of function to severe loss of function (discussed below). That TnC, and more specifically Ca^{2+} binding to the regulatory domain of TnC, is necessary and essential for muscle contraction has been established in numerous extraction/reconstitution/exchange studies of TnC in various skinned muscle preparations. Furthermore, knockout of the adult cardiac TnI in mice leads to loss of troponin in the heart and lethality around the time developmental cues turn off the expression of the neonatal TnI. On the other hand, continuous expression of the neonatal cardiac TnI in adult myocytes is a prime example of how a Tn subunit can alter cardiac biochemistry, physiology and improve pathophysiology (for example see [8, 9]). These examples, demonstrate that the Ca^{2+}-dependent interactions of the Tn subunits can influence the magnitude and speed of cardiac muscle contraction *in vitro* and *in vivo*. Thus, changes in or the modulation of different Tn subunit protein-protein interactions can alter cardiac contractile reserve and cardiac function.

The simplest interpretation for the Ca^{2+} dependent behavior of cardiac muscle mechanics relies on the Ca^{2+} binding properties of the Tn Ca^{2+} binding subunit, TnC. It is clear however that neither the apparent Ca^{2+} sensitivity nor the behavior of muscle mechanics can be adequately explained by the Ca^{2+} dependent properties and function of isolated TnC (in the absence of its partner Tn subunits and other myofilament interactions) [10]. Numerous biochemical and physiological studies have demonstrated that the qualitative and quantitative Ca^{2+} dependent behavior of the regulatory domain of TnC is dependent upon its direct and indirect interactions with several of its near (TnI, TnT, Tm and actin) and far (myosin, myosin-binding protein C) thin and thick filament protein neighbors. Ultimately, the Ca^{2+} sensitivity of cardiac muscle contraction entails not only the intrinsic binding of Ca^{2+} to TnC (and its modulation) but also all of the multiple protein-protein interactions (TnI-TnT, TnI-TnC, TnI-Tm, TnI-actin, TnT-Tm, Tm-actin, etc.) that lead to or block myosin's ability to interact with actin and generate force.

The Ca^{2+} binding properties of the thin filament (Ca^{2+} association and dissociation rates, and thus the Ca^{2+} sensitivity) and its influence on muscle function can be further tuned by the presence of mutations, truncations and/or post-translational modifications of several sarcomeric proteins that alter of the protein-protein interactions within the Tn complex [11]. Among these Tn protein-protein interactions, one of the strongest influences on Ca^{2+} binding to the regulatory domain of TnC is its ability to bind the switch peptide of TnI (see Figure 1). Both the steady-state and kinetic Ca^{2+} binding properties of TnC can be altered orders of magnitude by this binding of TnI to TnC [10]. In essence, the Ca^{2+} dependent behavior of TnC and ultimately the Ca^{2+} dependent mechanical function (myosin binding to actin, force development, speed of muscle contraction or shortening, ATPase activity, *in vitro* motility, etc.) is highly dependent on the system being studied and the states (phosphorylation, truncation, mutation, gene splicing, etc.) of the other sarcomeric proteins. One of the best characterized and most well accepted means for cardiac muscle to adjust its functional response to the Ca^{2+} signal is through phosphorylation of the N-terminal domain of TnI [12]. It is well established that the Ca^{2+} sensitivity of Tn and the Ca^{2+} sensitivity of force development are significantly decreased by the beta-adrenergic stimulation induced protein kinase A dependent phosphorylation of TnI. Interestingly, it appears that even this fundamental behavior of Tn can be ablated or potentiated by additional Tn modifications (described below). Thus, there appears to be several different mechanisms related to Tn structure/function that can work together or separately to alter the Ca^{2+} response of cardiac muscle and its systolic function.

Activation of Tn

Just prior to the initiation of systole, the free concentration of Ca^{2+} is at its lowest point in the myocyte, where minimal Ca^{2+} is thought to be bound to the regulatory domain of TnC (there may be substantial Ca^{2+} bound to the non-regulatory TnC C-terminal structural sites, depending on the heart rate or frequency of stimulation). At this time, the switch peptide of TnI is dissociated from TnC with the inhibitory peptide, and potentially other regions of the far C-terminal of TnI, bound to actin. This binding of TnI to actin, along with less clear TnT interactions [12a, 12b], helps to pin tropomyosin in a position on actin prohibiting strong, force producing binding of myosin. As the free Ca^{2+} rises in the space surrounding the myofibrils, it is thought that the regulatory domain of TnC can bind this Ca^{2+} with a kinetic that is nearly diffusion controlled, much like what has been observed for the behavior of the isolated TnC, Tn complex and reconstituted thin filament [10]. In the absence of Ca^{2+}, the switch peptide of TnI has a very low affinity for TnC, but once TnC binds Ca^{2+} the affinity of the TnI switch peptide for TnC is extremely high (~250 nM). This high affinity binding event drastically sensitizes Ca^{2+} binding to TnC primarily by slowing the Ca^{2+} dissociation rate from TnC (from greater than 1000/s to ~60/s at 37°C in ventricular myofibrils) [13]. The binding of TnI to TnC is therefore critical to stabilize the TnC-Ca^{2+} bound Tn state that together with other changes in Tn interactions (discussed in detail in the previous chapter) induces an activated state of the thin filament allowing contraction to ensue.

The speed at which the Tn system is normally turned on by Ca^{2+} to activate the thin filament precedes that of force development [4]. In other words, the amount of Tn activated by the Ca^{2+} will directly influence the amount of Tm movement that then sets the amount of myosin binding and ultimately the extent and rate of contraction critical to heart function. As mentioned above, it would appear that only a fraction of the Tn are activated during a normal heart beat. Changing the fraction of Tn activated by altering the Ca^{2+} binding properties of TnC or those protein-protein interactions within the Tn complex responsible for relaying the Ca^{2+} signal therefore not only alters the amount of Tm movement and extent of force, but also the rate at which cardiac muscle contracts (the higher the affinity, the faster the muscle contracts at submaximal concentrations of Ca^{2+}). Thus, altering the amount of Tn activated by Ca^{2+}, by either changing the Ca^{2+} signal itself and/or by altering the Ca^{2+} binding properties of the Tn can have a profound effect on the contraction of the heart.

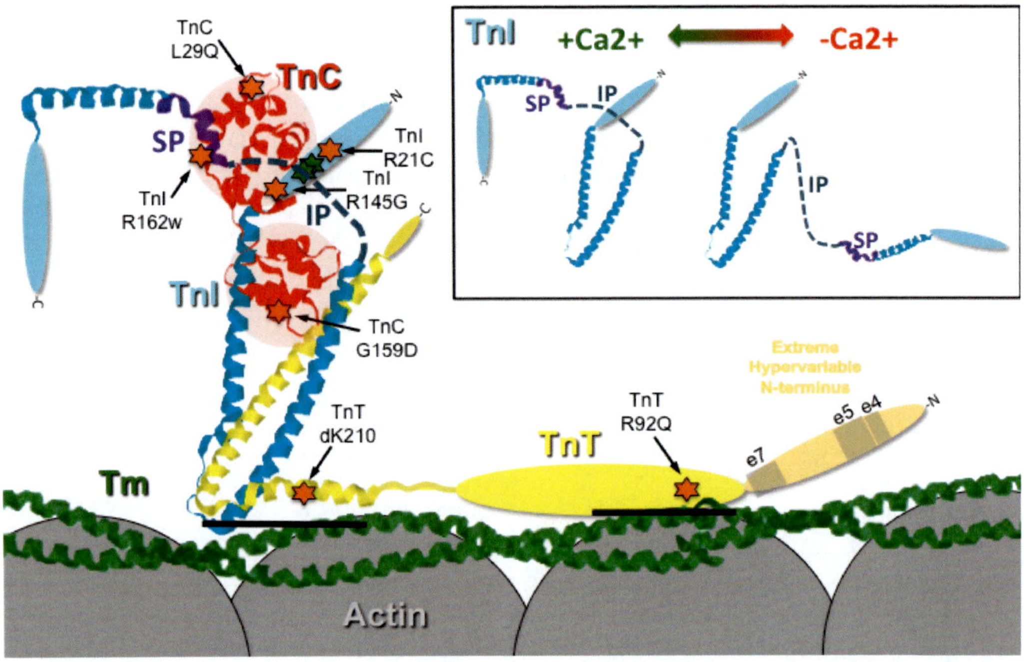

Figure 1. Cartoon representation demonstrating the interactions of a single Tn complex with the tropomyosin strand (Tm; green) and actin thin filament (grey circles) in the Ca^{2+} bound state. The Tn core structure is modified from that of the Ca^{2+} bound cardiac Tn crystal structure [68] with missing regions added based upon biochemical data (troponin C (TnC) red, troponin I (TnI) blue, troponin T (TnT) yellow). For clarity TnT is depicted on the same actin strand while in reality the TnT central and N-terminal regions likely reaches across to interact with the adjacent actin strand [14]. The partial representation of Tm strands and the Tm head-to-tail overlap is based upon the Tm crystal structure [122]. TnT sites of interaction with Tm are based upon biochemical data and are indicated by black bars [58-62]. The structural location of abnormal Tn discussed in the text are denoted by orange stars and TnI Ser-23/24 bis-phosphorylation location is denoted by green stars. The structural location of abnormally spliced TnT N-terminal exons discussed in the text are indicated by grey boxes. Insert demonstrating the potential change in TnI structure with and without Ca^{2+}. IP; TnI inhibitory peptide in purple, SP; TnI switch peptide in aqua.

Altering any of the multiple Tn complex protein-protein interactions involved in the regulation of myosin binding to actin can modulate the intrinsic binding of Ca^{2+} to TnC, the cardiac contractile properties and function of the heart. In addition to post-translational modification of the Tn proteins, several studies have demonstrated that disease associated abnormal Tn modifications can alter the Ca^{2+} sensitivity of Tn on the thin filament by altering both the Ca^{2+} association and dissociation rates of Tn [13a] [51, 52]. Based upon our and others studies we propose that several of the disease associated abnormal Tn modifications affect Ca^{2+} binding to TnC on the thin filament by primarily altering the interaction of TnI with TnC and/or actin. Essentially, there is a competition between actin and TnC for binding of the C-terminal of TnI. Processes (phosphorylation, mutation, alternate gene splicing, etc.) that increase the potential for TnC to bind TnI (increasing the affinity of TnC for TnI or decreasing the interaction of TnI with actin) increase the apparent Ca^{2+} sensitivity of the thin filament (for examples see below). Correspondingly, the opposite happens when this probability is decreased. While we believe this interaction of TnI with actin and TnC is a significant Tn protein-protein interaction in contractile modulation, other Tn protein-protein interactions are also critical to Tn regulation of myosin's interaction with actin and force production. How these other interactions affect Ca^{2+} regulated cardiac contraction is less clear.

As a first step to fully understand the mechanisms by which Ca^{2+} regulates muscle contraction through Tn, it is necessary to determine the role of the Tn protein-protein interactions in this regulation and their functional effects. Abnormal Tn can be used as tools to delineate the important Tn interactions involved in the Ca^{2+} regulation of cardiac contraction and lend insights into potential therapies to treat cardiomyopathies. In the following sections we discuss specific abnormal Tn modifications and their effects on Ca^{2+} regulation of cardiac contraction as models to develop the Tn protein-protein interactions critical to contractile regulation.

Abnormal Tn As Models of Structure-function

The Tn complex represents a critical link in the transmission of the Ca^{2+} activating signal into cardiac muscle contraction and is therefore a significant determinant of heart systolic function. Abnormal Tn are alterations of the Tn subunit amino acids directly linked to the development of cardiac disease. By definition this linkage to disease indentifies abnormal Tn as structural modifications that alter contractile function to cause cardiac disease. Many of these mutations do not destroy the regulatory function of Tn but rather modulate the ability of Tn to respond to Ca^{2+} or pass on this activating signal to actin and myosin. The study of these abnormal Tn can therefore provide biochemical and animal level information into specific Tn protein-protein interactions and the functionality they impart on the Tn Ca^{2+} regulation of cardiac contraction. Tn disease related abnormalities therefore serve as naturally occurring models to identify and probe regional Tn interactions important to modulate the regulation of heart function as well as towards pathological disease development.

Unlike amino acid polymorphisms, familial linked Tn modifications alter Tn structure-function to cause disease. The ultimate function of the Tn complex is to regulate the contraction of the cardiac muscle through transmission of the Ca^{2+} activating signal into the interaction of myosin with actin. This transmission of the Ca^{2+} signal involves a number of interactions between the three Tn subunits and the actin thin filament that summate in the Ca^{2+} responsiveness of force. Biochemical studies have identified Tn protein interactions critical to the Ca^{2+} regulation of contraction, however other Tn interactions are clearly involved with yet to be determined effects on contraction. As outlined above, Tn protein interactions involved in the Tn Ca^{2+} regulation of force include: Ca^{2+} binding to TnC, TnC's interactions with TnI, TnI's interaction with Actin/Tm, TnI's interaction with TnC, TnI's interaction with TnT and TnT's interaction with Tm. While an alteration in any one of these interactions may result in a similar Ca^{2+} sensitivity of force development response, the differential function each of these interactions imparts within Tn may result in the modification of other cardiac contractile functions (e.g. rate of contraction) in addition to Ca^{2+}

sensitivity and thus overall heart function. Cardiomyopathy linked abnormal Tn identify single amino acid residue or small regional modifications that cause disease as the direct result of their functional effects on critical interactions in the Tn Ca^{2+} regulation of contraction. Thus the study of cardiomyopathy linked Tn modifications refines our current mechanistic understanding of the basic biochemical Tn interactions and their role in the Tn regulation of Ca^{2+} sensitive force.

Abnormal Tn modifications structurally consist of amino acid residue mutation, single residue exclusion, alternate splicing or altered post-translational modification. Abnormal Tn have been linked to the development of hypertrophic cardiomyopathy (HCM), [14] dilated cardiomyopathy (DCM) [15] and restrictive cardiomyopathy (RCM) [16] clinical phenotypes. HCM and RCM linked abnormal Tn typically present clinically with diastolic dysfunction while DCM linked abnormal Tn present with systolic dysfunction. To date the mechanistic classification of abnormal Tn mutations has centered on their alteration of steady-state Ca^{2+} regulated force development (for a thorough review see [17, 18]). Abnormal Tn resulting in HCM and RCM typically demonstrate an increased Ca^{2+} sensitive force development, while DCM related Tn mutations typically result in a decreased sensitivity [19]. This Ca^{2+} sensitivity mechanism of disease development generally holds true, however exceptions to this classification exist. In addition, Tn abnormalities that exhibit similarly altered Ca^{2+} sensitivity can present with different phenotypic severity. These findings suggest the disease inducing abnormal Tn affect on Ca^{2+} sensitivity occurs through different mechanisms thus resulting in differential effects on heart function. While clearly important to cardiac phenotype, the classification of abnormal Tn by Ca^{2+} sensitivity alone only addresses the overall steady-state outcome of the multiple Tn interactions to regulate force. Classification by Ca^{2+} sensitivity ignores the role of these individual Tn interactions to affect dynamic contractile events also important to overall heart function as discussed above. The mechanistic classification of abnormal Tn by Ca^{2+} sensitivity may therefore be an over simplification of the Tn mechanisms causal to cardiomyopathy development because it only looks at one outcome of multiple Tn interactions. To date minimal mechanistic biochemical experiments have been conducted to understand the role of multiple Tn interactions on the overall Ca^{2+} regulation of force production.

The structural effects of abnormal Tn can alter contractile regulation through direct or indirect means. Direct effects of abnormal Tn include those that modify the intrinsic ability of the thin filament (Tn/Tm/Actin) to regulate contraction. Such effects include structural changes in Tn that directly or allosterically result in altered regulatory Ca^{2+} binding to TnC through modulation of the Tn regulatory interactions. Direct effects of abnormal Tn are typically measured as the ability of myosin to interact with actin in reconstituted thin filaments, by *in vitro* motility, myosin ATPase, or upon Tn exchange into skinned fibers. Indirect effects of Tn modifications function through altering various post-translational modifications impact on Tn Ca2+ regulation. Examples of indirect targets include stress induced myofilament post-translational modifications such as phosphorylation. The investigation of abnormal Tn indirect effects is difficult to study as it requires the functional investigation of modulatory, labile interactions that may or may not be present under normal conditions. To date most studies of abnormal Tn involve characterization of their primary structural effects.

By investigating the occurrence and location of abnormal Tn we can begin to identify the function of Tn interactions critical to the Ca^{2+} regulation of force. Tn regions that exhibit a high occurrence of multiple mutations identify critical Tn interactions susceptible to structural alteration such that they modulate but do not destroy the regulatory function of Tn. The absence, or low occurrence, of abnormal and variant Tn changes also provides information about the function of this Tn region in contractile regulation. Regions of Tn that lack Tn variants tell us that the interactions within this region either do not impart a functional effect on regulation or the interaction is so structurally critical that its modification is incompatible with regulatory function and life. In the first case it is possible that a region of Tn is sufficiently able to sustain structural modification without altering the regulatory function of Tn. Such non-functional Tn variants are naturally occurring genetic Tn polymorphisms and are not linked to the development of cardiomyopathy. On the other hand if the Tn variant induced structural alteration occurred in a region not amendable to structural change the variant would result in a non-functional Tn. Such a Tn variant would be incompatible with contractile regulation, life and would therefore not be observed. Unfortunately an accurate account of non-functional Tn polymorphisms and those regions of Tn devoid of mutations have not been determined. This data in itself would lend to a more complete structure/function understanding of the molecular mechanisms involved in Tn regulation of contraction. In the absence of data on variant Tn we focus our discussion on cardiomyopathy linked abnormal Tn to define the mechanistic interactions within Tn on contractile regulation.

Troponin C

TnC is the critical Ca^{2+} sensing subunit of the Tn complex that directly binds the contractile activating Ca^{2+} signal [12]. The TnC molecule consists of structural N- and regulatory C-terminal globular domains connected by an alpha-helical linker (see Figure 1). Biochemical data has demonstrated Ca^{2+} binding to a single TnC N-terminal domain is directly responsible for initiation of cardiac contraction. While the TnC C-terminal domain can modulate the Ca^{2+} binding function of the N-terminal domain, during normal cardiac function it is largely believed to be structural in nature.

Ca^{2+} binding to the TnC N-terminal domain is absolutely critical to the Tn regulation of cardiac contraction. Consistent with this critical nature of Ca^{2+} binding to TnC, TnC does not undergo contractile modulatory post-translational modification as do the other Tn subunits. Correspondingly the number of cardiomyopathy associated TnC mutations is the fewest of the three Tn subunits. Likewise the frequency of any particular TnC mutation is also low with the majority of cardiomyopathy linked mutations residing in a single family. This small number of mutations and their low occurrence compared to the other Tn subunits supports the critical nature of TnC Ca^{2+} binding and the inability of the Tn regulatory system to withstand significant TnC structural induced functional alterations.

Of the TnC cardiomyopathy linked mutations, only the TnC L29Q and G159D mutations have been reasonably studied. While the TnC L29Q mutation is located in the TnC N-terminal domain close to the inactive first TnC Ca^{2+} binding site, the TnC G159D mutation is located in the structural C-terminal domain of TnC. These two divergent TnC mutations only minimally alter primary intrinsic Tn Ca^{2+} regulation [20]. Both mutations exhibit a small Ca^{2+} sensitization [21], desensitization [22-25] or no effect at all [22, 24, 26, 27] dependant upon the system and conditions investigated. This discrepancy of the TnC mutations evaluated in varied experimental systems, themselves measuring different Tn interactions, suggests the need to further investigate the specific TnC interactions critical to transmission of the Ca^{2+} binding signal. Overall the lack of significant Ca^{2+} binding effects in these mutations and their location in completely different TnC functional regions suggest these two mutations cause disease through mechanisms other than primary effects on the intrinsic regulatory ability of Tn [28, 29]. Rather these two TnC mutations likely affect contractile function through their secondary extrinsic Tn effects (discussed below).

On the whole, the limited number of well characterized abnormal TnC mutations in limited locations does not provide a sufficient data set to draw insight into the role of specific TnC interactions and their regulation of contraction. From this limited data it is however clear that TnC is critical to Tn Ca^{2+} regulation of contraction and that this Tn subunit is likely unable to sustain large structural modification and maintain its ability to bind Ca^{2+}.

Troponin I

The primary role of the TnI subunit is to contribute to both the inhibition of contraction as well as the maintenance of the Ca^{2+} activating signal (for review see [12, 30, 31]). TnI fulfills this role by the toggling of its C-terminal region between actin bound *inhibition* and TnC bound *activation* (see discussion above and Figure 1). In the absence of Ca^{2+}, TnI directly contributes to contractile inactivation through it's inhibitory peptide (IP; residues 129 to 148) and C-terminal binding to actin-Tm [10, 32-34]. The binding of Ca^{2+} binding to TnC induces TnI switch peptide (SP; residues 149-164) [35] binding to TnC, releasing the IP from actin and resulting in stabilization of the TnC Ca^{2+} bound state. These interactions of TnI with actin and TnC are critical to Tn Ca^{2+} regulation of contraction [36]. Alteration of TnI's interactions between inhibitory IP binding to actin-Tm and activating SP binding to TnC can therefore directly modulate contractile regulation, as well as Ca^{2+} binding to TnC.

Significance of the TnI IP/SP interaction is demonstrated by abnormal TnI mutations. The majority of the TnI cardiomyopathy linked mutations present with a HCM or RCM phenotype and exhibit increased Ca^{2+} sensitive force production [17]. Survey of these mutations demonstrates a significant number of these mutations are located within the TnI IP, SP or C-terminal 2nd actin-Tm binding regions. Classical biochemical studies have developed the significance of these regions as critical to the TnI inhibitory/activation role of Tn regulation [33, 34]. Furthermore, the primary structure of the IP and SP TnI regions is highly conserved across species and muscle isoforms suggesting functional importance. The high occurrence of TnI mutations in the IP, SP and 2nd actin binding regions suggests a modulatory role of the interactions within these regions to the contractile regulatory function of Tn. Studying the effects of cardiomyopathy linked TnI IP and SP mutations can therefore shed light onto the structure-function role of these TnI interactions within the Ca^{2+} regulation of contraction.

The TnI Arg-145 residue represents a "hot spot" mutation located directly in the TnI IP. Mutations of TnI Arg-145 to Pro, Gly or Gln have been identified as linked with a HCM phenotype while mutation to Trp is linked to an RCM phenotype [37, 38]. Linkage of these multiple mutations suggests a critical modulatory role of the IP region on contractile regulation. The R145G TnI mutation represents the most thoroughly characterized of the Arg-

145 mutations and we therefore focus our discussion on this mutation pointing out differences in the other mutations where applicable. Mice transgenic for the expression of TnI R145G exhibit a HCM like phenotype [39] demonstrating linkage of the mutation to phenotype. Consistent with its HCM association, skinned cardiac fibers and papillary muscles from TnI R145G expressing mice demonstrate increased Ca^{2+} sensitive force production [39, 40], although Ca^{2+} sensitivity in single myofibrils was slightly decreased [41]. TnI R145G transgenic expression in the rabbit, expressing the β myosin heavy chain, also demonstrates increased Ca^{2+} sensitivity [42]. The R145G mutation sensitizes cardiac muscle to Ca2+ independent of the myosin isoform expression [43]. Increased Ca^{2+} sensitivity was also observed upon R145G Tn exchange into cardiac fibers [44-47], reconstituted ATPase [48, 49] and with short-term adenoviral expression [50], demonstrating the increased Ca^{2+} sensitivity is a primary effect of the mutation. In general TnI R145Q/W exhibit similar increases in Ca^{2+} sensitive regulation [40, 44, 46, 49]. The primary functional effect of the TnI R145G HCM linked mutation to increase intrinsic Ca^{2+} sensitive force production supports further investigation of the TnI IP interaction within Tn.

Mechanistic insight into TnI IP interactions involved in Tn regulation of contraction can be gained from biochemical investigations of the TnI R145G mutation. Similar to that observed in the muscle, the TnI R145G increased Ca^{2+} binding to TnC in reconstituted thin filaments, however TnC Ca^{2+} binding was not different in isolated Tn [51]. The lack of altered Ca^{2+} binding in isolated Tn demonstrates the modulatory effects of the R145G require additional interactions within the thin filament beyond those of isolated Tn. Furthermore, surface plasma resonance binding affinity of isolated TnC for wild-type or TnI R145G was not different [48]. The most likely source of the R145G filament interaction responsible for increased sensitivity would therefore involve a decreased interaction of the TnI IP or C-terminal binding with actin/Tm. Such decreased TnI IP binding to actin is supported by the R145G increased reconstituted ATPase in the absence of Ca^{2+} (decreased inhibition) [44-46, 49] and the BDM sensitive increase in force production at sub-activating [Ca^{2+}] [41, 43]. Further probing this mechanism the location of the TnI IP was investigated by Forster resonance energy transfer (FRET). FRET measurement of the distance between TnI containing R145G and TnC demonstrates a decrease in the Ca^{2+} dependent average distance of the SP to TnC [52]. These data suggest that the mutation of TnI R145G decreases IP binding to actin/Tm, decreasing the ability of the TnI IP to inhibit force production and making it easier for Ca^{2+} to turn on the filament. Overall these data demonstrate the TnI R145G mutation shifted the TnI IP/SP interaction away from actin and increased Ca^{2+} activation of Tn and force development. Thus TnI R145G changes in Ca^{2+} sensitive force development result from an effect of the mutation to shift the TnI-actin interaction away from inhibition.

Contrary to the TnI Arg-145 mutation located in the IP, the TnI Arg-162 mutation is located in the SP and provides an alternate model to investigate TnI IP/SP regulation of contraction. TnI Arg-162 mutations to Trp, Gln or Pro have been linked to the development of an HCM phenotype [38, 53]. Of these mutations only R162W has been functionally investigated. The reconstitution of thin filaments with Tn containing the TnI R162W and its exchange into skinned fibers both increased the Ca^{2+} sensitivity of activation consistent with the HCM phenotype [44, 54]. Biochemically, the R162W decreases the ability of the thin filament to inhibit myosin S-1 ATPase in the absence of Ca^{2+}, although to a lesser extent than the R145G [44, 48]. This decreased inhibition is consistent with FRET findings that the TnI

SP region moves closer to TnC upon the addition of Ca^{2+} compared to wild-type TnI [52]. Although the association of TnI with TnC in the presence of Ca^{2+} was unaltered, the disassociation of the TnI with TnC was dramatically decreased [48]. These data support a role for the TnI R162W mutation to increase Ca^{2+} regulated force development by increasing the SP interaction with TnC and shifting the TnI IP/SP from inhibitory actin binding to activating TnC binding. Thus in comparison to the TnI R145G mutation that increases Ca^{2+} sensitive force production by decreasing IP binding to actin, the R162W exhibits similar increased Ca^{2+} sensitivity effects by increasing SP binding to TnC. These two different mutations demonstrate that modifying different protein-protein interactions can lead to the same functional outcomes. Together these results present support for the balance of TnI IP-actin and the SP-TnC interactions in the Tn regulation of contraction.

The functional effects of the TnI IP R145G and SP R162W mutations demonstrate the significance of these TnI interactions in the Tn regulation of contraction. Any Tn modification resulting in weakening of IP binding to actin or strengthening of SP binding to TnC shifts the distribution of this TnI region towards TnC and overall contractile activation. However, as demonstrated by the differential decrease of inhibition in the absence of Ca^{2+} by R145G compared to R162W, the specific amino acid interactions involved may result in overall differences in Ca^{2+} regulation.

Troponin T

Similar to TnI, TnT is a critical allosteric protein in transmission of the TnC-Ca^{2+} binding signal to activate cardiac muscle contraction (for review see [12, 31, 55]). TnT further imparts an additional structural role within the Tn complex. TnT functions as the structural organizing protein of the Tn complex by anchoring the Tn complex to Tm and the actin thin filament. TnT also directly contributes to contractile regulation by modulating the position of Tm on actin in response to Ca^{2+}. This interaction of TnT with Tm is therefore critical to the Tn Ca^{2+} regulation of contraction.

Unlike the more globular structures of TnI and TnC, TnT is a largely extended protein consisting of its C-terminal domain within the Tn core and an extended N-terminus protruding outside of the core region [56] (see Figure 1). Contrary to the TnT C-terminus, the TnT N-terminal domain is extended and precise structural information is lacking. Although alpha-helical in nature, the structure of the N-terminal domain exhibits a high degree of flexibility necessary to maintain Tm inhibition [14, 57]. TnT contains two alpha-helical Tm binding domains, one C-terminal located adjacent to the Tn core (residues 205-229) and one N-terminal in the extended N-terminal domain (residues 88-126) [58-62]. The two TnT-Tm binding domains are connected by a long flexible linker such that the C-terminal domain interacts with Tm on one actin strand and the N-terminal domain interacts with Tm on the second adjacent actin strand [14]. The TnT N-terminal Tm domain binds Tm in a Ca^{2+} insensitive manner promoting Tm binding to actin and maintaining Tm localization on actin in the blocked position to inhibit the interaction of myosin with actin [63-65]. Contrary, the TnT C-terminal Tm binding domain binds Tm in a Ca^{2+} sensitive manner [66]. In the absence of Ca^{2+}, both the TnT N-terminal and C-terminal Tm binding interactions are strong contributing to an inhibition of contraction. In the presence of Ca^{2+}, the N-terminal domain activates the acto-myosin interaction in a TnI independent manner [67]. In addition to Tm binding, the TnT C-terminus forms the I-T arm of the Tn core through direct interactions with both TnI and TnC [68]. The TnT-TnC interaction has been demonstrated to occur in a Ca^{2+} sensitive manner and may also contribute to contractile regulation [66], [69-71]. The inhibitory/activating nature of the TnT-Tm interaction and the extended nature of the two TnT-Tm binding regions located on adjacent actin strands highlights the significance of TnT as an allosteric protein in the Tn Ca^{2+} regulation of Tm and therefore contraction.

The significance of the TnT-Tm interactions to contractile regulation are demonstrated in the regionally dependent structure-function resulting from abnormal cardiomyopathic related point mutations. TnT represents the protein containing the largest number of identified abnormal cardiomyopathy associated point mutations of the Tn subunits [17]. The location of these mutations are distributed across most regions of the molecule, however the occurrence and location of specific mutations has begun to unravel the molecular level interactions of this molecule within the Ca^{2+} regulation of the thin filament and contraction. Nearly 1/3 of the identified cardiomyopathy TnT mutations are located in the TnT N-terminal Tm binding region (residues 88-126) and are associated with a HCM phenotype. DCM related TnT mutations are all but absent from this N-terminal TnT-Tm binding region. Contrary to this, a large majority of the DCM related TnT mutations are located in the C-terminal TnT-Tm binding region (residues 205-229) with relatively few HCM linked TnT mutations identified in this region. The differential association of the HCM and DCM phenotypic presentation between the N- and C- terminal TnT-Tm binding regions suggests differential function of these sites in the Tn Ca^{2+} regulation of contraction.

The TnT Arg-92 residue represents a "hot spot" for TnT cardiomyopathy mutations within the N terminal TnT-Tm binding region. The mutation of Arg-92 imparts a phenotypic presentation consistent with a critical role of this TnT interaction in Tn regulation. TnT Arg-92 undergoes an HCM phenotype linked to 3 different residues including Trp, Gln and Leu [72-74]. Mice transgenic for expression of the TnT R92 mutations exhibit the development of a HCM, hypercontractile phenotype [75-77]. Consistent with this phenotype, skinned cardiac fibers from these mice demonstrate increased Ca^{2+} sensitivity of force development [78, 79]. Similarly the exchange of TnT R92 containing Tn into fibers also induces increased Ca^{2+} sensitive force production [80, 81]. These findings demonstrate the primary effect of the TnT R92Q mutation to increase Ca^{2+} sensitive dependant force production as causal to the disease phenotype.

The primary functional effect of the TnT Arg-92 cardiomyopathy linked mutation demonstrate the usefulness of this naturally occurring model to investigate the mechanistic role of the TnT N-terminal interaction in the Tn regulation of cardiac contraction. Biochemical investigations demonstrate the TnT N-terminal Tm binding region as significant to the maintenance of the Tm blocked state [63]. The effect of the TnT R92Q mutation to increase Ca^{2+} sensitized force production results from the direct increase in activating Ca^{2+} binding to TnC, however this sensitization of Ca^{2+} binding to TnC does not occur in the isolated Tn complex [51]. This lack of effect in isolated Tn demonstrates the necessity of TnT R92Q's interaction with the actin thin filament to Tn Ca^{2+} regulation of contraction. Although highly alpha-helical in structure, the TnT N-terminal Tm binding region requires a large degree of flexibility necessary to allow for the TnT-Tm binding interaction [14, 57, 82]. Structural flexibility of the TnT N-terminal binding domain is therefore critical to this TnT-Tm interaction. The TnT R92Q mutation increases the overall TnT stability [83] resulting from deceased bending forces at TnT residues 104 and 105 [77] to generate a less flexible TnT N-terminus. Consistent with this finding TnT R92Q decreases TnT binding to Tm, inhibits TnT's ability to stabilize the Tm overlap and is less effective in promoting Tm binding to actin [83]. The importance of TnT N-terminal structure is further supported by evidence that other N-terminal TnT-Tm binding mutations (R92L, R92W, R94L and F110I) that elicit similar effects on the TnT-Tm interaction, while TnT mutations outside this region (I79N, Delta160E and E163K) do not [83]. Together these data suggest the TnT Arg-92 mutations increase Ca^{2+} sensitive force production by inducing flexibility of the N-terminal TnT-Tm binding domain resulting in a decreased TnT-Tm interaction and an increased ability of TnT to influence Tm towards the active state. Correspondingly, mice transgenic for expression of the TnT R92Q mutation exhibit increased numbers of myosin force generating cross-bridges, increased sub-maximal k_{tr} and increased length dependent activation [78], while reconstituted ATPase experiments and Tn exchanged fibers exhibit a decreased ability to inhibit activation at low Ca^{2+} [19, 81]. Thus the effect of altered TnT R92Q structural flexibility disrupts the TnT-Tm interaction to increase the movement of Tm towards the active state resulting in increased Ca^{2+} dependant force generation through increased numbers of myosin heads binding to actin. Thus, this TnT modification represents another way to mechanistically influence Ca^{2+} binding to TnC.

Animal models of the TnT R92Q mutation further shed light into the muscle level effects of the N-terminal TnT-Tm binding region on cardiac contractile function. Consistent with an inability of the R92Q to inhibit force production, R92Q expression in single intact cardiac myocytes isolated from mice transgenic for the TnT R92Q mutation demonstrate shorter, BDM sensitive, sarcomere lengths with increased fractional shortening [76]. These cellular changes result in a hypercontractile state of the *ex vivo* working heart preparation evident by an increase in the rate of pressure development over time, time to peak contraction and left ventricular peak pressure [76]. Likewise the short-term inducible expression of the TnT R92Q in mice resulted in increased systolic function including elevated fractional shortening and left ventricular ejection fraction [75] demonstrating the nature of the R92Q induced TnT-Tm alterations to increased force production as primary to the development of the HCM phenotype. Overall this data demonstrates the significance of the TnT N-terminal Tm binding region as important to the regulation of the Tm state on actin, the activation of muscle force development at sub-maximal Ca^{2+} and overall cardiac contractility. Thus the detailed investigation of the abnormal TnT R92Q mutation as a model of the N-terminal TnT-Tm binding region expands our mechanistic understanding of this TnT region in the regulation of cardiac contractility and its role in contractile function.

Contrary to the large association of TnT HCM cardiomyopathy mutations located in the TnT N-terminal Tm binding region, TnT mutations located in the C-terminal Tm binding region predominantly exhibit DCM cardiomyopathy linked mutations. Unlike the N-terminal Tm binding region, the C-terminal TnT region not only binds Tm but is also intimately associated with the Tn core structure [55, 68] (see Figure 1). The differential expression of TnT HCM and DCM mutations in the two different TnT-Tm binding regions provides a unique opportunity to compare the functional differences of naturally occurring abnormal TnT models from these two regions. The deletion of TnT residue Lys-210 (dK210) is linked to the development of DCM and is located in the TnT C-terminal Tm binding region. Mice transgenic for the expression of the dK210 TnT exhibit a DCM phenotype recapitulating the human phenotype [84, 85]. Demembranated cardiac fibers from the dK210 mice exhibit decreased Ca^{2+} sensitive force production consistent with the DCM phenotype [84]. Similarly, the exchange of Tn containing the dK210 TnT into fibers also decreases Ca^{2+} sensitive force production demonstrating this effect on Ca^{2+} sensitivity is a primary effect of the mutation [13a] [51] [86-88]. The measurement of Ca^{2+} binding in reconstituted thin filaments demonstrates a similar desensitization of Ca^{2+} binding to TnC that is not evident in isolated Tn, indicating the functional effects of the dK210 requires interactions of TnT with the thin filament [51]. Consistent with this decrease in Ca^{2+} sensitivity, a transgenic knock-in model of the dK210 exhibited depressed systolic function evident by decreased fractional shortening and ejection fraction with trends to decrease the maximum rate of left ventricle pressure change and left ventricular end systolic pressure [84].

The TnT dK210 mutation located in the C-terminal TnT-Tm binding region exhibits significantly different functional effects compared to that of the R92Q N-terminal Tm binding region mutation. While the R92Q exhibits increased Ca^{2+} sensitivity of force and increased systolic function, the dK210 exhibits decreased Ca^{2+} sensitivity of force and decreased systolic function. These two TnT mutations also differ mechanistically. Unlike the TnT R92Q, the dK210 does not directly alter the interaction of TnT with Tm [86], but rather increases TnT-TnI binding [89] suggesting a dK210-TnI effect. Differential mechanisms are also observed in the dynamic binding of Ca^{2+} to TnC. While the R92Q mutation does not alter TnC Ca^{2+} biding kinetics, the dK210 exhibits increased Ca^{2+} disassociation and decreased association rates [51]. Correspondingly, the dK210 does not alter Tn inhibition of acto-myosin ATPase [88]. These findings highlight significantly different Tn regulatory mechanisms between the TnT N- and C- terminal Tm binding regions. While the TnT R92Q appears to primarily affect the TnT-Tm binding interactions, TnT dK210 results in allosteric modulation of primarily Tn core interactions. We propose that the TnT N-terminal Tm binding interaction dominates contractile inhibition through regulation of Tm position on actin, while the C-terminal Tm binding region dominates allosteric modulation of the neighboring Tn core interaction to affect dynamic Ca^{2+} binding to TnC. This differential function of the TnT-Tm binding regions may therefore account for the lack of Ca^{2+} sensitizing HCM mutations localized to the C-terminal TnT-Tm binding region.

Together the findings from cardiomyopathy linked abnormal Tn mutations located in the TnT N- and C- terminal Tm binding sites suggest these Tm binding regions function differentially in the Ca^{2+} regulation of contraction. Modulation of the N-terminal Tm binding interaction by R92Q increases regional stability resulting in decreased N-terminal TnT-Tm binding favoring the TnT activation role enhanced Tn Ca^{2+} sensitivity and contractile activation. On the other hand, modulation of the C-terminal Tm binding interaction by dK210 does not alter Tm binding but appears to function through altered TnT interactions in the Tn core to result in depressed Ca^{2+} and contractile activation. Direct detailed biochemical comparisons between these two regional interactions are currently lacking and would be beneficial to furthering the differential mechanistic roles of these two TnT-Tm interactions in contractile regulation.

Abnormal Tn Splicing

Normal alternative RNA splicing represents a regulated mechanism for the cell to modify protein structure-function through the expression of alternate protein isoforms. Of the cardiac Tn subunits only the TnT gene is alternately spliced. Normal cardiac TnT alternative splicing can potentially produce up to 8 different TnT isoforms depending on species [55, 90, 91]. In humans, and the majority of mammals, the cardiac TnT gene contains 14 constitutively expressed axons and 3 alternately spliced exons consisting of exons 4, 5, and 13 [92-95]. Alternative splicing of the C-terminal exon 13 does not appear to demonstrate altered regulated function and therefore likely does not play a large role in the modulation of contractile function. By comparison exons 4 and 5 are both located in the extreme N-terminal TnT hypervariable region. Unlike the remaining TnT molecule, the extreme N-terminus is not highly conserved, is hypervariable in length and charge between isoforms, does not contain cardiomyopathy linked mutations and does not directly interact with TnI, TnC or Tm [59].

The tight regulation and resulting charge effects of TnT N-terminal exon alternate splicing supports the importance of the TnT extreme N-terminus in Tn contractile regulatory function. In the normal mammalian heart, cardiac TnT exon 4 alternative splicing is regulated such that the expression of the exon 4 excluded TnT isoform occurs at a consistent level [90, 91]. Unlike exon 4, the alternative splicing of exon 5 is regulated in a developmental dependent manner. Prior to birth all TnT isoforms expressed in mammalian hearts contain exon 5. Shortly after birth TnT isoform expression switches to expression of the adult TnT isoforms from which exon 5 is excluded [90, 91]. For this reason exon 5 is termed the embryonic exon. Both exon 4 (consisting of 4-5 residues) and exon 5 (consisting of 9-10 residues) contain a large number of acidic amino acid residues that contribute significantly to both local and overall isoform charge and size. The inclusion of exon 4 introduces 2 additional negative charges in the TnT N-terminus. While exon 5 inclusion results in 5 additional negatively charged residues. The inclusion of these acidic residues not only alters local charge but also the charge of the overall TnT isoform. A direct correlation between overall TnT isoform charge, size and Ca^{2+} sensitivity has been demonstrated for the normal cardiac TnT isoforms [91, 96]. This tight developmentally regulated splicing and the correlation of TnT N-terminal modulated overall charge to functional regulation support multiple amino acid alterations within the extreme N-terminus as significant to TnT functional regulation of muscle contraction.

Significance of the extreme TnT N-terminus is further supported by aberrant N-terminal splicing. Turkeys have long been described to exhibit "round heart" disease with cardiac contractile dysfunction [97]. Selection for this dilated cardiomyopathy phenotype confirmed a genetic component of this contractile dysfunction [98, 99]. Western blot analysis of turkey ventricular tissue demonstrates the presence of an additional and unusually low molecular weight isoform. This low molecular weight isoform accounts for approximately 30% of the total TnT and is present at all stages of development [100]. Cloning identified the low molecular weight TnT isoform as resulting from the abnormal exclusion of TnT N-terminal exon 8 (mammalian exon 7, avians contain an extra TnT exon and therefore avian exon 8 correlates to mammalian exon 7). The linkage of this exon exclusion to disease development and its presence prior to the development of disease supports this abnormal spliced isoform as casual to cardiac contractile dysfunction [100]. Subsequently a number of mammalian species prone to spontaneous cardiac contractile dysfunction have also been identified to exhibit aberrant N-terminal splicing. Domestic cats, dogs and guinea pigs exhibit a high incidence of cardiomyopathy and also exhibit abnormal TnT N-terminal exon splicing [101-105]. Adult cats express aberrant spliced TnT variants retaining the embryonic exon 5 while adult dogs express aberrant TnT species for both abnormal exon 5 inclusion and exon 7 exclusion [106]. In addition to abnormal retention of exon 5, adult guinea pigs further exhibit abnormal exclusion of exon 6 [106]. The linked expression of these aberrant N-terminal spliced TnT variants in animals prone to cardiac dysfunction supports their role as causal to the development of cardiac contractile dysfunction.

Linkage of abnormal N-terminal spliced TnT isoforms to cardiac disease identifies abnormally spliced TnT as naturally occurring models to understand the structure-function effect of the TnT extreme N-terminal region on cardiac contraction. The cardiac TnT exon 5 is located at the extreme TnT hypervariable N-terminus. Aberrant inclusion of this acidic embryonic exon 5 in the adult induces local structural changes that are transmitted to the central TnT region resulting in altered binding to tropomyosin and TnI [107]. This allosteric regulation of the TnT regulatory interactions increases Ca^{2+} sensitivity of ATPase activation and muscle force development [96]. The TnT mammalian exon 7 is also located in the extreme TnT N-terminal tail domain in close proximity to the TnT N-terminal Tm binding site [58]. Similar to exon 5, the exclusion of exon 7 altered central TnT structure, binding to both Tm and TnI and increased the Ca^{2+} regulated ATPase activation [100]. The linkage of N-terminal exon splicing induced local change on global structure and altered Ca^{2+} sensitivity of force development has also been demonstrated for a number of alternately spliced skeletal Tn isoforms [91, 96]. TnT N-terminal alternate splicing induced changes in local charge therefore represent a long-range allosteric mechanism to modulate the direct TnT interactions responsible for the regulation of contractile function.

Transgenic mouse models expressing aberrant spliced TnT further elucidate the role of the extreme N-terminus to modulate cardiac contractility. Transgenic aberrant inclusion of exon 5 or exclusion of exon 7 both increase Ca^{2+} sensitivity *in vitro* and depressed ex vivo contractile function in hearts from these transgenic mouse models including depressed maximal left ventricular pressure development, +dp/dt and stroke volume [108]. Interestingly, isolated cardiac myocytes from mice transgenic for the inclusion of exon 5 demonstrated decreased shortening and re-lengthening velocities while mice transgenic for exon 7 exclusion were not different from that of wild-type [106]. This depressed myocardial function was observed at an early age before the development of cardiovascular disease demonstrating the aberrant TnT N-terminal splicing as casual to the contractile dysfunction. It is interesting that both Ca^{2+} sensitizing aberrant spliced TnT's exhibit depressed cardiac contractility. Furthermore, Ca^{2+} sensitized abnormal Tn usually exhibit a hypertrophic phenotype opposite to that of the "round heart" dilated cardiomyopathy observed in turkeys expressing exon 8 excluded TnT (mammalian exon 7 exclusion). Isolated cells from mice transgenic for exon 7 excluded TnT also exhibit depressed shortening upon isoproterenol beta-adrenergic stimulation while cells from mice transgenic for exon 5 inclusion exhibited a largely normal response [106]. These findings beg the question of whether the dilated cardiomyopathy observed in the turkeys resulted from a direct intrinsic result of the aberrant N-terminal spliced TnT to alter the Tn/Tm/Actin regulation or an indirect effect extrinsic to Tn regulation.

It is interesting to note both the large size and charge variation necessary in the extreme TnT N-terminus to affect regulatory function compared to other regions of TnT where single amino acid residue modifications exhibit contractile effects. This necessity likely results from the extended nature of the TnT molecule and supports that this region is relatively amendable to structural alteration. The linkage of TnT extreme N-terminal alternative splicing with the development of cardiovascular disease, its altered contraction and effect on Tn protein-protein interactions however clearly establish the role of the extreme TnT N-terminus structure as significant to allosteric regulation of contractile function.

Beyond Abnormal Tn Primary Function: Integrative Function

While many abnormal Tn exhibit Tn Ca^{2+} regulation consistent with the disease phenotype and severity, others present with a variable phenotype or a phenotype not accurately predicted by its Tn Ca^{2+} regulated force production. One area of study that may help to resolve this lack of genotype/phenotype correlation is a more detailed understanding of the multi-factorial effects that abnormal Tn modifications impart on the Tn regulation of contraction. To date the mechanism of abnormal Tn function has centered on the intrinsic regulatory interactions directly involved in the Tn/Tm/Actin Ca^{2+} regulation of contraction. It is clear that the heart has the ability to adjust this intrinsic "set point" of Tn/Tm/Actin regulation to match cardiac contractile demand through signaling induced post-translational modification. Recent data demonstrates that the effect of TnI Ser-23/24 bis-phosphorylation post-translational modification is functionally different depending upon the modification status of the myofilament proteins [27, 109-112]. These findings suggests that the aberrant effects of abnormal Tn are not limited to their direct intrinsic effects on Tn/Tm/Actin Ca^{2+} regulation but rather abnormal Tn can also affect Ca2+ regulation through other "extrinsic" Tn posr-translasional modifications [113]. Thus the focused mechanistic understanding of abnormal Tn structure/function effects based solely upon their direct intrinsic Tn Ca^{2+} regulatory effects without regard for their effect on extrinsic factors, such as dynamic post-translational modifications, does not present a comprehensive understanding of abnormal Tn on the overall integrated heart function and the development of disease.

The necessity to expand our mechanistic understanding to include indirect extrinsic effects is evident in the effects of abnormal Tn on β-adrenergic induced TnI phosphorylation. The β-adrenergic signaling induced bis-phosphorylation of TnI at Ser-23/24 by PKA represents a powerful Tn level contraction adaptation to match cardiac stress demands [114, 115]. TnI Ser-23/24 bis-phosphorylation functions at the level of Tn decreasing Ca^{2+} dependent force development, enhancing relaxation and pre-load at high heart rates resulting in improved cardiovascular performance through the Frank-Starling effect [12, 116]. The normal *in vivo* heart "set point" consists of approximately 40% Ser-23/24 bis-phosphorylated TnI [117, 118]. This basal TnI Ser-23/24 bis-phosphorlyation "set point" constitutes a desensitized myofilament state compared to the *ex vivo* non-phosphorylated Tn/Tm/Actin state. Thus modification in the ability of TnI Ser-23/24 bis-phosphorylation to induce Ca^{2+} desensitization of force can aberrantly shift Tn regulation of contraction from this basal "set point". It has long been noted that patients with HCM exhibit an altered response to β-adrenergic stimulation. While this altered response may lie in any component of the β-adrenergic response, the disruption of this signaling mechanism directly at the level of the Tn complex does result from certain abnormal Tn [27, 109]. Abnormal Tn disruption of TnI bis-phosphorylation desensitization can occur through at least 4 mechanisms: 1) an alteration of the upstream β-adrenergic signaling; 2) a structural modification of Tn that alters the ability of the kinase to phosphorylate TnI (decreased TnI Ser-23/24 bis-phosphorlyation); 3) an intact and normal TnI PKA bis-phosphorlyation function that is altered upon combination with an independent abnormal Tn; or 4) the direct abnormal Tn disruption of the PKA TnI bis-phosphorylation functional effect. While the abnormal alteration of upstream β-adrenergic signaling likely involves a complicated compensatory effect, the effect of an abnormal Tn on TnI PKA kinase phosphorylation and altered bis-phosphorlyation functional effects are Tn based Ca^{2+} regulatory effects. Investigation of abnormal Tn modifications to alter these indirect, extrinsic functional effects of TnI PKA bis-phosphorylation can serve as a model to shed light into the functional regions required for the transmission of this critical TnI phosphorylation Ca^{2+} desensitization effect. Ultimately understanding the overall structural effects of an abnormal Tn on muscle regulatory function must include its indirect ability to modify TnI PKA bis-phosphorylation Ca^{2+} desensitization through any of these mechanisms in addition to its direct, intrinsic Tn/Tm/Actin contractile regulatory effects. To this end we will discuss the abnormal Tn effects on TnI PKA bis-phosphorylation as a starting point in moving towards understanding the overall functional effects of abnormal Tn on heart function.

One indirect mechanism of abnormal Tn is a structural modification that disrupts the binding of PKA to TnI thereby abnormally decreasing TnI PKA bis-phosphorlyation and the Tn level effects of β-adrenergic signaling. This mechanism of the abnormal Tn may be independent or in addition to the abnormal Tn direct intrinsic effects. The consensus sequence for PKA binding (RRRSS) in TnI consists of residues number 19-24 [119]. The HCM linked abnormal Tn mutation of TnI Arg-21 to Cys is located directly within this TnI PKA binding sequence and is therefore a prime model to investigate indirect effects of abnormal Tn on PKA binding. While a large amount of investigation has not been conducted into the function of the TnI R21C mutation, it is clear that this mutation affects PKA binding to TnI. Skinned fibers from mice transgenic for expression of the TnI R21C mutant demonstrate Ca^{2+} sensitive force production similar to wild-type in the absence of TnI bis-phosphorylation (propranolol treatment) indicating no primary functional effect of the mutation [120]. Biochemical investigations have however demonstrated that the TnI R21C mutation decreases the ability of PKA to bis-phosphorylate TnI at Ser-23/24 [46] with R21C transgenic mice demonstrating decreased TnI bis-phosphorylation [120]. Together these data suggest that while the R21C mutation alone does not directly alter the intrinsic Tn/Tm/Actin contractile regulation, its secondary indirect effect to decrease PKA binding to TnI and induce Ser-23/24 bis-phosphorylation ultimately results in a state of increased Ca^{2+} sensitivity. Expression of the TnI R21C mutation therefore contributes to the development of HCM disease by its secondary indirect, extrinsic effect inhibiting the ability of PKA to phosphorylate TnI.

Another indirect mechanism of abnormal Tn is the combination of mechanistically independent changes in Ca^{2+} sensitivity of the abnormal Tn and TnI PKA Ser-23/24 bis-phosphorlyation to disrupt the overall Ca^{2+} sensitivity "set point". Early on it was demonstrated that thin filaments reconstituted with Tn containing the TnI R145G cardiomyopathy mutation exhibit altered filament regulation following PKA phosphorylation compared to those filaments containing wild-type TnI [47]. The direct intrinsic effect of the TnI R145G mutation is to increases Ca^{2+} sensitivity [39, 44, 48]. While PKA bis-phosphorylation of R145G Tn exchanged fibers decreases Ca^{2+} sensitivity, in the presence of the R145G increased Ca^{2+} sensitivity the resultant overall Ca^{2+} sensitivity remains increased compared to PKA phosphorylated WT fibers [47]. This combination of different Ca^{2+} sensitivities is also evident in TnI phosphorylation based post-translational modification where the increased Ca^{2+} sensitivity of TnI Ser-150 phosphorylation in combination with PKA Ser-23/24 bis-phosphorylation eliminates the Ser-23/24 Ca^{2+} desensitization of force production [110]. Together these results demonstrate modulation of the TnI IP to increase Ca^{2+} sensitivity as a TnI region critical to indirect blunting the TnI Ser-23/24 PKA phosphorylation Ca^{2+} desensitization.

A final indirect mechanism of abnormal Tn to modulate TnI bis-phosphorylation is the abnormal Tn induced alteration of the TnI PKA 23/24 bis-phosphorylation function in the absence of an abnormal Tn direct effect. The TnC G159D mutation has been linked to a DCM phenotype [121]. TnC Gly-159 is located in the C-terminal lobe of TnC in a region that structurally interacts with the N-terminus of TnI but does not directly bind the contractile initiating Ca^{2+} signal. As discussed above the TnC G159D mutation has no direct effect on the intrinsic Tn Ca^{2+} regulation of force development in skinned cardiac trabeculae such that the sensitivity of G159D Tn exchanged fibers is identical to that of wild-type [27]. This lack of a direct functional effect in the presence of DCM drove us to investigate the effect of the C-terminal TnC located G159D on its interaction with the TnI N-terminal Ser-23-24 bis-phosphorylation. Combination of the TnC G159D Tn with specific TnI Ser-23/24 PKA bis-phosphorylation (by employing the TnI S23/24D phosphorylation-mimetic) demonstrated that the TnC G159D mutation dramatically blunted the effect of TnI PKA bis-phosphorylation to decrease Ca^{2+} dependent force development [27]. Therefore the TnC G159D induces cardiomyopathy by disrupting the TnI Ser-23/24 bis-phosphorylation Ca^{2+} sensitivity "set point", effectively sensitizing the myofilament. This finding of the TnC G159D mutation to indirectly alter Tn regulation by altering the TnI PKA signaling effect demonstrates the structural significance of the TnC C-terminus interaction with TnI as critical to extrinsic TnI PKA dependent bis-phosphorylation desensitization regulation.

The heart is clearly a highly integrated and adaptive system. From these examples it is clear that it is not always sufficient to understand the mechanism(s) of abnormal Tn contractile (mis-)regulation based upon its independent functional effects alone, but rather that abnormal Tn regulatory effects must ultimately be understood as part of the integrated regulatory system in which they reside. While the potential combined effects of abnormal Tn are extensive we have only begun to explore these extrinsic effects. Such integrated data on abnormal Tn function is ultimately necessary to fully understand their role in Tn regulation and disease.

Conclusion

The heart functions to pump blood through the contraction of its muscular walls. Both the extent of force development and its speed are significant Ca^{2+} dependent parameters that determine this systolic contractile performance of the heart. Tn plays a critical regulatory role in controlling the extent and speed of isometric force and the velocity of isotonic contraction by transmitting the Ca^{2+} signal into cardiac contraction. It is clear that multiple interactions between all three Tn subunits, Tm and actin help to tune the contractile and thus systolic performance of the heart. Disease linked mutations in Tn have been useful in delineating the significant protein-protein interactions that regulate systolic function of the heart. Many of these interactions can be further tuned by key post-translational modifications within Tn.

Clearly more work needs to be done to characterize the existing and constantly expanding database of disease linked mutations to gain a full understanding of the Tn interactions critical to is contractile regulation. This work must be multidisciplinary in scope (biochemical, biophysical, physiological and medical) to delineate the full behavior of Tn from molecule-to-man. Such data will ultimately provide the necessary understanding to design future targeted therapies for treatment of the different cardiomyopathies. In addition to disease associated mutations, the compilation of a polymorphisms database would be helpful to further identify significant Tn interactions modulatory to the structure-function of Tn. Finally, it will be important and necessary to determine how these different mutations cross-talk with the various regulatory post-translational modifications occurring in Tn to fully understand the integrated effects of Tn interaction modifications. While we have begun to elucidate key Tn protein-protein interactions responsible for the regulation of cardiac contraction there is still much to be learned about how the heart adapts to altered demand and the specific role of Tn in this regulated function.

References

[1] Solaro, R. J. (2011) *Regulation of Cardiac Contractility.* in *I*ntegrated Systems Physiology: From Molecule to Function to Disease, Morgan & Claypool Life Sciences, San Rafael. pp.

[2] McDonald, K. S., and Moss, R. L. (2000) *Circ Res* 87, 768-773.

[3] Bers, D. M. (2001) *Excitation-Contraction Coupling and Cardiac Contractile Force* 2nd ed., Kluwer Academic Publishers, Norwell, MA.

[4] Gordon, A. M., Homsher, E., and Regnier, M. (2000) *Physiol Rev* 80, 853-924.

[5] Stehle, R., Solzin, J., Iorga, B., and Poggesi, C. (2009) *Pflugers Arch* 458, 337-357.

[6] Gao, W. D., Backx, P. H., Azan-Backx, M., and Marban, E. (1994) *Circ Res* 74, 408-415.

[7] Venkataraman, R., Baldo, M. P., Hwang, H. S., Veltri, T., Pinto, J. R., Baudenbacher, F. J., and Knollmann, B. C. (2013) *J Mol Cell Cardiol* 60, 8-15.

[8] Wattanapermpool, J., Reiser, P. J., and Solaro, R. J. (1995) *Am J Physiol* 268, C323-330.

[9] Arteaga, G. M., Warren, C. M., Milutinovic, S., Martin, A. F., and Solaro, R. J. (2005) *Am J Physiol Heart Circ Physiol* 289, H2183-2192.

[10] Davis, J. P., and Tikunova, S. B. (2008) *Cardiovasc Res* 77, 619-626.

[11] Metzger, J. M., and Westfall, M. V. (2004) *Circ Res* 94, 146-158.

[12] Kobayashi, T., and Solaro, R. J. (2005) *Annu Rev Physiol* 67, 39-67.

[12a] Mamidi R, Michael JJ, Muthuchamy M, Chandra M. Interplay between the overlapping ends of tropomyosin and the N terminus of cardiac troponin T affects tropomyosin states on actin. *FASEB J.* 2013;27(9):3848-3859.

[12b] Mamidi R, Mallampalli SL, Wieczorek DF, Chandra M. Identification of two new regions in the N-terminus of cardiac troponin T that have divergent effects on cardiac contractile function. *J Physiol.* 2013;591(Pt 5):1217-1234.

[13] Little, S. C., Biesiadecki, B. J., Kilic, A., Higgins, R. S., Janssen, P. M., and Davis, J. P. (2012) *J Biol Chem* 287, 27930-27940.

[13a] Liu B, Tikunova SB, Kline KP, Siddiqui JK, Davis JP. Disease-related cardiac troponins alter thin filament Ca2+ association and dissociation rates. *PLoS One.* 2012;7(6):e38259.

[14] Tardiff, J. C. (2011) *Circ Res* 108, 765-782.

[15] Chang, A. N., and Potter, J. D. (2005) *Heart Fail Rev* 10, 225-235.

[16] Parvatiyar, M. S., Pinto, J. R., Dweck, D., and Potter, J. D. (2010) *J Biomed Biotechnol* 2010, 350706.

[17] Willott, R. H., Gomes, A. V., Chang, A. N., Parvatiyar, M. S., Pinto, J. R., and Potter, J. D. (2010) *J Mol Cell Cardiol* 48, 882-892.

[18] Ohtsuki, I., and Morimoto, S. (2008) *Biochem Biophys Res Commun* 369, 62-73.

[19] Robinson, P., Griffiths, P. J., Watkins, H., and Redwood, C. S. (2007) *Circ Res* 101, 1266-1273.

[20] Dong, W. J., Xing, J., Ouyang, Y., An, J., and Cheung, H. C. (2008) *J Biol Chem* 283, 3424-3432.

[21] Dyer, E. C., Jacques, A. M., Hoskins, A. C., Ward, D. G., Gallon, C. E., Messer, A. E., Kaski, J. P., Burch, M., Kentish, J. C., and Marston, S. B. (2009) *Circ Heart Fail* 2, 456-464.

[22] Dweck, D., Hus, N., and Potter, J. D. (2008) *J Biol Chem* 283, 33119-33128.

[23] Schmidtmann, A., Lindow, C., Villard, S., Heuser, A., Mugge, A., Gessner, R., Granier, C., and Jaquet, K. (2005) *FEBS J* 272, 6087-6097.

[24] Gollapudi, S. K., and Chandra, M. (2012) *Biochem Res Int* 2012, 824068.

[25] Liang, B., Chung, F., Qu, Y., Pavlov, D., Gillis, T. E., Tikunova, S. B., Davis, J. P., and Tibbits, G. F. (2008) *Physiol Genomics* 33, 257-266.

[26] Neulen, A., Stehle, R., and Pfitzer, G. (2009) *Basic Res Cardiol* 104, 751-760.

[27] Biesiadecki, B. J., Kobayashi, T., Walker, J. S., Solaro, R. J., and de Tombe, P. P. (2007) *Circ Res* 100, 1486-1493.

[28] Baryshnikova, O. K., Li, M. X., and Sykes, B. D. (2008) *J Mol Biol* 375, 735-751.

[29] Baryshnikova, O. K., Robertson, I. M., Mercier, P., and Sykes, B. D. (2008) *Biochemistry* 47, 10950-10960.

[30] Perry, S. V. (1999) *Mol Cell Biochem* 190, 9-32.

[31] Tobacman, L. S. (1996) *Annu Rev Physiol* 58, 447-481.

[32] Tripet, B., Van Eyk, J. E., and Hodges, R. S. (1997) *J Mol Biol* 271, 728-750.

[33] van Eyk, J. E., Thomas, L. T., Tripet, B., Wiesner, R. J., Pearlstone, J. R., Farah, C. S., Reinach, F. C., and Hodges, R. S. (1997) *J Biol Chem* 272, 10529-10537.

[34] Syska, H., Wilkinson, J. M., Grand, R. J., and Perry, S. V. (1976) *Biochem J* 153, 375-387.

[35] Li, M. X., Spyracopoulos, L., and Sykes, B. D. (1999) *Biochemistry* 38, 8289-8298.

[36] Hoffman, R. M., Blumenschein, T. M., and Sykes, B. D. (2006) *J Mol Biol* 361, 625-633.

[37] Mogensen, J., Kubo, T., Duque, M., Uribe, W., Shaw, A., Murphy, R., Gimeno, J. R., Elliott, P., and McKenna, W. J. (2003) *J Clin Invest* 111, 209-216.

[38] Niimura, H., Patton, K. K., McKenna, W. J., Soults, J., Maron, B. J., Seidman, J. G., and Seidman, C. E. (2002) *Circulation* 105, 446-451.

[39] James, J., Zhang, Y., Osinska, H., Sanbe, A., Klevitsky, R., Hewett, T. E., and Robbins, J. (2000) *Circ Res* 87, 805-811.

[40] Wen, Y., Xu, Y., Wang, Y., Pinto, J. R., Potter, J. D., and Kerrick, W. G. (2009) *J Mol Biol* 392, 1158-1167.

[41] Kruger, M., Zittrich, S., Redwood, C., Blaudeck, N., James, J., Robbins, J., Pfitzer, G., and Stehle, R. (2005) *J Physiol* 564, 347-357.

[42] Sanbe, A., James, J., Tuzcu, V., Nas, S., Martin, L., Gulick, J., Osinska, H., Sakthivel, S., Klevitsky, R., Ginsburg, K. S., Bers, D. M., Zinman, B., Lakatta, E. G., and Robbins, J. (2005) *Circulation* 111, 2330-2338.

[43] Wen, Y., Pinto, J. R., Gomes, A. V., Xu, Y., Wang, Y., Potter, J. D., and Kerrick, W. G. (2008) *J Biol Chem* 283, 20484-20494.

[44] Takahashi-Yanaga, F., Morimoto, S., Harada, K., Minakami, R., Shiraishi, F., Ohta, M., Lu, Q. W., Sasaguri, T., and Ohtsuki, I. (2001) *J Mol Cell Cardiol* 33, 2095-2107.

[45] Lang, R., Gomes, A. V., Zhao, J., Housmans, P. R., Miller, T., and Potter, J. D. (2002) *J Biol Chem* 277, 11670-11678.

[46] Gomes, A. V., Liang, J., and Potter, J. D. (2005) *J Biol Chem* 280, 30909-30915.

[47] Deng, Y., Schmidtmann, A., Redlich, A., Westerdorf, B., Jaquet, K., and Thieleczek, R. (2001) *Biochemistry* 40, 14593-14602.

[48] Elliott, K., Watkins, H., and Redwood, C. S. (2000) *J Biol Chem* 275, 22069-22074.

[49] Kobayashi, T., and Solaro, R. J. (2006) *J Biol Chem* 281, 13471-13477.

[50] Westfall, M. V., Borton, A. R., Albayya, F. P., and Metzger, J. M. (2002) *Circ Res* 91, 525-531.

[51] Liu, B., Lee, R. S., Biesiadecki, B. J., Tikunova, S. B., and Davis, J. P. (2012) *J Biol Chem* 287, 20027-20036.

[52] Zhou, Z., Rieck, D., Li, K. L., Ouyang, Y., and Dong, W. J. (2012) *Arch Biochem Biophys*.

[53] Kimura, A., Harada, H., Park, J. E., Nishi, H., Satoh, M., Takahashi, M., Hiroi, S., Sasaoka, T., Ohbuchi, N., Nakamura, T., Koyanagi, T., Hwang, T. H., Choo, J. A., Chung, K. S., Hasegawa, A., Nagai, R., Okazaki, O., Nakamura, H., Matsuzaki, M., Sakamoto, T., Toshima, H., Koga, Y., Imaizumi, T., and Sasazuki, T. (1997) *Nat Genet* 16, 379-382.

[54] Yumoto, F., Lu, Q. W., Morimoto, S., Tanaka, H., Kono, N., Nagata, K., Ojima, T., Takahashi-Yanaga, F., Miwa, Y., Sasaguri, T., Nishita, K., Tanokura, M., and Ohtsuki, I. (2005) *Biochem Biophys Res Commun* 338, 1519-1526.

[55] Perry, S. V. (1998) *J Muscle Res Cell Motil* 19, 575-602.

[56] Cabral-Lilly, D., Tobacman, L. S., Mehegan, J. P., and Cohen, C. (1997) *Biophys J* 73, 1763-1770.

[57] Greenfield, N. J., Palm, T., and Hitchcock-de Gregori, S. E. (2002) *Biophys J* 83, 2754-2766.

[58] Jin, J. P., and Chong, S. M. (2010) *Arch Biochem Biophys* 500, 144-150.

[59] Pearlstone, J. R., and Smillie, L. B. (1982) *J Biol Chem* 257, 10587-10592.
[60] Morris, E. P., and Lehrer, S. S. (1984) *Biochemistry* 23, 2214-2220.
[61] Mak, A. S., and Smillie, L. B. (1981) *J Mol Biol* 149, 541-550.
[62] Pato, M. D., Mak, A. S., and Smillie, L. B. (1981) *J Biol Chem* 256, 602-607.
[63] Hinkle, A., Goranson, A., Butters, C. A., and Tobacman, L. S. (1999) *J Biol Chem* 274, 7157-7164.
[64] Xu, C., Craig, R., Tobacman, L., Horowitz, R., and Lehman, W. (1999) *Biophys J* 77, 985-992.
[65] Tobacman, L. S., Nihli, M., Butters, C., Heller, M., Hatch, V., Craig, R., Lehman, W., and Homsher, E. (2002) *J Biol Chem* 277, 27636-27642.
[66] Pearlstone, J. R., and Smillie, L. B. (1983) *J Biol Chem* 258, 2534-2542.
[67] Potter, J. D., Sheng, Z., Pan, B. S., and Zhao, J. (1995) *J Biol Chem* 270, 2557-2562.
[68] Takeda, S., Yamashita, A., Maeda, K., and Maeda, Y. (2003) *Nature* 424, 35-41.
[69] Heeley, D. H., Golosinska, K., and Smillie, L. B. (1987) *J Biol Chem* 262, 9971-9978.
[70] Pearlstone, J. R., and Smillie, L. B. (1978) *Can J Biochem* 56, 521-527.
[71] Zot, A. S., and Potter, J. D. (1987) *Annu Rev Biophys Biophys Chem* 16, 535-559.
[72] Moolman, J. C., Corfield, V. A., Posen, B., Ngumbela, K., Seidman, C., Brink, P. A., and Watkins, H. (1997) *J Am Coll Cardiol* 29, 549-555.
[73] Varnava, A. M., Elliott, P. M., Baboonian, C., Davison, F., Davies, M. J., and McKenna, W. J. (2001) *Circulation* 104, 1380-1384.
[74] Watkins, H., McKenna, W. J., Thierfelder, L., Suk, H. J., Anan, R., O'Donoghue, A., Spirito, P., Matsumori, A., Moravec, C. S., Seidman, J. G., and et al. (1995) *N Engl J Med* 332, 1058-1064.
[75] Lutucuta, S., Tsybouleva, N., Ishiyama, M., Defreitas, G., Wei, L., Carabello, B., and Marian, A. J. (2004) *J Am Coll Cardiol* 44, 2221-2230.
[76] Tardiff, J. C., Hewett, T. E., Palmer, B. M., Olsson, C., Factor, S. M., Moore, R. L., Robbins, J., and Leinwand, L. A. (1999) *J Clin Invest* 104, 469-481.
[77] Ertz-Berger, B. R., He, H., Dowell, C., Factor, S. M., Haim, T. E., Nunez, S., Schwartz, S. D., Ingwall, J. S., and Tardiff, J. C. (2005) *Proc Natl Acad Sci USA* 102, 18219-18224.
[78] Chandra, M., Rundell, V. L., Tardiff, J. C., Leinwand, L. A., De Tombe, P. P., and Solaro, R. J. (2001) *Am J Physiol Heart Circ Physiol* 280, H705-713.
[79] Montgomery, D. E., Tardiff, J. C., and Chandra, M. (2001) *J Physiol* 536, 583-592.
[80] Chandra, M., Tschirgi, M. L., and Tardiff, J. C. (2005) *Am J Physiol Heart Circ Physiol* 289, H2112-2119.
[81] Szczesna, D., Zhang, R., Zhao, J., Jones, M., Guzman, G., and Potter, J. D. (2000) *J Biol Chem* 275, 624-630.
[82] Hitchcock-DeGregori, S. E., Song, Y., and Greenfield, N. J. (2002) *Biochemistry* 41, 15036-15044.
[83] Palm, T., Graboski, S., Hitchcock-DeGregori, S. E., and Greenfield, N. J. (2001) *Biophys J* 81, 2827-2837.
[84] Du, C. K., Morimoto, S., Nishii, K., Minakami, R., Ohta, M., Tadano, N., Lu, Q. W., Wang, Y. Y., Zhan, D. Y., Mochizuki, M., Kita, S., Miwa, Y., Takahashi-Yanaga, F., Iwamoto, T., Ohtsuki, I., and Sasaguri, T. (2007) *Circ Res* 101, 185-194.
[85] Zhan, D. Y., Morimoto, S., Du, C. K., Wang, Y. Y., Lu, Q. W., Tanaka, A., Ide, T., Miwa, Y., Takahashi-Yanaga, F., and Sasaguri, T. (2009) *Cardiovasc Res* 84, 64-71.

[86] Lu, Q. W., Morimoto, S., Harada, K., Du, C. K., Takahashi-Yanaga, F., Miwa, Y., Sasaguri, T., and Ohtsuki, I. (2003) *J Mol Cell Cardiol* 35, 1421-1427.
[87] Morimoto, S., Lu, Q. W., Harada, K., Takahashi-Yanaga, F., Minakami, R., Ohta, M., Sasaguri, T., and Ohtsuki, I. (2002) *Proc Natl Acad Sci U S A* 99, 913-918.
[88] Venkatraman, G., Harada, K., Gomes, A. V., Kerrick, W. G., and Potter, J. D. (2003) *J Biol Chem* 278, 41670-41676.
[89] Sfichi-Duke, L., Garcia-Cazarin, M. L., Sumandea, C. A., Sievert, G. A., Balke, C. W., Zhan, D. Y., Morimoto, S., and Sumandea, M. P. (2010) *J Mol Cell Cardiol* 48, 934-942.
[90] Wei, B., and Jin, J. P. (2011) *Arch Biochem Biophys* 505, 144-154.
[91] Jin, J. P., Chen, A., Ogut, O., and Huang, Q. Q. (2000) *Am J Physiol Cell Physiol* 279, C1067-1077.
[92] Jin, J. P., and Lin, J. J. (1989) *J Biol Chem* 264, 14471-14477.
[93] Jin, J. P., and Lin, J. J. (1988) *J Biol Chem* 263, 7309-7315.
[94] Jin, J. P., Huang, Q. Q., Yeh, H. I., and Lin, J. J. (1992) *J Mol Biol* 227, 1269-1276.
[95] Jin, J. P., Wang, J., and Zhang, J. (1996) *Gene* 168, 217-221.
[96] Gomes, A. V., Guzman, G., Zhao, J., and Potter, J. D. (2002) *J Biol Chem* 277, 35341-35349.
[97] Czarnecki, C. M. (1984) *Comp Biochem Physiol A Comp Physiol* 77, 591-598.
[98] Jankus, A. S., N. A. Noren, G. R. (1973) Turkey round heart disease: A model cardiomyopathy. in *Research Animals in Medicine* (Harmison, L. T. ed.). pp 53-60.
[99] Noren, G. R., Staley, N. A., Jankus, E. F., and Stevenson, J. E. (1971) *Virchows Arch A Pathol Pathol Anat* 352, 285-295.
[100] Biesiadecki, B. J., and Jin, J. P. (2002) *J Biol Chem* 277, 18459-18468.
[101] Lawler, D. F., Templeton, A. J., and Monti, K. L. (1993) *J Vet Intern Med* 7, 383-387.
[102] Tilley, L. P., Liu, S. K., Gilbertson, S. R., Wagner, B. M., and Lord, P. F. (1977) *Am J Pathol* 86, 493-522.
[103] Liu, S. K., and Tilley, L. P. (1980) *Yale J Biol Med* 53, 191-211.
[104] Liu, S. K., Maron, B. J., and Tilley, L. P. (1979) *J Am Vet Med Assoc* 174, 708-713.
[105] Liu, S. K., Maron, B. J., and Tilley, L. P. (1979) *Am J Pathol* 94, 497-508.
[106] Biesiadecki, B. J., Elder, B. D., Yu, Z. B., and Jin, J. P. (2002) *J Biol Chem* 277, 50275-50285.
[107] Biesiadecki, B. J., Chong, S. M., Nosek, T. M., and Jin, J. P. (2007) *Biochemistry* 46, 1368-1379.
[108] Feng, H. Z., and Jin, J. P. (2010) *Am J Physiol Heart Circ Physiol* 299, H97-H105.
[109] Memo, M., Leung, M. C., Ward, D. G., Dos Remedios, C., Morimoto, S., Zhang, L., Ravenscroft, G., McNamara, E., Nowak, K. J., Marston, S. B., and Messer, A. E. (2013) *Cardiovasc Res*.
[110] Nixon, B. R., Thawornkaiwong, A., Jin, J., Brundage, E. A., Little, S. C., Davis, J. P., Solaro, R. J., and Biesiadecki, B. J. (2012) *J Biol Chem* 287, 19136-19147.
[111] Boontje, N. M., Merkus, D., Zaremba, R., Versteilen, A., de Waard, M. C., Mearini, G., de Beer, V. J., Carrier, L., Walker, L. A., Niessen, H. W., Dobrev, D., Stienen, G. J., Duncker, D. J., and van der Velden, J. (2011) *J Mol Cell Cardiol* 50, 487-499.
[112] Kooij, V., Saes, M., Jaquet, K., Zaremba, R., Foster, D. B., Murphy, A. M., Dos Remedios, C., van der Velden, J., and Stienen, G. J. (2010) *J Mol Cell Cardiol* 48, 954-963.

[113] Marston, S. B. (2011) *J Cardiovasc Transl Res* 4, 245-255.
[114] Layland, J., Solaro, R. J., and Shah, A. M. (2005) *Cardiovasc Res* 66, 12-21.
[115] Kranias, E. G., and Solaro, R. J. (1982) *Nature* 298, 182-184.
[116] Solaro, R. J., Moir, A. J., and Perry, S. V. (1976) *Nature* 262, 615-617.
[117] Ayaz-Guner, S., Zhang, J., Li, L., Walker, J. W., and Ge, Y. (2009) *Biochemistry* 48, 8161-8170.
[118] Zhang, J., Guy, M. J., Norman, H. S., Chen, Y. C., Xu, Q., Dong, X., Guner, H., Wang, S., Kohmoto, T., Young, K. H., Moss, R. L., and Ge, Y. (2011) *J Proteome Res* 10, 4054-4065.
[119] Kennelly, P. J., and Krebs, E. G. (1991) *J Biol Chem* 266, 15555-15558.
[120] Wang, Y., Pinto, J. R., Solis, R. S., Dweck, D., Liang, J., Diaz-Perez, Z., Ge, Y., Walker, J. W., and Potter, J. D. (2012) *J Biol Chem* 287, 2156-2167.
[121] Mogensen, J., Murphy, R. T., Shaw, T., Bahl, A., Redwood, C., Watkins, H., Burke, M., Elliott, P. M., and McKenna, W. J. (2004) *J Am Coll Cardiol* 44, 2033-2040.
[122] Whitby, F. G., and Phillips, G. N., Jr. (2000) *Proteins* 38, 49-59.

In: Troponin
Editor: J.-P. Jin

ISBN: 978-1-62948-320-7

Chapter XI

Cardiac Troponin Mutations: Diastolic Dysfunction and Therapeutic Options

Pierre-Yves Jean-Charles, PhD[1], Changlong Nan, PhD[1], Lei Zhang, MD[1,2], Jie Tian, MD[2] and Xupei Huang, MD, PhD[1,2,*]

[1]Charles E. Schmidt College of Medicine,
Florida Atlantic University, Boca Raton, FL, US
[2]Division of Cardiology, Children's Hospital,
Chongqing Medical University, Chongqing, China

Abstract

Cardiac troponin I (cTnI) plays a critical role in regulation of cardiac function. The physiological effect of cTnI, as an inhibitory subunit of troponin complex, is to prevent the interaction between myosin heavy chain heads and actins, i.e., the cross-bridge formation, and to ensure a proper relaxation of cardiac myofilaments. In pathological conditions, the deficiency of cTnI or mutations in cTnI especially in the C-terminus of cTnI is associated with diastolic dysfunction caused by myofibril hypersensitivity to Ca^{2+}. In this chapter, we summarize the recent studies from our laboratory and the others in exploring the mechanisms underlying diastolic dysfunction and heart failure caused by cTnI mutations in cell-based assays and in transgenic animal models. These studies provide us with useful information in searching for therapeutic strategies and target-oriented medication for the treatment of diastolic dysfunction and heart failure.

* Corresponding author: Dr. Xupei Huang, Charles E. Schmidt College of Medicine, Center of Molecular Biology and Biotechnology, Florida Atlantic University, 777 Glades Road, Boca Raton, FL 33431, USA. Tel: 1-561-2972443. Fax: 1-561-2972221. Email: xhuang@fau.edu.

1. Introduction

Cardiomyopathies represent a group of diseases that primarily affect the cardiac muscle. On the ground of cardiac morphology and pathophysiology, cardiomyopathies may be classified into three major types: hypertrophic cardiomyopathy (HCM), dilated cardiomyopathy (DCM) and restrictive cardiomyopathy (RCM). HCM and DCM are characterized, as their name indicates, by a hypertrophied heart or a dilated ventricle, whereas RCM is defined by the restricted diastolic filling that results from an abnormally stiffened myocardium. In the past, most cardiomyopathy cases are described as idiopathic cardiomyopathies, i. e. etiology is unknown (Siegel R. J. 1984 Aug) (Ammash N. M. 2000 May 30). Now we know that most cardiomyopathy cases are heritable and caused by single gene mutations (Braunwald E. 2008). Cardiac troponin I (cTnI) mutations have been found in these major cardiomyopathies causing severe cardiac dysfunction and early death. Most of the disease-causing cTnI mutations are located in the COOH terminal of that protein which is highly conserved and essential for the proper function of the molecule. Determining how cTnI mutations disrupt myofilament dynamics and cause cardiac dysfunction is critical for developing therapeutic strategies for the treatment of the myocardial diseases. Functional studies in normal and mutant cTnI revealed that alteration of myofilament Ca^{2+} sensitivity was the main physiological anomaly in diseased myofibers presenting contractile dysfunction. This leads to investigate whether restoration of the myofilament Ca^{2+} sensitivity is a potential therapeutic option for cardiomyopathies due to myofibril protein mutations. In this article, the effect of cTnI mutations in the development of cardiomyopathies will be summarized and the therapeutic implications based on the recent discoveries from biochemical and transgenic animal studies will be discussed.

2. Cardiac Troponin I and cTnI C-Terminus

2.1. Regulation of Cardiac Function by cTnI

Cardiac troponin I plays an essential role in the regulation of cardiac muscle contraction. This cardiac specific protein forms the contraction regulatory complex, troponin, with the Ca^{2+} binding protein cardiac troponin C (cTnC) and the tropomyosin binding troponin subunit cardiac troponin T (cTnT). As the inhibitory subunit of that complex, cTnI prevents actomyosin cross-bridge formation by binding to actin in the absence of Ca^{2+}. This inhibitory action is extended over 7 actin monomers by tropomyosin which blocks additional myosin binding sites in the actin filament. The displacement of cTnI from its actin site will release its inhibitory effect and induce contraction. This occurs upon binding of Ca^{2+} to TnC which causes a conformational change in TnC increasing the affinity of cTnI to that molecule. Stronger interaction of cTnC with cTnI leads the latter to leave its actin binding site, changing the conformation of the troponin complex and causing the dislodgment of tropomyosin to open myosin binding sites and allow muscle contraction. Thus, the affinity of cTnI to actin and TnC seem to be a decisive factor behind the induction of contraction or relaxation. Targeting such affinity to modify myofilament dynamics may help resolve the myofibrillar mechanical dysfunction that characterizes the cardiomyopathies associated with cTnI mutations.

2.2. Functional Domains in the N Terminal of cTnI

Extensive research has been done to locate cTnI's interacting sites with its associated proteins and to identify the major domains in its sequence that are necessary for its inhibitory function. The sequence of cTnI is highly conserved between species and possesses at least 60% homology with its isoforms slow skeletal troponin I (ssTnI) and fast skeletal troponin I (fsTnI) (Wilkinson J. M. 1978 Jan 5). A major difference between the cardiac isoform of troponin I and its fetal and skeletal isoforms is a 32 amino acid extension at the NH2 terminal of that protein. This NH2-terminal extension is known to interact with the C terminal of TnC and contains two serines residues 23 and 24 which can be phosphorylated by cAMP-dependent protein kinase (PKA) under beta adrenergic stimulation. The phosphorylation of these residues decreases the affinity of cTnI to TnC and accelerates cardiac relaxation. The adjacent region covering residues 33 to 80 is very conserved and known to bind to the COOH terminus of cTnC (Krudy G. A. 1994 Sep 23) (Kleerekoper Q. 1995 Oct 17). The residues 33 to 53 were found to be necessary for Ca^{2+}-dependent activation of cardiac myofilaments (Rarick H. M. 1999 Feb). Residues 90 to 136 cTnI correspond to a coiled-coil interface between cTnI and cTnT in the troponin complex and is a result essential for the transmission of the Ca^{2+} signal to tropomyosin (Takeda S. 2003 Jul 3) (Parmacek M. S. 2004 Nov-Dec) (Vinogradova M. V. 2005 Apr 5). Cardiomyopathy causing mutations in the NH2 terminal half of cTnI are rare. Most cTnI mutations cluster in the COOH terminal half of that protein, particularly in exon 7 and 8 of the *TNNI3* gene.

2.3. Functional Domains in the C Terminal of cTnI

The C-terminal half of cTnI is more conserved than the N-terminal region of the protein (Wilkinson J. M. 1978 Jan 5). It is particularly known to play an important role in its inhibitory action. It contains specific regions that are crucial for the normal activity of the protein and cardiac relaxation (Zhou Z. 2012 Mar 2). Among them is the inhibitory region, which is the minimum sequence necessary for inhibition of actomyosin ATPase activity. This domain spans from residue 138 to 149 and binds to actin and the C terminal domain of TnC in a non-concurrent manner that depends on the binding of Ca^{2+} to the C-terminal of TnC (Talbot J. A. 1981 Mar 25) (Rarick H. M. 1997 Oct 24). Next to it is a regulatory region that comprehends residues 151 to 167. Such region also termed the second TnC binding site binds to the N-terminal domain of TnC following Ca^{2+} activation of muscle contraction (Dong W. J. 1999 Oct 29) (Tripet B. 1997 Sep 5) (Li M. X. 1999 Jun 29). This regulatory region enhances cTnI's interaction with TnC upon Ca^{2+} activation, playing a critical role for the switch of cTnI from actin to TnC. Downstream to the regulatory region is a second actin-binding site located at residues 168 to 188 of cTnI. This domain binds specifically to the actin-tropomyosin filament and is known to contribute to the inhibitory activity of cTnI (Tripet B. 1997 Sep 5). The remaining C-terminal domain, 192 to 210, while not fully characterized, was found to play a role in the stabilization of tropomyosin in the actin filament upon Ca^{2+} activation of muscle contraction (Galińska A. 2010 Mar 51). Mutations in the various domains in the C terminus of cTnI disrupt the molecule, altering the affinity of cTnI to actin or TnC, thus the response of the troponin complex to Ca^{2+} stimulation which will affect cardiac function.

The integrity of the cTnI molecule is essential for proper conformation of the troponin complex in the myofilament and the inhibition of actomyosin ATPase activity. The presence of cTnI degradation products in serum is often an indicator of myocardial damage (Antman E. M. 1996 Oct 31). While proteolysis of the NH2 terminal extension of cTnI has been observed in physiological and pathological conditions including following simulation of microgravity, such modification was found to be beneficial for the diastolic function of the heart (Barbato J. C. 2005 Feb 25). On the other hand, proteolysis of the carboxyl end of cTnI coincides with the development of human ischemic myocardial diseases such as myocardial stunning and acute pressure overload (McDonough J. L. 2001 Jan 2) (Feng J. 2001 Apr 24). Numerous mutations in the carboxyl half of the protein are associated with the development of cardiomyopathies further confirming the importance of the C terminal domains of cTnI for proper regulation of cardiac contraction. Diastolic dysfunction is the main anomaly observed either from ablation, deficiency, proteolysis or mutations of cTnI (Huang X. 1999 Jan 8-22) (Liu J. 2007 Aug) (Huang X. P. 2004 Dec).

3. Diastolic Dysfunction in Human Patients Caused by cTnI Mutations

Unlike DCM which is characterized by damage of cardiac contractility or systolic dysfunction, HCM and RCM share a similar key feature characterized by a defect in cardiac diastolic function, i. e. diastolic dysfunction (Maron B. J. 2006 Apr 11). Diastolic dysfunction is an important clinical problem since up to 40% of heart failure cases are attributed to this cardiac condition (Bonow R. O. 1992 Sep 15) (Chen H. H. 2002 Oct) (Senni M. 2001 Nov 1). Therapies for diastolic dysfunction are limited especially in HCM and RCM which are often associated with a poor prognosis and sudden death. The role that cTnI plays in regulating cardiac relaxation and the different myocardial diseases or cardiomyopathies that result from its mutation make it a particularly interesting molecule for deciphering the mechanisms behind the impaired myofilament kinetics in the disease myocardium and designing therapies aimed at restoring the mechanical attributes of the myofilament. In this article, we will summarize the diastolic dysfunction caused by cTnI mutations in human patients and in animal studies.

3.1. cTnI Mutations in Patients with Hypertrophic Cardiomyopathy

Hypertrophic cardiomyopathy is one of the most common cases of heart failure and a common cause of sudden death in young athletes (Go A. S. 2013). It is characterized by diastolic dysfunction, increased ventricular wall and septal dimensions accompanied with fibrosis and myocardial disarrays. The systolic function is usually preserved. Most mutations in familial HCM are autosomal dominant and sarcomeric protein gene mutations account for most of the cases of familial HCM (Kaski J. P. 2009 Oct). The MH7 and MYBPC3 genes coding respectively for β myosin heavy chain and cardiac myosin binding protein C accounts for most of the HCM mutations (Richard P. and Project 2003 May 6). The majority of cTnI mutations have been linked to HCM. However, the prevalence of cTnI mutations in HCM is low, ranging from 2 to 9% (Mogensen J. 2004 Dec 21) (Curila K. 2012 Feb). Clinically, cTnI mutations are often associated with a more malignant course (Doolan A. 2005 Feb). Other sarcomeric proteins linked with HCM include ventricular myosin essential light chain, ventricular myosin regulatory light chain, cardiac alpha actin, alpha tropomyosin, cTnT, and titin.

Kimura et al. were the first to identify cTnI mutations in HCM (Kimura A. 1997 Aug). They reported six mutations, R145G, R145Q, R162W, K183del, G203S, K206Q in different families presenting mild to severe symptoms. Normal hypertrophy was observed in the probands carrying the R145G, R145Q, K206Q mutations, while the R162W and G203S mutations induced hypertrophy only in the apex. The carrier of the K183del mutation in that study presented apical HCM while his son who had the same mutation presented typical HCM. This confirmed that both typical and atypical HCM are part of the clinical manifestation of cTnI mutations. Currently, more than 40 HCM linked cTnI mutations have been identified (Willott R. H. 2010 May). The majority of the mutations occur in the C terminal half of the molecule, specifically in exons 7 and 8 of the *TNNI3* gene which correspond with the inhibitory region, regulatory domain, second actin tropomyosin binding site and the C terminal domain of the cTnI protein. A rare case of mutation in the N terminal of that protein occurred in residue 21, substituting arginine with cysteine (Gomes A. V. 2005 Nov). The R21C substitution is located close to the serine 23 and 24 residues in the N-terminal extension of cTnI which may affect their phosphorylation state and the response of the molecule to β adrenergic stimulation.

Significant clinical heterogeneity is noticeable in HCM-linked sarcomeric protein mutations. Mutations were classified as benign or malignant depending on whether they induced mild symptoms or caused heart failure and sudden death. Nevertheless, some benign mutations were found to cause severe symptoms and some malignant mutations were found silent in other cases. Thus, while the type and location of the mutations are important determinants of disease presentation and severity, they are not the only causative factors since symptoms and prognosis for the same mutation may also vary across and within families (Girolami F. 2010 Apr 6). This leads to consider additional causes (genetic or environmental) that may influence the clinical manifestation of the disease. The correlation between the dosage of mutant proteins in carriers of monogenic protein mutation and disease phenotype while observed in transgenic animal models of HCM is not well established in clinical studies. This is due to the low availability of cardiac samples from the biopsy and the difficulty in differentiating mutant from intact proteins. Also, multiple mutations in the same and different genes are considered to play a dosage effect in HCM. Patients carrying double or compound mutations in sarcomeric genes present more severe symptoms and are at greater risk of sudden death (Maron B. J. 2012 Jan).

3.2. cTnI Mutation in Patients with Restrictive Cardiomyopathy

RCM is highly linked with cTnI mutations. This myocardial disease is characterized by cardiac relaxation and ventricular filling abnormalities due to an increase of ventricular stiffness. Cardiac dysfunction in RCM is not accompanied with ventricular morphological changes as opposed to HCM and DCM (Maron B. J. 2006 Apr 11). RCM is differentiated from a resembling disease, constrictive pericarditis, by the absence of pericardial thickening (Schoenfeld M. H. 1990 Nov) (Zwas D. R. 2012 Sep). Bi-atrial enlargement often occurs in RCM due to chronic increase of atrial pressure. The disease when confined to the cardiac muscle is classified as primary RCM. The secondary RCM develops from extrinsic factors such as infiltrative disorders (amyloidosis and sarcoidosis), storage diseases (haemochromatosis, glycogen storage disease and Fabry disease), and inflammatory diseases like in the cases of Loeffler cardiomyopathy, endomyocardial fibrosis, and eosinophilic endomyocardial disease (Kushwaha S. S. 1997 Jan). Primary RCM can be idiopathic or from genetic causes. The prevalence of RCM is considerably less than HCM and DCM as it accounts for approximately 2-5% of all cardiomyopathies. Nevertheless, RCM is the most lethal and least treatable cardiomyopathy having the worst prognosis (Malcić I. 2002 May) (Lipshultz S. E. 2003 Apr 24) (Nugent A. W. 2003 Apr 24) (Russo L. M. 2005 Sep) (Rivenes S. M. 2000 Aug 22). The incidence of mortality is very high in RCM pediatric patients with 50% of deaths occurring within 2 years after diagnosis (Katritsis D. 1991 Nov 1). Diastolic heart failure and sudden cardiac death are common and cardiac transplantation is highly recommended for RCM patients (Rivenes S. M. 2000 Aug 22) (Nugent A. W. 2003 Apr 24).

Familial RCM was associated with mutations and accumulation of the intermediate filament desmin (Arbustini E. 2006 Aug) (Pruszczyk P. 2007 Apr 25). Skeletal myopathy and atrioventricular block often accompany desmin-related cardiomyopathies. Most of the RCM genetic causes are attributed to mutations in sarcomeric protein genes. cTnI represents more than half of the RCM mutations. Mogensen et al. were the first to find a connection between cTnI mutations and RCM (Mogensen J. 2003). They reported six missense mutations (L144Q, R145W, A171T, K178E, D190G, and R192H) in exons 7 or 8 of the *TNNI3* gene (Mogensen J. 2003). Two de novo mutations in that study, cTnI R192H and K178E, were associated with the worst clinical phenotype and sudden death. Additional RCM clinical cases involving cTnI and other myofibril proteins were later discovered thanks to genetic screening of different individuals and families suffering from that disease (van den Wijngaard A. 2011 May 2). To date, at least 10 mutations in the *TNNI3* gene have been associated with RCM, including two deletions causing frame-shift mutations and premature stop codons in the cTnI protein. All the RCM mutations reported are located in the C terminal half of cTnI (Kaski J. P. 2008 Nov) (van den Wijngaard A. 2011 May 2).

RCM-causing cTnI mutations are very lethal with early onset and worst clinical presentation. Nevertheless, it is worth noting that different mutations at the same position in cTnI may cause different types of cardiomyopathies. For example, the substitutions of arginine 145 with glycine and glutamine (R145G and R145Q) are associated with HCM while its replacement by tryptophan (R145W) is linked with RCM (Takahashi-Yanaga F. 2001 Dec). HCM and RCM phenotypes may develop from the same mutations as well. This can be seen by the presence of left ventricular hypertrophy and ECGs typical for HCM in different families carrying the R145W mutations (van den Wijngaard A. 2011 May 2). Other cTnI mutations observed in both HCM and RCM cases include the RCM-causing R192H mutation which was detected in a patient presenting hypertrophy with restrictive physiology while his father and sister died of restrictive cardiomyopathy (Kaski J. P. 2008 Nov) (Rai T. S. 2009 Nov). Restrictive physiology is usually associated with poor outcome in hypertrophic cardiomyopathy (Pinamonti B. 2010 Sep) (Maskatia S. A. 2012 Jan). Also, a de novo mutation, R204H, which was previously associated with HCM, was observed in two individual RCM cases (Doolan A. 2005 Feb). The first case involved a young female presenting pure RCM with myofibrillar disarray and absence of hypertrophy (Gambarin F. I. 2008 Oct). The second case presented RCM in combination with ventricular septal defect (Yang S. W. 2010 Oct). In Mogensen's study, 12 patients of the same family carrying the cTnI D190G mutation showed a clinical phenotype of mixed appearance of RCM and HCM (Mogensen J. 2003). Patients with HCM/RCM phenotype usually have better transplant-free survival than those presenting pure RCM (Webber S. A. and Investigators 2012 Sep 4). The clinical variety of the cTnI mutations raises some questions about the mechanism behind the development and progression of the resulting myocardial diseases. Determining the factors that cause the distinct features in the cardiomyopathies will allow better management of the diseases and the development of specific therapies.

4. Diastolic Dysfunction Caused by cTnI Mutations Is Confirmed in Animal Studies

Using transgenic techniques to create disease mouse models is a valuable tool for confirming myofibril protein mutation-caused cardiac disorders in human patients and exploring the mechanisms of the diseases. Various HCM transgenic animal models with cTnI mutations have been developed. The first transgenic mouse model for cTnI mutation (cTnI R146G) was generated in 2000 (James J. 2000 Oct 27). This mutation is located in the inhibitory domain of that protein which is essential for the inhibitory function of cTnI. The 146Gly mice presented damaged diastolic function accompanied with myocardial disarray and interstitial fibrosis. Myocyte hypertrophy, myofiber disarray and interstitial fibrosis were also reported in HCM transgenic mice containing the cTnI-G203S mutation located in the C terminal domain of cTnI (Tsoutsman T. 2006 Oct). A Lys184 deletion located in the second actin-tropomyosin binding site of cTnI did not induce hypertrophy in the transgenic mice as opposed to the human patients carrying the same mutation (Kimura A. 1997 Aug) (Kokado H. 2000 Aug 8). Nevertheless cardiac relaxation kinetics was impaired in these animals (Iorga B. 2008 Mar 1). HCM transgenic mice containing the R21C mutation in the N-terminal extension of cTnI developed cardiac hypertrophy and fibrosis associated with activation of the fetal gene program with increased expression of atrial and brain natriuretic peptides and β myosin heavy chain (Wang Y. 2012 Jan 13). The phenotype of cTnI transgenic mice is consistent with the manifestation and the pathology of HCM in clinical patients and in other HCM animal models containing the disease-causing mutations in different sarcomeric proteins.

RCM causing cTnI mutations have overt effects on cardiac diastolic function. Transgenic animal models of RCM present diastolic abnormalities with normal systolic function. Our laboratory has been among the first to begin to define the effect of the RCM cTnI mutation on the development of diastolic dysfunction. We have created transgenic mice expressing human RCM mutation cTnI R192H (cTnI R193H in mouse sequence) and cTnI K178E (cTnI K179E in mouse sequence) in the heart (Du J. 2006 Dec 15) (Du J. 2008) (Li Y. 2010 Sep) (Figure 1). The most extensively studied RCM animal model, $cTnI^{193His}$, contains the cTnI R193H mutation and displayed cardiac diastolic dysfunction (Du J. 2006 Dec 15). These $cTnI^{193His}$

mice presented no cardiac hypertrophy or ventricular dilation. Bi-atrial enlargement was the main morphological change in the animals which is consistent with the clinical characteristics of the disease (Figure 2). Diastolic function was manifested by a reduced left ventricular end diastolic dimension and increased isovolumetric relaxation time (IVRT). The prolonged IVRT was the earliest parameter change in the animals and was followed at a later stage by an increased deceleration time and reduced E to A ratio in Doppler echocardiography. Systolic dysfunction while absent for the most part, became apparent at the end stage of the disease as the animals presented signs of congestive heart failure and died prematurely (Du J. 2008). Another mouse model developed by Wen et al. contained the RCM causing R145W cTnI mutation which is located at the same residue as HCM causing R145G and R145Q mutations (Wen Y. 2009 Oct 9) (Kimura A. 1997 Aug).

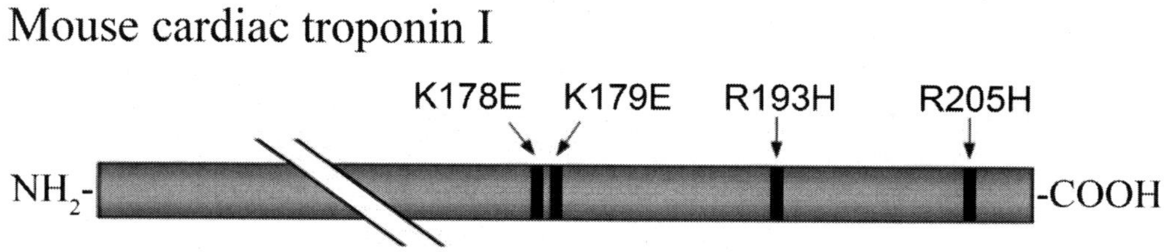

Figure 1. Cardiac troponin I C-terminal mutations studied in animal models. Red columns represent four different mutations on C-terminus of mouse cardiac troponin I generated in our laboratory. Among the four mutations on mouse cTnI, K179E, R193H and R205H are expressed corresponding to human RCM cTnI mutations K178E, E192H and R204H, respectively. Mouse cTnI K178E is generated due to its similarity with K179E mutation since both 178 and 179 amino acids are lysine and possibly play a similar role in forming protein structure.

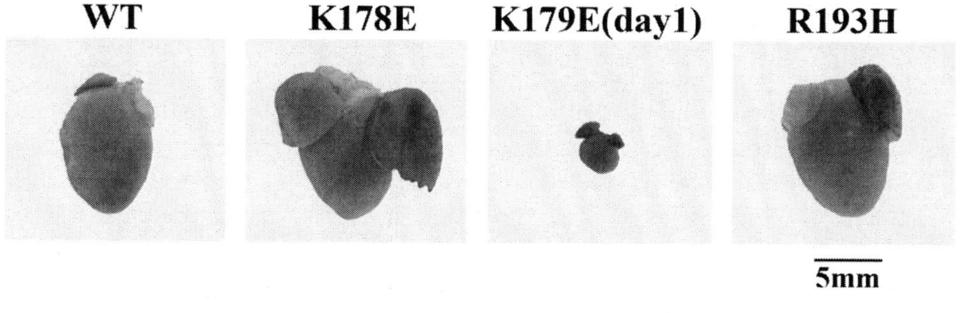

Figure 2. Morphological changes in the hearts from our RCM cTnI mutation mouse models. Hearts from 2 months old K178E and R193H transgenic mice or day 1 K179E mouse were harvested for morphological assessment. A dramatic bi-atrial enlargement was observed in the hearts of K178E and R193H transgenic mice. Heart from K179E at age of day 1 also shows an enlargement of atria.

Our laboratory developed as well two additional transgenic mice containing the RCM cTnI K178E mutation reported in Mogensen's study and a K177E mutation in the preceding codon in a different animal as a comparison (Mogensen J. 2003). As opposed to the K178E mutation, the K177E substitution is not found in human patient but was induced in transgenic animals to determine the specificity of the residues and mutation in the development of RCM. Both transgenic animals (cTnI K178E and K179E in the mouse genome) presented drastic bi-atrial enlargement in the absence of ventricular hypertrophy and dilation. They presented similar hemodynamic characteristics to the $cTnI^{193His}$ animals in our laboratory confirming the development of RCM as a consequence of cTnI mutation. The cardiac dysfunction was severe in the animals as most of them underwent a premature death. Our laboratory is also developing transgenic RCM mice containing the mutation R205H in the mouse genome corresponding to cTnI R204H mutation discovered in human RCM patients (Gambarin F. I. 2008 Oct) (Yang S. W. 2010 Oct). Analyzing these animals will allow determining the mechanism behind the restrictive physiology and designing treatment for the diseased animals.

While some cTnI mutations are associated with both HCM and RCM in clinical patients, the cardiomyopathy observed in the transgenic animals developed so far is very specific. No sign of hypertrophy is observed in the RCM animals and their cardiac dysfunction is very severe. This indicates that HCM and RCM are distinct diseases, although both HCM and RCM share a common cardiac dysfunction which is an impaired relaxation due to increased ventricular stiffness. The factors that account for presence or lack of ventricular morphological changes in HCM and RCM respectively are not well understood. It can be presumed that the dysfunction is more severe in RCM since it is associated with the worst clinical phenotype. Determining the functional changes induced by the mutations and the mechanisms underlying the development of the mutation-caused cardiomyopathies will allow the design of a common therapy for these myocardial diseases.

5. Mechanism Studies: Effect of cTnI Mutations on Myofibril Sensitivity for Ca^{2+}

Mutations in cTnI associated with cardiomyopathies were found to alter Ca^{2+} sensitivity of force production. Morimoto first suggested that Ca^{2+} desensitization of myofilament contraction was a primary mechanism for DCM by functional measurement of the deletion delta K210 in cTnT (Morimoto S. 2002 Jan 22). The reduced myofilament Ca^{2+} sensitivity in DCM was observed in myocytes containing various mutated sarcomeric proteins such as cTnT, TnC, and alpha-tropomyosin (Mirza M. 2005 Aug 5) (Karam C. N. 2011 Mar) (Lakdawala N. K. 2010 Jan 26). Also, analysis of the DCM-causing cTnI mutations K36Q and N185K confirmed that the mutant proteins conferred lower maximum ATPase rates and lower Ca^{2+} sensitivity of force production (Carballo S. 2009 Aug 14). Nevertheless it is worth noting that some DCM causing mutations induce conflicting results regarding the relation between reduced myofilament Ca^{2+} sensitivity and DCM (Venkatraman G. 2003 Oct 24). This indicates that such aberration is not the only mechanism behind the systolic dysfunction that characterizes the disease. Such irregularity has not been reported yet in DCM causing-cTnI mutations.

Myofilament Ca^{2+} hypersensitivity is a major feature of HCM and RCM-linked myofibril mutations. The HCM mutations R145G, R145Q, R162W, DeltaK183 and K206Q were found to increase the Ca^{2+}-sensitivity of myofibrillar ATPase activity and force generation in skinned muscle fibers. The G203S mutation while showing a tendency to increase the Ca^{2+} sensitivity in both myofibrils and skinned muscle fibers, did not present statistically significant difference compared to wild-type cTnI (Takahashi-Yanaga F. 2001 Dec). The Ca^{2+} sensitizing effect of HCM associated mutations is observed in various sarcomeric protein mutants confirming that this functional alteration is characteristic of the disease (Morimoto S. 1998 Jul) (Chang A. N. 2005 Oct 7) (Fraysse B. 2012 Jun). Myofilament Ca^{2+} hypersensitivity implies a greater myofilament response to Ca^{2+} binding to TnC. Therefore not as much cytosolic Ca^{2+} is necessary to stimulate contraction and generate force. This increases the probability of the myofilament to become contracted while limiting relaxation. This physiological aberration coincides with the hyper-contractility and diastolic dysfunction observed in the HCM heart.

Drastic myofilament Ca^{2+} hyper-sensitization was observed on assembled human and mice RCM myofibers containing cTnI mutations. The RCM-associated mutations are unable to properly inhibit actomyosin ATPase activity and have a greater calcium sensitivity of force development (Yumoto F. 2005 Dec) (Gomes A. V. 2005 Sep 2). This feature is characteristic of RCM and observed in various sarcomeric protein mutations (Pinto J. R. 2008 Jan 25). Ca^{2+} hypersensitivity observed in RCM is a feature that is shared with HCM and is opposite to that found in DCM which often displays reduced myofibril Ca^{2+} sensitivity (Morimoto S. 2002 Jan 22) (Robinson P. 2002 Oct 25) (Lu Q. W. 2003 Dec) (Mirza M. 2005 Aug 5) (Du C. K. 2007 Jul 20). Nevertheless, RCM causing cTnI mutations are found to confer a greater Ca^{2+} sensitivity on the myofilament than the sarcomeric protein mutations associated with HCM (Yumoto F. 2005 Dec) (Wen Y. 2009 Oct 9). This may account for the more severe cardiac dysfunction and worst prognosis associated with RCM.

The increase in Ca^{2+} sensitivity in RCM is also linked with greater affinity of TnC to Ca^{2+} in assembled RCM myofibers as reported (Kobayashi T. 2006 May 12). While the affinity of TnC to Ca^{2+} was increased in both HCM and RCM myofibers, no significant difference was observed between them (Kobayashi T. 2006 May 12). A recent study demonstrated that the changes in Ca^{2+} sensitivity were related with modifications in binding kinetics of Ca^{2+} to TnC. In RCM myofilaments, the dissociation rate of Ca^{2+} from TnC was decreased while the Ca^{2+} association rate to TnC was increased (Liu B. 2012). Such attributes indicate that, in RCM, Ca^{2+} has a tendency to bind to and remain longer on TnC. This increase of TnC affinity to Ca^{2+} may account for the myofilament hypersensitivity to Ca^{2+} observed in HCM and RCM myocardium. Thus, the cardiomyopathy causing mutations, by altering the affinity of cTnI for actin favor interactions between cTnI and TnC and maintain the myofilament in the Ca^{2+} activated state which will facilitate myocardial contraction. The inhibitory function of cTnI is compromised as a result and relaxation becomes minimal.

The increase of Ca^{2+} sensitivity was reflected at the cellular level with changes in Ca^{2+} cycling kinetics and sarcomere relaxation rate and amplitude. In various studies, HCM and RCM cardiomyocytes present shorter end diastolic dimension, slower sarcomere relaxation rates and delayed Ca^{2+} transient decay. They generally show normal sarcomere shortening or peak systolic Ca^{2+} amplitudes indicating no systolic dysfunction (Knollmann B. C. 2003 Mar 7) (Davis J. 2008 May) (Li Y. 2010 Sep). The drastic changes in myofilaments kinetics and Ca^{2+} dynamics in HCM and RCM associated with the increased myofilament Ca^{2+} sensitivity indicates that Ca^{2+} remains longer in the myofilament, inducing actomyosin cross-bridge formation. Thus, therapies for diastolic dysfunction in HCM and RCM should aim at modifying the Ca^{2+} sensitivity in the diseased hearts in order to improve cardiac relaxation. The restoration of the myofilament Ca^{2+} sensitivity in the mutant myofibers will set in motion physiological changes that will lead to the correction of the dramatic aberrations in cardiac function observed in the cardiomyopathic heart.

6. Alteration of Cardiomyofibril Ca^{2+} Sensitivity As a Therapeutic Option for Cardiomyopathies Due to Myofibril Protein Mutations

6.1. Compounds That Increase Myofilament Ca^{2+} Sensitivity

Systolic dysfunction in DCM is linked with decreased myofilament Ca^{2+} sensitivity while a myofilament hypersensitivity to Ca^{2+} accounts for the diastolic dysfunction in HCM and RCM. Therefore, treatment for these cardiomyopathies should aim at restoring the altered myofilament Ca^{2+} sensitivity since this functional aberration is a major characteristic of these diseases. This approach is more direct and therefore potentially more efficient than affecting Ca^{2+} homeostasis through drugs like catecholamines and Ca^{2+} channel antagonists that target Ca^{2+} handling proteins. Different components of the sarcomere (thin and thick filament proteins) can be targeted to correct the Ca^{2+} sensitivity and cardiac function. The troponin complex offers a lot of advantages due to the crucial role that its subunits play in the regulation of cardiac contraction and relaxation. The structure and conformation of the troponin complex has been extensively studied and compounds affecting Ca^{2+} sensitivity by their interaction with cTnC have been developed. By affecting calcium affinity to TnC or the interactions of cTnI with its associated molecules, myofilament cross-bridge kinetics and cardiac function can be modified.

Compounds that directly or indirectly modify cTnI-TnC interactions are being explored for therapeutic development. Those that enhance myofilaments' response to Ca^{2+} binding (Ca^{2+} sensitizers) may be used for the treatment of systolic dysfunction, while those that reduce the myofibril sensitivity for Ca^{2+} (Ca^{2+} desensitizers) will be beneficial for diastolic function. Ca^{2+} sensitizers can operate by increasing the binding affinity of TnC to Ca^{2+}, stabilizing the Ca^{2+} bound conformation of TnC, or facilitating actomyosin cross-bridge formation (Endoh M. 2008 Dec). Ca^{2+} desensitizers are believed to have the opposite effect which will reduce the affinity of TnC to Ca^{2+}, limit cTnI-TnC interactions or inhibit actomyosin cross-bridges. Levosimendan, the most studied Ca^{2+} sensitizer, is known to improve contractile function in cases of ischemia/ reperfusion, myocardial stunning, and acute heart failure (García González M. J. 2006). This compound binds to the hydrophobic pocket of the N-terminal domain of cTnC in a Ca^{2+} dependent manner (Pollesello P. 1994 Nov 18) (Robertson I. M. 2008 Jul 15). While, it has not been tested in transgenic animal models of DCM, another Ca^{2+} sensitizer, pimobendan that also has a selective phosphodiesterase III inhibitory action, was shown to rescue a DCM animal model containing cTnT K210del mutation (Morimoto S. 2002 Jan 22). More studies on the actions of Ca^{2+} sensitizers on different sarcomeric mutations should be performed to establish their therapeutic effect for DCM.

6.2. Biological Factors That Decrease Myofilament Sensitivity for Ca^{2+}

While different sensitizing agents are being considered for therapeutic use, Ca^{2+} desensitizers that improve diastolic dysfunction are less established. It is known that cTnI phosphorylation at residues 23 and 24 under β adrenergic stimulation facilitates cardiac muscle relaxation and reduces myofilament Ca^{2+} sensitivity in rodents (Kentish J. C. 2001 May 25) (Burkart E. M. 2003). Such action is confirmed in transgenic mice containing rat cTnI with aspartate replacing the ser 23 & 24 to mimic phosphorylation. These animals show reduced myofilament Ca^{2+} sensitivity while other mouse models, in which the serine residues are replaced with non-phosphorylable alanine, display neither reduction of calcium sensitivity nor positive lusotropic response under beta adrenergic stimulation (Sakthivel S. 2005) (Ramirez-Correa G. A. 2010 May). Functional studies in RCM cardiomyocytes transduced with mutant cTnI showed that the diseased myofilament can still be desensitized by PKA mediated serine phosphorylation and their relaxation enhanced (Davis J. 2008 May). Clinical drugs such as β adrenoceptor agonists, phosphodiesterase 3 inhibitors mediate the phosphorylation of these serines residues inducing a lusotropic effect (Robertson S. P. 1982 Jan 10) (Zhang R. 1995 Jun) (Kranias E. G. 1985 Feb 15). Nevertheless, their inotropic and chronotropic effects secondary to their actions on Ca^{2+} handling proteins increase the cytosolic Ca^{2+} content and the workload of the heart (Endoh 2003 May). Therefore, they are detrimental for the treatment of diastolic dysfunction.

6.3. Ca^{2+} Desensitization via Alteration of Myofilament Proteins

Several attempts have been done to improve diastolic dysfunction by using Ca^{2+} desensitizing proteins to replace the Ca^{2+} hyper-sensitizing mutant proteins. At least these trials are feasible so far in experimental animals which not only confirm the key pathological feature of Ca^{2+} hypersensitivity in diastolic dysfunction and heart failure, but also explore a novel gene therapeutic option for the future in the treatment of myocardial disorders. Reduction of myofilament Ca^{2+} sensitivity was achieved in a HCM mice model carrying a tropomyosin mutation. The HCM mice presenting increased Ca^{2+} sensitivity and diastolic dysfunction were mated with chimeric α-/β-tropomyosin transgenic mice displaying decreased myofilament Ca^{2+} sensitivity. The myofilament Ca^{2+} sensitivity was restored in these double transgenic mice accompanied with improved cardiac function, decreased fibrosis and reduced myocardial disarray (Jagatheesan G. 2007 Aug). In another study, cross-breeding of transgenic mice presenting diastolic dysfunction due to a cTnI mutation with a cTnT animal model with depressed systolic function led to the mutual rescue of their phenotype (Wei B. 2010 Sep 3). These show that alteration of the dynamics of the proteins in the cTnI complex had a potential therapeutic role for cardiac dysfunction.

A modified cTnI molecule was found to possess Ca^{2+} desensitizing properties. The molecule lacked the N-terminal extension of cTnI that is specific for the cardiac isoform of that protein (Barbato J. C. 2005 Feb 25). Proteolysis of the N terminal of cTnI can occur as a functional adaptation in simulated microgravity and also under physiological and stress conditions (Yu Z. B. 2001 May 11) (McConnell B. K. 2009 Jan 16) (Biesiadecki B. J. 2010 Jun 18). The phosphorylable serine residues 23 and 24 are eliminated because of the truncation and the interaction of cTnI with TnC altered. The NH2 terminal truncated cTnI presented similar properties as wild type (WT) cTnI when submitted to β adrenergic stimulation (Barbato J. C. 2005 Feb 25). The myofilament Ca^{2+} sensitivity was decreased in the presence of that molecule and diastolic function improved.

The therapeutic effect of this cTnI desensitizing molecule was investigated in our RCM $cTnI^{193His}$ mice presenting damaged diastolic function. Li et al. developed double transgenic mice (Double-TG) expressing both cTnI-ND and the RCM causing transgenic protein ($cTnI^{193His}$) in a wild type null background (Li Y. 2010 Sep). The expression of the RCM-causing cTnI mutant molecule was the same in transgenic and double transgenic mice but the cTnI-ND desensitizing molecule completely replaced the wild type cTnI protein in the Double-TG. Ca^{2+} sensitivity was restored in the animals accompagnied with improved diastolic function and increased lifespan. Such findings confirm that calcium desensitization is a promising alternative to restore diastolic function in RCM hearts and highlight the need to further explore this therapeutic option. While cTnI-ND provides a good gene therapy option for RCM, the search for chemical Ca^{2+} desensitizers is not obsolete since they may help compensate for the limitations in gene therapy techniques and the dependency of cTnI-ND therapeutic effect on its incorporation rate on the myofilament.

6.4. Small Molecule and Chemical Ca^{2+} Desensitizer

Finding Ca^{2+} desensitizers that primarily affect myofilament dynamics is necessary. So far, compound with such properties are very scarce. Myosin inhibitors such as blebbistatin and 2, 3-butanedione monoxime (BDM) may alter myofilament sensitivity to Ca^{2+} via their inhibitory effect on actomyosin cross-bridge formations. These myosin ATPase inhibitors while useful in functional studies in *vitro* and ex *vivo*, are too toxic for therapeutic use (Gwathmey J. K. 1991 Nov) (Kettlewell S. 2004 Mar) (Dou Y. 2007 Sep). There is a great need to develop or find small molecules and chemical Ca^{2+} desensitizers that can be used to alter myofibril sensitivity for Ca^{2+} and myofilament sliding kinetics. A great concern about biological agents is their non-toxicity and their good bioavailability but there are few Ca^{2+} desensitizers possessing such qualities. The catechin, (-)-epigallocatechin-3-gallate (EGCg) has been newly found to possess Ca^{2+} desensitizing abilities via its interaction with cTnC (Liou Y. M. 2008 Aug) (Robertson I. M. 2009 Aug 21). This compound is the most abundant catechin in green tea and is credited for the numerous health benefits attributed to green tea consumption (Higdon J. V. 2003). EGCg desensitizes myofilament to Ca^{2+} by forming a ternary complex with the C-terminal domain of troponin C and the anchoring region of cTnI (Robertson I. M. 2009 Aug 21). The affinity of TnC to Ca^{2+} is reduced as a result which facilitates cardiac relaxation. The ability of EGCg to correct myofilament Ca^{2+} hypersensitivity and diastolic dysfunction has been demonstrated in a HCM mouse model and in our RCM $cTnI^{193His}$ mice confirming the therapeutic potential of that compound for diastolic dysfunction (Tadano N. 2010 Nov) (Zhang L. 2012 June 2-5).

Conclusion

Mutations in conserved regions in the C terminal of cTnI induce changes in the conformation of the troponin complex, compromising myofilaments cross-bridge kinetics, resulting in cardiac dysfunction. The different cardiomyopathies that develop from these mutations present specific functional aberrations attributed to damaged myofilament Ca^{2+}

sensitivity. Decreased Ca^{2+} sensitivity or increased Ca^{2+} sensitivity caused by different troponin mutations are the mechanical feature for systolic dysfunction or diastolic dysfunction, which has been confirmed in clinical patients and in experimental animal studies. At this regard, using Ca^{2+} sensitizers and desensitizers that interact with specific myofilament proteins can help restore the cardiac dysfunction found in DCM, HCM and RCM. The TnC-cTnI interface which is the target of different cardiotonic agents highlights the unique role of the troponin complex in the regulation of muscle contraction and relaxation. The use of Ca^{2+} desensitizers for the treatment of HCM and RCM is a particularly novel therapeutic approach since both diseases share a common pathology of the increased myofilament Ca^{2+} sensitivity and diastolic dysfunction. Since the therapies for these myocardial diseases are very limited, such therapeutic strategy that takes into account a common physiological mechanism behind the development of the diseases deserves more investigation. Ca^{2+} desensitizers, even though they cannot correct the genetic mutations that cause these cardiomyopathies, may alter the myofilament response to Ca^{2+} and therefore cardiac contractility. It is needed to set up a high throughput screening test for small molecules and chemical desensitizers. Then trial of these agents on animal models presenting various cardiomyopathies will permit determining their therapeutic potentials and the development of new drugs based on the basic structure of these chemicals.

Acknowledgment

This work was supported by grants from the NIH (HL112130-01), NSFC (31271218) and the American Heart Association (11GRNT7860030) to X. P. Huang. P. Y. Jean-Charles is a recipient of William Townsend Porter Pre-doctoral Fellowship from the American Physiological Society.

References

Ammash N. M., Seward J. B., Bailey K. R., Edwards W. D., Tajik A. J. Clinical profile and outcome of idiopathic restrictive cardiomyopathy. *Circulation*, 2000 May 30: 101(21): 2490-6.

Antman E. M., Tanasijevic M. J., Thompson B., Schactman M., McCabe C. H., Cannon C. P., Fischer G. A., Fung A. Y., Thompson C., Wybenga D., Braunwald E. Cardiac-specific troponin I levels to predict the risk of mortality in patients with acute coronary syndromes. *N Engl J Med*, 1996 Oct 31: 335(18):1342-9.

Arbustini E., Pasotti M., Pilotto A., Pellegrini C., Grasso M., Previtali S., Repetto A., Bellini O., Azan G., Scaffino M., Campana C., Piccolo G., Viganò M., Tavazzi L. Desmin accumulation restrictive cardiomyopathy and atrioventricular block associated with desmin gene defects. *Eur J Heart Fail*, 2006 Aug: ;8(5):477-83.

Barbato J. C., Huang Q. Q., Hossain M. M., Bond M., Jin J. P. Proteolytic N-terminal truncation of cardiac troponin I enhances ventricular diastolic function. *J Biol Chem.,* 2005 Feb 25: 280(8):6602-9.

Biesiadecki B. J., Tachampa K., Yuan C., Jin J. P., de Tombe P. P., Solaro R. J. Removal of the cardiac troponin I N-terminal extension improves cardiac function in aged mice. *J Biol Chem.,* 2010 Jun 18: 285(25):19688-98.

Bonow R. O., Udelson J. E. Left ventricular diastolic dysfunction as a cause of congestive heart failure. Mechanisms and management. *Ann Intern Med*, 1992 Sep 15: 117(6):502-10.

Braunwald E. The management of heart failure: the past, the present, and the future. *Circulation*, 2008: 1(1): 58-62.

Burkart E. M., Sumandea M. P., Kobayashi T., Nili M., Martin A. F., Homsher E., Solaro R. J. Phosphorylation or glutamic acid substitution at protein kinase C sites on cardiac troponin I differentially depress myofilament tension and shortening velocity. *J Biol Chem.,* 2003: 278(13):11265-72.

Carballo S., Robinson P., Otway R., Fatkin D., Jongbloed J. D., de Jonge N., Blair E., van Tintelen J. P., Redwood C., Watkins H. Identification and functional characterization of cardiac troponin I as a novel disease gene in autosomal dominant dilated cardiomyopathy. *Circ Res*, 2009 Aug 14: 105(4):375-82.

Chang A. N., Harada K., Ackerman M. J., Potter J. D. Functional consequences of hypertrophic and dilated cardiomyopathy-causing mutations in alpha-tropomyosin. *J Biol Chem*, 2005 Oct 7: 280(40):34343-9.

Chen H. H., Lainchbury J. G., Senni M., Bailey K. R., Redfield M. M. Diastolic heart failure in the community: clinical profile, natural history, therapy, and impact of proposed diagnostic criteria. *J Card Fail*, 2002 Oct: 8(5):279-87.

Curila K., Benesova L., Penicka M., Minarik M., Zemanek D., Veselka J., Widimsky P., Gregor P. Spectrum and clinical manifestations of mutations in genes responsible for hypertrophic cardiomyopathy. *Acta Cardiol*, 2012 Feb: 67(1):23-9.

Davis J., Wen H., Edwards T., Metzger J. M. Allele and species dependent contractile defects by restrictive and hypertrophic cardiomyopathy-linked troponin I mutants. *J Mol Cell Cardiol*, 2008 May: 44(5):891-904.

Dong W. J., Xing J., Villain M., Hellinger M., Robinson J. M., Chandra M., Solaro R. J., Umeda P. K., Cheung H. C. Conformation of the regulatory domain of cardiac muscle troponin C in its complex with cardiac troponin I. *J Biol Chem*, 1999 Oct 29: 274(44): 31382-90.

Doolan A., Tebo M., Ingles J., Nguyen L., Tsoutsman T., Lam L., Chiu C., Chung J., Weintraub R. G., Semsarian C. Cardiac troponin I mutations in Australian families with hypertrophic cardiomyopathy: clinical, genetic and functional consequences. *J Mol Cell Cardiol.,* 2005 Feb: 38(2):387-93.

Dou Y., Arlock P., Arner A. Blebbistatin specifically inhibits actin-myosin interaction in mouse cardiac muscle. *Am J Physiol Cell Physiol*, 2007 Sep: 293(3):C1148-53.

Du C. K., Morimoto S., Nishii K., Minakami R., Ohta M., Tadano N., Lu Q. W., Wang Y. Y., Zhan D. Y., Mochizuki M., Kita S., Miwa Y., Takahashi-Yanaga F., Iwamoto T., Ohtsuki I., Sasaguri T. Knock-in mouse model of dilated cardiomyopathy caused by troponin mutation. *Circ Res*, 2007 Jul 20: 101(2):185-94.

Du J., Liu J., Feng H. Z., Hossain M. M., Gobara N., Zhang C., Li Y., Jean-Charles P. Y., Jin J. P., Huang X. P. Impaired relaxation is the main manifestation in transgenic mice expressing a restrictive cardiomyopathy mutation, R193H, in cardiac TnI. *Am J Physiol Heart Circ Physiol.,* 2008: 294(6):H2604-13.

Du J., Zhang C., Liu J., Sidky C., Huang X. P. A point mutation (R192H) in the C-terminus of human cardiac troponin I causes diastolic dysfunction in transgenic mice. *Arch Biochem Biophys.,* 2006 Dec 15: 456(2):143-50.

Endoh M. Cardiac Ca^{2+} signaling and Ca^{2+} sensitizers. *Circ J*, 2008 Dec: 72(12):1915-25.

Endoh M. The therapeutic potential of novel cardiotonic agents. *Expert Opin Investig Drugs*, 2003 May: 12(5):735-50.

Feng J., Schaus B. J., Fallavollita J. A., Lee T. C., Canty J. M. Jr. Preload induces troponin I degradation independently of myocardial ischemia. *Circulation*, 2001 Apr 24: 103(16): 2035-7.

Fraysse B., Weinberger F., Bardswell S. C., Cuello F., Vignier N., Geertz B., Starbatty J., Krämer E., Coirault C., Eschenhagen T., Kentish J. C., Avkiran M., Carrier L. Increased myofilament Ca^{2+} sensitivity and diastolic dysfunction as early consequences of Mybpc3 mutation in heterozygous knock-in mice. *Mol Cell Cardiol*, 2012 Jun: 52(6):1299-307.

Galińska A., Hatch V., Craig R., Murphy A. M., van Eyk J. E., Wang C. L., Lehman W., Foster D. B. The C terminus of cardiac troponin I stabilizes the Ca^{2+}-activated state of tropomyosin on actin filaments. *Circ Res*, 2010 Mar 51: 106(4):705-11.

Gambarin F. I., Tagliani M., Arbustini E. Pure restrictive cardiomyopathy associated with cardiac troponin I gene mutation: mismatch between the lack of hypertrophy and the presence of disarray. *Heart*, 2008 Oct: 94(10):1257.

García González M. J., Domínguez Rodríguez A. Pharmacologic treatment of heart failure due to ventricular dysfunction by myocardial stunning: potential role of levosimendan. *Am J Cardiovasc Drugs*, 2006: 6(2):69-75.

Girolami F., Ho C. Y., Semsarian C., Baldi M., Will M. L., Baldini K., Torricelli F., Yeates L., Cecchi F., Ackerman M. J., Olivotto I. Clinical features and outcome of hypertrophic cardiomyopathy associated with triple sarcomere protein gene mutations. *J Am Coll Cardiol*, 2010 Apr 6: 55(14):1444-53.

Go A. S., Mozaffarian D., Roger V. L., Benjamin E. J., Berry J. D., Borden W. B., Bravata D. M., Dai S., Ford E. S., Fox C. S., Franco S., Fullerton H. J., Gillespie C., Hailpern S. M., Heit J. A., Howard V. J., Huffman M. D., Kissela B. M., Kittner S. J., Lackland D. T., Lichtman J. H., Lisabeth L. D., Magid D. Heart Disease and Stroke Statistics--2013 Update A Report From the American Heart Association. *Circulation*, 2013: 127: e6-e245.

Gomes A. V., Harada K., Potter J. D. A mutation in the N-terminus of troponin I that is associated with hypertrophic cardiomyopathy affects the $Ca(^{2+})$-sensitivity, phosphorylation kinetics and proteolytic susceptibility of troponin. *J Mol Cell Cardiol*, 2005 Nov: 39(5):754-65.

Gomes A. V., Liang J., Potter J. D. Mutations in Human Cardiac Troponin I That Are Associated with Restrictive Cardiomyopathy Affect Basal ATPase Activity and the Calcium Sensitivity of Force Development. *J Biol Chem.,* 2005 Sep 2: 280(35):30909-15.

Gwathmey J. K., Hajjar R. J., Solaro R. J. Contractile deactivation and uncoupling of crossbridges. Effects of 2,3-butanedione monoxime on mammalian myocardium. *Circ Res*, 1991 Nov: 69(5):1280-92.

Higdon J. V., Frei B. Tea catechins and polyphenols: health effects, metabolism, and antioxidant functions. *Crit Rev Food Sci Nutr*, 2003: 43(1):89-143.

Huang X., Pi Y., Lee K. J., Henkel A. S., Gregg R. G., Powers P. A., Walker J. W. Cardiac troponin I gene knockout: a mouse model of myocardial troponin I deficiency. *Circ Res*, 1999 Jan 8-22: 84(1):1-8.

Huang X. P., Du J. F. Troponin I, cardiac diastolic dysfunction and restrictive cardiomyopathy. *Acta Pharmacol Sin*, 2004 Dec: 25(12):1569-75.

Iorga B., Blaudeck N., Solzin J., Neulen A., Stehle I., Lopez Davila A. J., Pfitzer G., Stehle R. Lys184 deletion in troponin I impairs relaxation kinetics and induces hypercontractility in murine cardiac myofibrils. *Cardiovasc Res*, 2008 Mar 1: 77(4):676-86.

Jagatheesan G., Rajan S., Petrashevskaya N., Schwartz A., Boivin G., Arteaga G. M., Solaro R. J., Liggett S. B., Wieczorek D. F. Rescue of tropomyosin-induced familial hypertrophic cardiomyopathy mice by transgenesis. *Am J Physiol Heart Circ Physiol*, 2007 Aug: 293(2):H949-58.

James J., Zhang Y., Osinska H., Sanbe A., Klevitsky R., Hewett T. E., Robbins J. Transgenic modeling of a cardiac troponin I mutation linked to familial hypertrophic cardiomyopathy. *Circ Res*, 2000 Oct 27: 87(9):805-11.

Karam C. N., Warren C. M., Rajan S., de Tombe P. P., Wieczorek D. F., Solaro R. J. Expression of tropomyosin κ induces dilated cardiomyopathy and depresses cardiac myofilament tension by mechanisms involving cross-bridge dependent activation and altered tropomyosin phosphorylation. *J Muscle Res Cell Motil*, 2011 Mar: 31(5-6):315-22.

Kaski J. P., Syrris P., Burch M., Tomé-Esteban M. T., Fenton M., Christiansen M., Andersen P. S., Sebire N., Ashworth M., Deanfield J. E., McKenna W. J., Elliott P. M. Idiopathic restrictive cardiomyopathy in children is caused by mutations in cardiac sarcomere protein genes. *Heart*, 2008 Nov: 94(11):1478-8.

Kaski J. P., Syrris P., Esteban M. T., Jenkins S., Pantazis A., Deanfield J. E., McKenna W. J., Elliott P. M. Prevalence of sarcomere protein gene mutations in preadolescent children with hypertrophic cardiomyopathy. *Circ Cardiovasc Genet*, 2009 Oct: 2(5):436-41.

Katritsis D., Wilmshurst P. T., Wendon J. A., Davies M. J., Webb-Peploe M. M. Primary restrictive cardiomyopathy: clinical and pathologic characteristics. *J Am Coll Cardiol.*, 1991 Nov 1: 18(5):1230-5.

Kentish J. C., McCloskey D. T., Layland J., Palmer S., Leiden J. M., Martin A. F., Solaro R. J. Phosphorylation of troponin I by protein kinase A accelerates relaxation and crossbridge cycle kinetics in mouse ventricular muscle. *Circ Res*, 2001 May 25;88(10): 1059-65.

Kettlewell S., Walker N. L., Cobbe S. M., Burton F. L., Smith G. L. The electrophysiological and mechanical effects of 2,3-butane-dione monoxime and cytochalasin-D in the Langendorff perfused rabbit heart. *Exp Physiol*, 2004 Mar: 89(2):163-72.

Kimura A., Harada H., Park J. E., Nishi H., Satoh M., Takahashi M., Hiroi S., Sasaoka T., Ohbuchi N., Nakamura T., Koyanagi T., Hwang T. H., Choo J. A., Chung K. S., Hasegawa A., Nagai R., Okazaki O., Nakamura H., Matsuzaki M., Sakamoto T., Toshima H., Koga Y., Imaizumi T., Sasazuki T. Mutations in the cardiac troponin I gene associated with hypertrophic cardiomyopathy. *Nat Genet*, 1997 Aug: 16(4):379-82.

Kleerekoper Q., Howarth J. W., Guo X., Solaro R. J., Rosevear P. R. Cardiac troponin I induced conformational changes in cardiac troponin C as monitored by NMR using site-directed spin and isotope labeling. *Biochemistry*, 1995 Oct 17: 34(41):13343-52.

Knollmann B. C., Kirchhof P., Sirenko S. G., Degen H., Greene A. E., Schober T., Mackow J. C., Fabritz L., Potter J. D., Morad M. Familial hypertrophic cardiomyopathy-linked mutant troponin T causes stress-induced ventricular tachycardia and Ca^{2+}-dependent action potential remodeling. *Circ Res*, 2003 Mar 7: 92(4):428-36.

Kobayashi T., Solaro R. J. Increased Ca^{2+} affinity of cardiac thin filaments reconstituted with cardiomyopathy-related mutant cardiac troponin I. *J Biol Chem.,* 2006 May 12:281(19): 13471-.

Kokado H., Shimizu M., Yoshio H., Ino H., Okeie K., Emoto Y., Matsuyama T., Yamaguchi M., Yasuda T., Fujino N., Ito H., Mabuchi H. Clinical features of hypertrophic cardiomyopathy caused by a Lys183 deletion mutation in the cardiac troponin I gene. *Circulation*, 2000 Aug 8: 102(6):663-9.

Kranias E. G., Garvey J. L., Srivastava R. D., Solaro R. J., Solaro R. J. Phosphorylation and functional modification of sarcoplasmic reticulum and myofibrils in isolated rabbit heart stimulated with isoprenaline. *Biochem J.,* 1985 Feb 15: 226(1):113-21.

Krudy G. A., Kleerekoper Q., Guo X., Howarth J. W., Solaro R. J., Rosevear P. R. NMR studies delineating spatial relationships within the cardiac troponin I-troponin C complex. *J Biol Chem*, 1994 Sep 23: 269(38):23731-5.

Kushwaha S. S., Fallon J. T., Fuster V. *Restrictive cardiomyopathy.* 1997 Jan 23: 336(4):267-76, N Engl J Med.

Lakdawala N. K., Dellefave L., Redwood C. S., Sparks E., Cirino A. L., Depalma S., Colan S. D., Funke B., Zimmerman R. S., Robinson P., Watkins H., Seidman C. E., Seidman J. G., McNally E. M., Ho C. Y. Familial dilated cardiomyopathy caused by an alpha-tropomyosin mutation: the distinctive natural history of sarcomeric dilated cardiomyopathy. *J Am Coll Cardiol*, 2010 Jan 26: 55(4):320-9.

Li M. X., Spyracopoulos L., Sykes B. D. Binding of cardiac troponin-I147-163 induces a structural opening in human cardiac troponin-C. *Biochemistry*, 1999 Jun 29: 38(26):8289-98.

Li Y., Charles P. Y., Nan C., Pinto J. R., Wang Y., Liang J., Wu G., Tian J., Feng H. Z., Potter J. D., Jin J. P., Huang X. Correcting diastolic dysfunction by Ca^{2+} desensitizing troponin in a transgenic mouse model of restrictive cardiomyopathy. *J Mol Cell Cardiol.,* 2010 Sep: 49(3):402-11.

Liou Y. M., Kuo S. C., Hsieh S. R. Differential effects of a green tea-derived polyphenol (-)-epigallocatechin-3-gallate on the acidosis-induced decrease in the Ca(2+) sensitivity of cardiac and skeletal muscle. *Pflugers Arch*, 2008 Aug: 456(5):787-800.

Lipshultz S. E., Sleeper L. A., Towbin J. A., Lowe A. M., Orav E. J., Cox G. F., Lurie P. R., McCoy K. L., McDonald M. A., Messere J. E., Colan S. D. The incidence of pediatric cardiomyopathy in two regions of the United States. *N Engl J Med.,* 2003 Apr 24:348 (17):1647-55.

Liu B., Tikunova S. B., Kline K. P., Siddiqui J. K., Davis J. P. Disease-related cardiac troponins alter thin filament Ca^{2+} association and dissociation rates. *PLoS One*, 2012: 7(6):e38259.

Liu J., Du J., Zhang C., Walker J. W., Huang X. Progressive troponin I loss impairs cardiac relaxation and causes heart failure in mice. *Am J Physiol Heart Circ Physiol*, 2007 Aug: 293(2):H1273-81.

Lu Q. W., Morimoto S., Harada K., Du C. K., Takahashi-Yanaga F., Miwa Y., Sasaguri T., Ohtsuki I. Cardiac troponin T mutation R141W found in dilated cardiomyopathy stabilizes the troponin T-tropomyosin interaction and causes a Ca^{2+} desensitization. *J Mol Cell Cardiol*, 2003 Dec: 35(12):1421-7.

Malcić I., Jelusić M., Kniewald H., Barisić N., Jelasić D., Bozikov J. Epidemiology of cardiomyopathies in children and adolescents: a retrospective study over the last 10 years. *Cardiol Young.*, 2002 May: 12(3):253-9.

Maron B. J., Maron M. S., Semsarian C. Double or compound sarcomere mutations in hypertrophic cardiomyopathy: a potential link to sudden death in the absence of conventional risk factors. *Heart Rhythm*, 2012 Jan: 9(1):57-63.

Maron B. J., Towbin J. A., Thiene G., Antzelevitch C., Corrado D., Arnett D., Moss A. J., Seidman C. E., Young J. B. Contemporary definitions and classification of the cardiomyopathies: an American Heart Association Scientific Statement from the Council on Clinical Cardiology, Heart Failure and Transplantation Committee; Quality of Care and Outcomes Research and Functio. *Circulation*, 2006 Apr 11: 113(14):1807-16.

Maskatia S. A., Decker J. A., Spinner J. A., Kim J. J., Price J. F., Jefferies J. L., Dreyer W. J, Smith E. O., Rossano J. W., Denfield S. W. Restrictive physiology is associated with poor outcomes in children with hypertrophic cardiomyopathy. *Pediatr Cardiol*, 2012 Jan: 33(1):141-9.

McConnell B. K., Popovic Z., Mal N., Lee K., Bautista J., Forudi F., Schwartzman R., Jin J. P., Penn M., Bond M. Disruption of protein kinase A interaction with A-kinase-anchoring proteins in the heart *in vivo:* effects on cardiac contractility, protein kinase A phosphorylation, and troponin I proteolysis. *J Biol Chem.*, 2009 Jan 16: 284(3):1583-92.

McDonough J. L., Labugger R., Pickett W., Tse M. Y., MacKenzie S., Pang S. C., Atar D., Ropchan G., van Eyk J. E. Cardiac troponin I is modified in the myocardium of bypass patients. *Circulation*, 2001 Jan 2: 103(1):58-64.

Mirza M., Marston S., Willott R., Ashley C., Mogensen J., McKenna W., Robinson P., Redwood C., Watkins H. Dilated cardiomyopathy mutations in three thin filament regulatory proteins result in a common functional phenotype. *J Biol Chem*, 2005 Aug 5: 280(31):28498-506.

Mogensen J., Kubo T., Duque M., Uribe W., Shaw A., Murphy R., Gimeno J. R., Elliott P., McKenna W. J. Idiopathic restrictive cardiomyopathy is part of the clinical expression of cardiac troponin I mutations. *J Clin Invest.*, 2003: 111(2):209-16.

Mogensen J., Murphy R. T., Kubo T., Bahl A., Moon J. C., Klausen I. C., Elliott P. M., McKenna W. J. Frequency and clinical expression of cardiac troponin I mutations in 748 consecutive families with hypertrophic cardiomyopathy. *J Am Coll Cardiol*, 2004 Dec 21:44(12):2315-25.

Morimoto S., Lu Q. W., Harada K., Takahashi-Yanaga F., Minakami R., Ohta M., Sasaguri T., Ohtsuki I. Ca$^{(2+)}$-desensitizing effect of a deletion mutation Delta K210 in cardiac troponin T that causes familial dilated cardiomyopathy. *Proc Natl Acad Sci U S A*, 2002 Jan 22: 99(2):913-8.

Morimoto S., Yanaga F., Minakami R., Ohtsuki I. Ca^{2+} sensitizing effects of the mutations at Ile-79 and Arg-92 of troponin T in hypertrophic cardiomyopathy. *Am J Physiol*, 1998 Jul: 275(1 Pt 1):C200-7.

Nozaki, S., Kobayashi, K., Katayama, E. & Muramatsu, I. (1980) *Chem. Lett.* 3, 345–348. Synthetic studies on troponin I active site. Preparation of a pentadecapeptide with inhibitory activity toward actomyosin adenosine triphosphatase. *Chem. Lett.*, 1980: 3, 345–348.

Nugent A. W., Daubeney P. E., Chondros P., Carlin J. B., Cheung M., Wilkinson L. C., Davis A. M., Kahler S. G., Chow C. W., Wilkinson J. L., Weintraub R. G. The epidemiology of childhood cardiomyopathy in Australia. *N Engl J Med.,* 2003 Apr 24: 348(17):1639-46.

Parmacek M. S., Solaro R. J. Biology of the troponin complex in cardiac myocytes. *Prog Cardiovasc Dis*, 2004 Nov-Dec: 47(3):159-76.

Pinamonti B., Merlo M., Nangah R., Korcova R., di Lenarda A., Barbati G., Sinagra G. The progression of left ventricular systolic and diastolic dysfunctions in hypertrophic cardiomyopathy: clinical and prognostic significance. *J Cardiovasc Med (Hagerstown),* 2010 Sep: 11(9):669-77.

Pinto J. R., Parvatiyar M. S., Jones M. A., Liang J., Potter J. D. A Troponin T Mutation That Causes Infantile Restrictive Cardiomyopathy Increases Ca^{2+} Sensitivity of Force Development and Impairs the Inhibitory Properties of Troponin. *J Biol Chem.,* 2008 Jan 25: 283(4):2156-66.

Pollesello P., Ovaska M., Kaivola J., Tilgmann C., Lundström K., Kalkkinen N., Ulmanen I., Nissinen E., Taskinen J. Binding of a new Ca^{2+} sensitizer, levosimendan, to recombinant human cardiac troponin C. A molecular modelling, fluorescence probe, and proton nuclear magnetic resonance study. *J Biol Chem*, 1994 Nov 18: 269(46):28584-90.

Pruszczyk P., Kostera-Pruszczyk A., Shatunov A., Goudeau B., Dramiñska A., Takeda K., Sambuughin N., Vicart P., Strelkov S. V., Goldfarb L. G., Kamiñska A. Restrictive cardiomyopathy with atrioventricular conduction block resulting from a desmin mutation. *Int J Cardiol*, 2007 Apr 25: 117(2):244-53.

Rai T. S., Ahmad S., Ahluwalia T. S., Ahuja M., Bahl A., Saikia U. N., Singh B., Talwar K. K., Khullar M. Genetic and clinical profile of Indian patients of idiopathic restrictive cardiomyopathy with and without hypertrophy. *Mol Cell Biochem.,* 2009 Nov: 331(1-2): 187-92.

Ramirez-Correa G. A., Cortassa S., Stanley B., Gao W. D., Murphy A. M. Calcium sensitivity, force frequency relationship and cardiac troponin I: critical role of PKA and PKC phosphorylation sites. *J Mol Cell Cardiol*, 2010 May: 48(5):943-53.

Rarick H. M., Tang H. P., Guo X. D., Martin A. F., Solaro R. J. Interactions at the NH2-terminal interface of cardiac troponin I modulate myofilament activation. *J Mol Cell Cardiol*, 1999 Feb: 31(2):363-75.

Rarick H. M., Tu X. H., Solaro R. J., Martin A. F. The C terminus of cardiac troponin I is essential for full inhibitory activity and Ca^{2+} sensitivity of rat myofibrils. *J Biol Chem*, 1997 Oct 24: 272(43):26887-92.

Richard P., Charron P., Carrier L., Ledeuil C., Cheav T., Pichereau C., Benaiche A., Isnard R., Dubourg O., Burban M., Gueffet J. P., Millaire A., Desnos M., Schwartz K., Hainque B., Komajda M., and EUROGENE Heart Failure Project. "Hypertrophic cardiomyopathy: distribution of disease genes, spectrum of mutations, and implications for a molecular diagnosis strategy. *Circulation*, 2003 May 6: 107(17):2227-32.

Rivenes S. M., Kearney D. L., Smith E. O., Towbin J. A., Denfield S. W. Sudden death and cardiovascular collapse in children with restrictive cardiomyopathy. *Circulation*, 2000 Aug 22: 102(8):876-82.

Robertson I. M., Baryshnikova O. K., Li M. X., Sykes B. D. Defining the binding site of levosimendan and its analogues in a regulatory cardiac troponin C-troponin I complex. *Biochemistry*, 2008 Jul 15: 47(28):7485-95.

Robertson I. M., Li M. X., Sykes B. D. Solution structure of human cardiac troponin C in complex with the green tea polyphenol, (-)-epigallocatechin 3-gallate. *J Biol Chem*, 2009 Aug 21: 284(34):23012-23.

Robertson S. P., Johnson J. D., Holroyde M. J., Kranias E. G., Potter J. D., Solaro R. J. The effect of troponin I phosphorylation on the Ca^{2+}-binding properties of the Ca^{2+}-regulatory site of bovine cardiac troponin. *J Biol Chem.,* 1982 Jan 10: 257(1):260-3.

Robinson P., Mirza M., Knott A., Abdulrazzak H., Willott R., Marston S., Watkins H., Redwood C. Alterations in thin filament regulation induced by a human cardiac troponin T mutant that causes dilated cardiomyopathy are distinct from those induced by troponin T mutants that cause hypertrophic cardiomyopathy. *J Biol Chem*, 2002 Oct 25: 277(43): 40710-6. .

Russo L. M., Webber S. A. Idiopathic restrictive cardiomyopathy in children. *Heart*, 2005 Sep: 91(9):1199-202.

Sakthivel S., Finley N. L., Rosevear P. R., Lorenz J. N., Gulick J., Kim S., van Buren P., Martin L. A., Robbins J. *In Vivo* and *in Vitro* Analysis of Cardiac Troponin I Phosphorylation. *J Biol Chem.,* 2005 : 280(1):703-14.

Schoenfeld M. H. The differentiation of restrictive cardiomyopathy from constrictive pericarditis. *Cardiol Clin*, 1990 Nov: 8(4):663-71.

Senni M., Redfield M. M. Heart failure with preserved systolic function. A different natural history? *J Am Coll Cardiol*, 2001 Nov 1: 38(5):1277-82.

Siegel R. J., Shah P. K., Fishbein M. C. Idiopathic restrictive cardiomyopathy. *Circulation*, 1984 Aug: 70(2):165-9.

Tadano N., Du C. K., Yumoto F., Morimoto S., Ohta M., Xie M. F., Nagata K., Zhan D. Y., Lu Q. W., Miwa Y., Takahashi-Yanaga F., Tanokura M., Ohtsuki I., Sasaguri T. Biological actions of green tea catechins on cardiac troponin C. *Br J Pharmacol*, 2010 Nov: 161(5):1034-43.

Takahashi-Yanaga F., Morimoto S., Harada K., Minakami R., Shiraishi F., Ohta M., Lu Q. W, Sasaguri T, Ohtsuki I. Functional consequences of the mutations in human cardiac troponin I gene found in familial hypertrophic cardiomyopathy. *J Mol Cell Cardiol.,* 2001 Dec: 33(12):2095-107.

Takeda S., Yamashita A., Maeda K., Maéda Y. Structure of the core domain of human cardiac troponin in the Ca($^{2+}$)-saturated form. *Nature*, 2003 Jul 3: 424(6944):35-4.

Talbot J. A., Hodges R. S. Synthetic studies on the inhibitory region of rabbit skeletal troponin I. Relationship of amino acid sequence to biological activity. *J Biol Chem*, 1981 Mar 25: 256(6):2798-802.

Tripet B., van Eyk J. E., Hodges R. S. Mapping of a second actin-tropomyosin and a second troponin C binding site within the C terminus of troponin I, and their importance in the Ca^{2+}-dependent regulation of muscle contraction. *J Mol Biol*, 1997 Sep 5: 271(5):728-50.

Tsoutsman T., Chung J., Doolan A., Nguyen L., Williams I. A., Tu E., Lam L., Bailey C. G., Rasko J. E., Allen D. G., Semsarian C. Molecular insights from a novel cardiac troponin I mouse model of familial hypertrophic cardiomyopathy. *J Mol Cell Cardiol*, 2006 Oct: 41(4):623-32.

van den Wijngaard A., Volders P., van Tintelen J. P., Jongbloed J. D., van den Berg M. P., Lekanne Deprez R. H., Mannens M. M., Hofmann N., Slegtenhorst M., Dooijes D., Michels M., Arens Y., Jongbloed R., Smeets B. J. Recurrent and founder mutations in the Netherlands: cardiac Troponin I (TNNI3) gene mutations as a cause of severe forms of hypertrophic and restrictive cardiomyopathy. *Neth Heart J.,* 2011 May 2: [Epub ahead of print].

Venkatraman G., Harada K., Gomes A. V., Kerrick W. G., Potter J. D. Different functional properties of troponin T mutants that cause dilated cardiomyopathy. *J Biol Chem*, 2003 Oct 24: 278(43):41670-6.

Vinogradova M. V., Stone D. B., Malanina G. G., Karatzaferi C., Cooke R., Mendelson R. A., Fletterick R. J. Ca($^{2+}$)-regulated structural changes in troponin. *Proc Natl Acad Sci U S A*, 2005 Apr 5: 102(14):5038-43.

Wang Y., Pinto J. R., Solis R. S., Dweck D., Liang J., Diaz-Perez Z., Ge Y., Walker J. W., Potter J. D. Generation and functional characterization of knock-in mice harboring the cardiac troponin I-R21C mutation associated with hypertrophic cardiomyopathy. *J Biol Chem*, 2012 Jan 13: 287(3):2156-67.

Webber S. A., Lipshultz S. E., Sleeper L. A., Lu M., Wilkinson J. D., Addonizio L. J., Canter C. E., Colan S. D., Everitt M. D., Jefferies J. L., Kantor P. F., Lamour J. M., Margossian R., Pahl E., Rusconi P. G., Towbin J. A., and Pediatric Cardiomyopathy Registry Investigators. Outcomes of restrictive cardiomyopathy in childhood and the influence of phenotype: a report from the Pediatric Cardiomyopathy Registry. *Circulation*, 2012 Sep 4: 126(10):1237-44.

Wei B., Gao J., Huang X. P., Jin J. P. Mutual rescues between two dominant negative mutations in cardiac troponin I and cardiac troponin T. *J Biol Chem*, 2010 Sep 3: 285(36):27806-16.

Wen Y., Xu Y., Wang Y., Pinto J. R., Potter J. D., Kerrick W. G. Functional effects of a restrictive-cardiomyopathy-linked cardiac troponin I mutation (R145W) in transgenic mice. *J Mol Biol.,* 2009 Oct 9: 392(5):1158-67.

Wilkinson J. M., Grand R. J. Comparison of amino acid sequence of troponin I from different striated muscles. *Nature*, 1978 Jan 5: 271(5640):31-5.

Willott R. H., Gomes A. V., Chang A. N., Parvatiyar M. S., Pinto J. R., Potter J. D. Mutations in Troponin that cause HCM, DCM AND RCM: what can we learn about thin filament function? *J Mol Cell Cardiol*, 2010 May: 48(5):882-92.

Yang S. W., Hitz M. P., Andelfinger G. Ventricular septal defect and restrictive cardiomyopathy in a paediatric TNNI3 mutation carrier. *Cardiol Young.,* 2010 Oct: 20(5):574-6.

Yu Z. B., Zhang L. F., Jin J. P. A proteolytic NH2-terminal truncation of cardiac troponin I that is up-regulated in simulated microgravity. *J Biol Chem.,* 2001 May 11: 276(19): 15753-60.

Yumoto F., Lu Q. W., Morimoto S., Tanaka H., Kono N., Nagata K., Ojima T., Takahashi-Yanaga F., Miwa Y., Sasaguri T., Nishita K., Tanokura M., Ohtsuki I. Drastic Ca^{2+} sensitization of myofilament associated with a small structural change in troponin I in inherited restrictive cardiomyopathy. *Biochemical and Biophysical Research Communications*, 2005: 338(3):1519-26.

Zhang L., Jean-Charles P. Y., Nan C., Huang X. Effects of green tea catechins on diastolic function in mice with restrictive cardiomyopathy. Myofilament proteins as structural scaffolds and mediators of function. Madison, WI, June 2-5, 2012.

Zhang R., Zhao J., Mandveno A., Potter J. D. Cardiac troponin I phosphorylation increases the rate of cardiac muscle relaxation. *Circ Res.,* 1995 Jun: 76(6):1028-35.

Zhou Z., Li K. L., Rieck D., Ouyang Y., Chandra M., Dong W. J. Structural dynamics of C-domain of cardiac troponin I protein in reconstituted thin filament. *J Biol Chem*, 2012 Mar 2: 287(10):7661-74.

Zwas D. R., Gotsman I., Admon D., Keren A. Advances in the differentiation of constrictive pericarditis and restrictive cardiomyopathy. *Herz*, 2012 Sep: 37(6):664-73.

Acknowledgment

I would like to thank the Medical Research Council of Canada, Heart and Stroke Foundation of Canada, Margaret Gunn Endowment for Animal Research, Doberman Foundation of America, Midwest Eye Bank, the Berlin Cardiovascular Research Fund, Wayne State University Vice President for Research Incubator Program, American Heart Association, March of Dimes, National Aeronautics and Space Administration, and National Institutes of Health for supporting the research in my laboratory, and Dr. Douglas Root at University of North Texas for helpful comments and suggestions.

Index

A

B

C

D

E

F

G

H

I

J

K

L

M

N

O

P

Q

R

S

T

U

V

W

X